NO MAN'S LAND

NO MAN'S LAND

REGINALD HILL

St. Martin's Press
New York

Permission to quote from the works of A. B. Paterson is gratefully
received from Retusa Pty Ltd., Australia.

NO MAN'S LAND. Copyright © 1985 by Reginald Hill. All rights reserved.
Printed in the United States of America. No part of this book may be
used or reproduced in any manner whatsoever without written permission
except in the case of brief quotations embodied in critical articles or
reviews. For information, address St. Martin's Press, 175 Fifth Avenue,
New York, N.Y. 10010.

Library of Congress Cataloging in Publication Data

Hill, Reginald.
 No man's land.

 1. World War, 1914-1918—Fiction. I. Title.
PR6058.I448N6 1986   823'.914   85-25081
ISBN 0-312-57670-6

First published in Great Britain by William Collins Sons & Co. Ltd.
First U.S. Edition
10 9 8 7 6 5 4 3 2 1

PREFACE

No Man's Land has been a long time in the making and passed through many changes. 'So what?' says the reader. 'Let's get on with the story.'

Quite right, but first I must offer brief but heartfelt thanks: to Caradoc King of A. P. Watt, who sowed the seed and helped winnow many harvests; to Pat, my wife, who typed draft after draft with many helpful comments and comparatively little complaint; and to Marjory Chapman of Collins, who came late to the book and was able to spot where familiarity had bred obscurity and reconstruction awkwardness.

'And now can we have the story?'

Not quite yet. At this point I could hold you up with a long bibliography, but it has always seemed to me that fiction should authenticate itself. In any case, though I acknowledge a tremendous debt to all those historians and memoirists whose books brought me into at times unbearably close contact with the Great War, nowhere did I come across any concern with my central theme, the fate of those men who, for whatever reason, walked away from the War, and didn't get caught. Paul Fussell in *The Great War and Modern Memory* (OUP 1975) refers briefly to the legend of a wild gang of deserters living in the waste land of the old Somme battlefield, but the truth behind the legend, the real story of what became of these men both during and after the war, must still be locked in individual minds and family tradition. I would be fascinated to hear from anyone who can turn the key.

Meanwhile here, at last, is my fiction.

DONCASTER   R.H.
1984

PROLOGUE

THE HORSEMEN (1)

Britain declared war on Germany on August 4th, 1914. The British Army killed its first German soldier on August 22nd near Mons. Seventeen days later on September 8th, near the Marne, the British Army killed its first British soldier, a 19-year-old private condemned for desertion. Thereafter, for the duration of the war, it continued to condemn them to death at the rate of 60 per month and execute them at the rate of 6 per month. These figures are approximate. Some months were better than others.

The boy heard the horses while they were still a quarter of a mile away, but he did not waken the man. This desolate landscape was a garden to the country of his mind and he was completely indifferent to its sights and sounds.

Even when the man awoke, the boy said nothing. By now the tread of many hooves had steadied to the direct approach of a single animal while the others fanned out. All this was perfectly clear to the boy's country-sharp ears, but the man's senses were still dulled by sleep. It was hunger that had awoken him; hunger, thirst and the need to piss.

Dully he wondered why a body which had taken in no more than a few mouthfuls of dirty ditchwater in twenty-four hours should still need to piss.

The man stood up and stretched his limbs, stiff beneath the ill-fitting khaki uniform with its single lance-corporal's stripe. Soon in the west the sun would set. Soon in the east the war would rise. Here in this ghastly desolation, they were safe from that at least.

Then, freezing in mid-yawn, he saw the approaching rider.

His mind registered that he was a lieutenant of cavalry riding a big grey. He was less than two hundred yards away, picking his way steadily across the broken terrain.

The man reached down and seized the boy's hand.

'Josh, *up*!' he commanded.

The boy obeyed. Told what to do, he would do it, no more, no less.

Still grasping his hand, the man said, 'Come!'

They had rested on the western slope of a low ridge where the ravaged earth had been partially repaired by some patches of spring greenery. The man set off down hill, partly because his weary limbs needed all the help they could get, but mainly because this had once been a defended height and at the foot of the slope lay a huge embuscade of rusting barbed wire. Get this between them and their pursuer and there was hope.

Then came the sound of more hooves, and trotting smartly forward on the far side of the wire he saw two more horsemen.

'Back up the hill, Josh!' he cried.

Panting for breath, the fugitives staggered back up the slope. Even now things did not look hopeless. The officer was still advancing very slowly, apparently caring more about his horse's welfare than their capture. And the coils of wire still lay between the two newcomers and the slope. Though sunset was still a little way off, the sun was about to take an early night by slipping behind a low bank of dark cloud on the horizon. And once dusk set in, the Desolation would become a strange place of shifting sounds and stealthy movement which put the odds on the side of the pursued rather than the pursuers.

But when they reached the crest of the ridge, the odds shifted firmly back.

There were two more horsemen making their way up the relatively easy and unencumbered eastern slope.

One wore a corporal's stripes; the other, a small moustachioed man, let out a wild hunting cry as soon as he saw them and spurred his horse into a gallop, waving a fierce-looking sabre in the air.

'Run, Josh, run!' screamed the man.

Hand in hand they sprinted along the ridge, tripping and stumbling on the uneven surface. It was a vain and pointless flight. The first pair of horsemen had hurried on till they found a gap in the wire and were now on a line to cut them off ahead. The fierce little trooper was coming up at a full gallop on the other side, with the corporal following more sedately. And a backward glance showed the officer still a good way behind but proceeding with a steady certainty which was unnerving.

The same glance brought their downfall, literally.

The man put his foot in a hole and went tumbling forward beyond recovery. He managed to let go of Josh's hand, not wanting to bring the boy falling with him. The loss of contact was as effective as switching off a machine, for when the man had done a head-cracking somersault, finishing upright against a charred tree-stump, he saw Josh standing completely still a few yards away.

It was the stillness of indifference, but the moustachioed trooper, arriving a moment later, was seeing things through a distorting glass of long-frustrated blood lust. He saw the boy's stillness as defiance; the fugitives had turned to make a fight of it! Thrusting his sabre straight out before him in the classic charge position, he rode straight at the unmoving figure.

The man was helpless. Half-stunned by the fall, he had let go of his rifle and it lay out of reach a few feet away. The trooper let out a cry of rage, anger, hatred, directed at he knew not what, as he prepared to let his sabre taste its first blood in this or any war.

Nearby a rifle cracked three times, or perhaps it was three rifles. The trooper's scream spiralled out of the range of human hearing, his short thick neck spouted blood like a punched wine-cask and the unchristened sabre was spattered close to the hilt before it fell from his lifeless hand and vibrated point down in the earth close to Josh's head.

The man staggered to his feet. He was on the highest point

of the ridge and had a clear view in all directions. The trooper, his foot caught in the stirrup, was being dragged along behind his panicking horse. The poor animal was eager to join his two fellows who were also riderless. This was no terrain for wild flight, however. One was already caught in the barbed wire, bucking and plunging ever deeper with terrible screams that sounded more human than the trooper's death cry. The other was in full flight when it stumbled in a weed-overgrown trench and cartwheeled to the ground with a backbreaking crash.

Their riders were down some way further back, one lying still, the other in the prone position, firing his rifle at a group of men running forward, crouched low, in the trench system which advanced along the crest of the ridge. It must have been one of these who had shot the man with the moustaches. Nor had it been a lucky bullet. One of them popped up out of the trench now, taking a snap shot at the firing trooper who jerked convulsively and lay still.

And now the newcomers came out of the trenches and began advancing at a low crouching run.

The man looked round. Half way up the far slope, the cavalry corporal had reined his horse to a halt and was peering anxiously upwards into the gathering gloom. But the sound of gunfire had had the opposite effect on the officer. Far from slowing down, he was spurring his horse from a cautious trot into a full-blooded gallop. He had drawn his pistol, and though its puny crack was hardly audible above the thundering hooves, the man saw quite clearly the spurt of muzzle-flame.

'Down, you fuckers! Get down!'

The screamed command came from behind. The man realized that he and Josh were in a direct line between the cavalry officers and their anonymous rescuers. Grabbing the boy's arms, he pulled him to the ground. Instantly a fusillade of rifle shots rang out. The horse, hit in the chest and head, collapsed in full stride. The rider flew out of the saddle, hit the ground with a bone-jarring thump and lay quite still. His pistol skittered across the dry clay surface and came to rest a foot away from the man. He reached out his hand to grasp its butt. A boot crashed down on the barrel, forcing the weapon from his grip.

'That's mine, sport,' said a deep growling voice, and a huge hand plucked the pistol from the earth.

There was still the crackle of rifle fire. The man raised his head to see the cavalry corporal galloping away full tilt, crouched low against his horse's neck.

'Bastard's away, Viney,' said someone.

'All right. Save your bullets. Going at that speed, he'll probably break his neck anyhow.'

'This one's alive, Viney,' said a man kneeling by the officer's body. He was almost as big as the one called Viney, but much flabbier. He had a knife in his hand.

'Shall I finish him?' he asked.

The man called Viney hesitated.

'Naw,' he said finally. 'Later, may be. On your feet, friends.'

The man stood upright and helped Josh to his feet too. The recent excitements seemed to have touched the boy not at all. His wide clear eyes regarded their saviours with an indifference his companion could not share. They were a wild-looking gang, in dirty ragged clothing and with unkempt hair and unshaven faces. They were also very well armed. Only one thing was clear about them; in stillness or in action there could be no doubt who was their leader.

He addressed himself to the huge, muscular man standing slightly apart with a stillness which matched Josh's, but which possessed a brooding, menacing quality completely absent from the boy's.

'Who are you?' he said.

The man registered the question with a flicker of his hard green eyes but his only reply was to say to the fat man with the knife, 'Bring them,' and walk away.

Immediately, the man's arms were seized and he found himself being half dragged, half carried, over ground too rough and dark for safe progress but which his captors seemed to treat as a sunlit pavement. Turning his head, he glimpsed two similar trios, one supporting the officer, the other Josh. He called the boy's name reassuringly a couple of times and received in reply the back of a hand across his mouth with such force that his teeth dug into his upper lip and the salty taste of blood bloomed on his tongue.

They travelled, he guessed, for rather more than a quarter of a mile, then halted. But the pause had not been long enough for him to get any bearings before he was forced forward again, this time only a couple of paces, when the ground opened up beneath his feet and he went crashing and sliding down what felt like a flight of rough-hewn steps. No time to get his breath. He was dragged and pushed along a narrow corridor, down another short flight of steps, through a doorway into what felt like a large chamber. From the sound of breathing and the stench of unwashed bodies, it was fully occupied. A door was closed. The darkness was complete.

'All right?' said a voice.

'All right.'

There was the sputter of a match which grew into the large flame of a hurricane lamp, at first flickering fitfully, but soon settling down to cast a dim, even light. Around the chamber, which was perhaps sixty feet by thirty, other lamps were lit, and the fingers of flame conjured fierce, lupine faces out of the dark, between twenty and thirty of them. They regarded the newcomers with hostile curiosity.

'You promised us food, not more mouths, Viney,' growled one man.

'Well, I fancy a slice of this poncy officer,' said another. 'He should be nice and tender considering what them bastards feed themselves!'

The speaker, a slight wild-eyed man with his right cheek one broad suppurating scab, began to posture before the lieutenant, making exaggerated salutes and crying in a strangulated voice, 'Yessir! Nosir! How'd you like to be cooked, sir? Stewed, fried, or baked in a pie, sir?'

'Stow it, Foxy,' growled Viney. 'Someone take this joker away and keep him safe till I want him. Not you, Foxy. Taff, you look after him.'

The lieutenant, now conscious enough to look alarmed,

was dragged from the chamber by a small dark man.

Viney slowly stripped off his battledress tunic. It was growing hot in the crowded room. Beneath it he wore a grubby singlet against which his pectoral muscles strained. Across the biceps of his right arm was tattooed a brown and gold butterfly, strangely delicate in such a place on such a person. The boy Josh could not take his eyes off it.

Someone tossed Viney a long beer bottle with a screw-top. He twisted it off, put the bottle to his lips and took a long suck.

'Thirsty, son?' he said to Josh. 'Try some of this.'

He passed the bottle over. Josh drank.

'Your mate too,' said Viney. 'He looks as if he could do with it.'

Josh passed the bottle to the man, who drank deep. It was good beer.

'Now you've wet your whistles, let's hear you use them,' said Viney. 'Introductions first. I'm Viney. I'm in charge here till someone proves different. Now it's your turn, sports. Name, rank, number to start with. No bullshit. Jildi!'

A huge finger pointed at the man in the lance-corporal's tunic.

He shook himself free of the restraining hands and said in a formal military voice, 'I am Lothar von Seeberg, Feldwebel, which is, I think, sergeant of field artillery . . .'

'A Boche?' cried someone incredulously. 'You're a Boche?'

'I am German, yes,' agreed Lothar von Seeberg.

'In British uniform? You Hun bastard!'

The voice rose to a scream of rage and suddenly the wild-eyed man with the scabby cheek flung himself on to Lothar, long-taloned fingers scrabbling for the throat. The German went backwards with his attacker locked astride him. From the ring of onlookers there rose an animal cry of encouragement and expectation.

'Stop it!' cried Josh. 'Stop it!'

Lothar had no time to register pleasure at this rare sign of independent thought from Josh. All his strength of body and mind was needed to ward off this new attack and he was so drained by his recent struggles that he did not know if it would suffice.

Josh was too firmly held to offer any physical help and he could only stand and watch with sudden tears streaming down his face. But Viney stepped forward and seized the scab-faced man by his hair, dragging him upright.

'All right, Foxy, turn it in!' he cried. 'No one fights till I say fight! Hold still, I tell you, or you'll end up bald as a baboon's arse and only half as pretty.'

The man called Foxy stopped struggling but snarled, 'He's a sodding Boche, you heard him, Viney.'

'You want to kill Germans, you should've stayed in the line,' said Viney. 'Not that I've anything against killing Germans, but not in the way of patriotism, you follow me? You better had!'

So saying, he contemptuously thrust the scab-faced man away from him and turned his attention to Josh, who as the threat to Lothar faded was relapsing into his near-catatonic indifference.

'Are you a Hun too, sonny?' enquired Viney in a kindly tone.

The boy did not speak but continued to stare at the tattooed butterfly. Lothar from the ground gasped, 'No. He is English.'

'A Pom, is it? All right then. Why're you getting so het up about a Kraut, son?'

'Perhaps they're man and wife. He's a pretty little thing,' quipped a voice from the onlookers.

Viney turned a terrible face on the speaker.

'That's plenty of that, Taylor,' he snarled. 'I've told you before about that. You follow me, sport?'

There was a silence, unbroken except by the hard breathing of the separated combatants.

'I said, do you follow me?' repeated Viney.

'It was just a joke.'

'Mebbe. There's some jokes I don't much like, Taylor. I won't say it again. Either you shape up or ship out.'

There was no answer, though the man called Taylor, a

square-faced gingery man, looked bitterly resentful. Viney returned to Lothar.

'Right, Fritz. Time for explanations. Never hang a man till you've heard what he's got to say, that's the Aussie way. You seem to speak pretty fair English, better than half of these jokers, I'd say.'

'I have relatives in England. Cousins. I have spent holidays with them.'

'Spying! I told you he was a fucking spy!' cried an unlikely-looking patriot.

Viney sighed and said, 'Stick a bung in it, for Christ's sake! Sorry about that, Fritz. You were explaining yourself before we hanged you.'

Emboldened by the albeit grisly humour of the man's tone, Lothar said, 'And you, will you then explain yourselves to me?'

There was more angry muttering but Viney only grinned.

'Fair do's, is that it? I don't rightly know if I could start explaining myself. How about you, Blackie? You're usually right handy with explanations.'

The fat man who had wanted to slit the officer's throat smiled broadly without altering by a shade the steady cruelty of his eyes.

'We're what you might call Viney's Volunteers, friend. Only we're a bit careful about who we let volunteer.'

'That's the strength of it,' agreed Viney. 'You'll mebbe learn some more if it seems worth the breath, which don't seem likely the way things look. So I'd start explaining, Fritz.'

'Very well,' said Lothar. 'But first, about the boy, he is not well . . .'

'The boy will talk for himself,' said Viney softly. 'Right, son?'

To Lothar's surprise, Josh nodded slowly. Once again tears filled his eyes.

'Hey, take it easy, son,' said Viney. 'There's nothing to cry about down here. You're among friends. Look around. What do you see? Lads like yourself, that's what. We're all mates here, you'll see.'

Slowly the tearful eyes moved round the ring of shadowy faces as the boy turned a full circle till he faced the big Australian again. Now he opened his mouth and spoke inaudibly.

'Sorry, son,' said Viney. 'Didn't catch that.'

'Wilf,' murmured the boy.

Then, throwing his head back, he let out a scream which made even these desperate, horror-sated men start with fear.

'*Wilf!*'

BOOK THE FIRST

Dissolution

PART ONE

SOMME

The Allied Somme offensive began on July 1st, 1916. On that first day, British casualties were 57,540, including 20,000 dead. The minimum weight of the infantryman's equipment on that day was 66 lbs. Some carried more, including rolls of barbed wire on their backs. These proved to be redundant.

1

Shortly after midnight, a rum ration was doled out.

Sergeant Renton took care of it himself, squeezing along the narrow rain-sodden trench to make sure his platoon got their fair share.

Wilf Routledge downed his in a single draught and belched appreciatively.

'Hey, Sarge,' he said. 'If them old legs of yours fold up today, just give us Outerdale lads a shout and we'll take turns carrying you piggy-back!'

Before the sergeant, a tough old professional, could reply, another voice came out of the dark.

'I hope you fight as hard as you talk, Routledge. Empty vessels make most noise is what I was taught at school.'

This was Lieutenant Maiden, the platoon commander. He had been a bank clerk before the war, and this was going to be his first experience under fire too, facts well known to Wilf.

'At school, sir?' he said. 'Was that Eton College then, sir?'

Maiden's sallow face flushed and he said sharply, 'I'll keep a close eye on you Routledge,' and moved off.

Behind him Wilf laughed and joked to the sergeant,

'Temporary gent; permanent chicken. What say you, Sarge?'

Sergeant Renton said sourly, 'I say he's likely as frightened as you ought to be, Routledge, but mebbe not as much as you will be. As for piggy-back, just make sure this lot of sheep-shaggers here get over the plonk with all their gear on, and I'll be well satisfied.'

Wilf laughed and made a rude gesture at Renton's back. Josh Routledge regarded his elder brother with affection and pride. He was a hero, a prince among men. It would be almost a pity if there wasn't any real fighting so that Wilf could prove his worth.

Almost. But in his heart Josh prayed that the barrage which had been hurled across the skies for over a week now would have had the promised effect.

'Jerry's either dead or running,' Wilf had assured them. 'Haven't you heard? They're bringing buses up the line to take us through the gaps with the cavalry. We'll all be back in Outerdale, shearing sheep, come Michaelmas!'

Now here they were, in the line, and so far there'd been no sign of buses. Nor of cavalry either.

But at least the group was still together. Wilf was responsible for that. There were five of them from Outerdale. They'd not fallen over themselves to sign up as soon as the war began. Up there in the Cumberland fells, men didn't rush into such things. So they'd watched Wilf and waited. And when he made the move, they'd readily followed: Jimmy Todhunter, squat and square as a mountain boulder; his young brother, 'laal' Jockey, who was Josh's best friend; Ed Birkett, tall, spare, taciturn; and youngest of them all, so young he'd had to lie about his age, Josh himself.

Wilf, six foot, athletically muscled, golden-haired, had taken on the Army single-handed, making sure they stayed together throughout training, in transit, at the Bull Ring, and here in the line.

But now as the minutes trickled away, all his powers of unification and inspiration were being called upon. Ed Birkett was the most in need. Even during training, separation from Outerdale had seemed to cut him off more than the others from some source of vital nourishment. Now he stood aside, indifferent as an old horse to the huge burden of equipment he had to bear, completely still except for his right hand scrabbling at his lips from time to time as if there was something in his mouth it wished to pluck out.

Laal Jockey too had gone very quiet, while his brother Jimmy, who normally used words like half-sovereigns, became almost voluble. When dawn came and passed, confirming what they'd already been told but found hard to believe, that the attack was timed for seven-thirty, Jimmy burst out indignantly, 'It'll be broad fucking daylight, clear enough to shoot a gnat's cock off at half a mile!'

'They'll need to be crack shots to hit thine!' laughed Wilf. But laughter couldn't soothe the unease caused by this fresh example of High Command stupidity.

Jockey, crouched next to Josh, said, 'If there's no Germans left to shoot, it'll not matter if it's dark or light, will it?'

The little man was shaking uncontrollably. Josh too was suffering from intermittent attacks of the shakes and it was a not altogether unselfish gesture for him to put his arm around Jockey's shoulders.

'We'll be all right, lad, you'll see,' he whispered. 'It'll be a quiet walk. This time tomorrow we'll be supping plinketty-plonk in Bapaume, waiting for the peace to begin.'

After a while he felt his friend's trembling subside, and his own too. Best of all, Wilf gave him an approving glance.

Shortly after dawn there was a slight shower of rain, but it didn't last, and soon the rising mist showed that the sun's heat was beginning to penetrate the damp ground. As the warm rays carded the fleecy vapour, drawing it up in curls and threads, the lines of chalk thrown up when the Germans dug their trenches became quite clear. There had been little real fighting here for almost two years and often the chalk was embroidered with vivid yellow flowers, while in no-man's land, which was a strip of pasture rising in four undulations to the enemy line, the uncropped grasses of two summers rose high, wreathed with weeds, smudged with poppies and flecked with the delicate hues of wild flowers.

There was no sign of activity in the enemy trenches and, better still, the only wire visible to Josh and his friends was that of their own defences through which paths had been cut in the dark of the night. Hope began to rise that perhaps their officers' optimism was right. But not all the hope in the world could still the rising pulse of fear or slacken the tight racking of the nerves as the minutes oozed by.

There had been some shelling during the night but relatively things had been quiet. Then at six twenty-five, the British batteries exploded into their final outburst of fury. The effect was devastating, on many of the British troops at least. Here with the shells screaming low overhead to explode only two hundred yards away, Josh felt the noise was tangible, like the weight of his mother's cheese-press being screwed tighter and together on his yielding brain, squeezing his essence out to trickle down into the sump of the trench to join the mingled rainwaters and urine stagnating there. He felt himself being pushed lower and lower down the muddy wall against which he leaned. Others were suffering the same reaction, Ed Birkett with his hands clasped tight over his ears, and laal Jockey with fingers clinging desperately to Josh's sleeve in search of a comfort he did not have the strength to give.

Then his shoulder was grasped and he felt himself being dragged upright.

'Look at this, Josh!' whooped Wilf excitedly. 'This'll finish the bastards off good and proper. Oh my, give it to 'em, my lovely boys!'

A ragged noise ran along the trench. It was a mingling of cheers and gleeful laughter. Men were pulling themselves up on to the parapet to view the devastation being done to the enemy's front line. Like spectators at a boxing match, they roared their approval as black, white, yellow and ochre fists of smoke marked where shell after shell punched destructively against the German trenches.

As Josh looked, he felt the weight easing from his mind. This noise, this violence, was his Friend. The enemy was being destroyed before his very eyes. No one could survive such an onslaught, or very few, and they must surely have no will to resist. He found that he had started cheering too. Beside him, laal Jockey was on his feet now, shouting and laughing. Only Ed Birkett remained crouching at the foot of the trench, hands on ears, wide, unblinking eyes fixed on some landscape of the mind.

And now the minutes began to trickle, and then to run. The mood of euphoria caused by the commencement of the barrage evaporated as rapidly as it had erupted. The German guns had been stung into life and though nothing came very near their sector of trench, this reminder that resistance could still be met was like a breath of frost on too-early shoots.

Just before seven another rum ration was issued and the thick sweet liquid brought a warm glow of comfort for a little while. Some men became inordinately cheerful, laughing and making ribald jokes and threatening to go over the top in advance of the rest to do the dirty work for them. Others relapsed into introspection or sometimes half-audible prayer. Next to Josh, Jockey had started shaking again. Ed Birkett was leaning like a felled tree against the parapet. Jimmy was standing next to him, silent now. And even Wilf was curiously quiet, with an uncharacteristically inward-looking expression on his face. It was going to be all right, Josh assured himself. Just one quick rush – no, not even a quick rush. They'd been drilled to form up in lines which would advance in waves a hundred yards apart at a sedate pace of no more than two miles an hour. They hadn't got to cheer or shout in case they warned the enemy, though as the enemy were notionally mainly dead, it seemed an odd prohibition. So, no quick rush, nothing at all like the glorious cavalry charges which had thundered through his boyish imagination. A quiet stroll across the grass, like walking with Wilf over the fields down to the lake after their evening meal. At the end of it, a pipe of tobacco for Wilf and a quiet chat, or simply a companionable silence. That was how it would be, Josh assured himself. That was how it was going to be.

At 7.28 the earth suddenly tremored and rippled beneath their feet. Men staggered against each other, the drunk startled out of their merriment and the pious out of their prayer.

'Mines,' said Wilf, suddenly his old self again. 'They've set off the mines. Another little surprise for Fritz. We'll be off over the plonk any time now!'

Two minutes later, the guns fell silent. There was a moment of complete peace, broken only by a brave bird which chattered its pleasure that the cloudless sky had been given back to its rightful owners once more. It seemed to Josh that all they had to do was hold their breath and this peaceful moment could be made to stretch forever.

Then whistles began to shriek, orders were shouted, and the British guns which had been having their sights adjusted from the German front line to his rear defences drowned the little snatch of birdsong again.

Officers and NCOs ran along the parapet urging the men out of the trenches. Sergeant Renton's voice was calmly authoritative, but Lieutenant Maiden, his face flushed with effort or perhaps with drink, his revolver in his hand, was having difficulty in remaining sub-soprano.

'Out, out, out!' he screamed, gesturing wildly with his gun. 'You there, Routledge, get a move on, man! No malingerers today or, by Christ, we know how to treat them!'

Wilf, who was giving Ed Birkett a bunk up out of the trench, looked at the officer with contempt but said nothing.

Soon they were all above ground and filing through a gap in their own wire to form the assault lines on the other side. They were in the first wave, and as they waited for the order to advance, Josh glanced to his left where Wilf stood, his handsome young face touched with a faint smile, his athletic body making the burden of his heavy equipment seem negligible.

A whistle blew a double blast. Maiden and Renton confirmed its meaning with commands. The first wave began to move forward. The attack had begun.

2

Feldwebel Lothar von Seeberg dreamt, but knew he was dreaming. His cousin Sylvie knelt astride him with his penis held deep inside her body, and rose and fell in an accelerating rhythm, her face flushed with desire and her mouth gaped wide to let out hoarse screams of ecstasy.

He forced himself awake; forced himself back to awareness that Sylvie was now his sister-in-law, married to his twin, Willi; back to awareness that he was lying on a narrow truckle-bed, sixty feet below the ground, sixty feet below the air, sixty feet below the war.

Then he turned and his swollen glans penis rubbed against the rough blanket and that was enough. For a moment love and shame and depth and darkness and the whole damned war were blanked out in a magnesium flare of ecstasy. Then it was gone, and war, darkness and shame crowded back.

He swung his legs over the edge of his truckle-bed and started cleaning himself up. He could have switched on the electric bulb but he didn't want to wake the other man in this narrow underground chamber.

It was a vain stealth. The light came on and Artillery Captain Dieter Loewenhardt blinked sleepily at him.

'So that's why you've stayed a sergeant,' he said. 'Once you're commissioned, they stop you having wet dreams.'

Lothar smiled at his captain's rare attempt at humour. It was a sign of the easy relationship which, after a stuttering start, had developed between them.

The two men were of an age, twenty-six, but Loewenhardt looked older. Stocky and dark, he was the youngest son of an impoverished Junker family who had joined the artillery because he wanted to be kept by the Army rather than having to subsidize it, as happened in the smarter regiments.

When Lothar von Seeberg had been posted to his section, his reaction had been one of resentment and suspicion. The officers' mess was full of von Seeberg stories, amused, envious, or plain contemptuous, and Loewenhardt found it hard to understand why a man who seemed born to be one of the most glittering ornaments of his class should be slumming it in the ranks instead of following his brother into the crack cavalry regiment of which their uncle was colonel.

His first inclination was to regard his new sergeant as a pain in the arse, but gradually, reluctantly, he had had to admit that he knew his job and was good with the men. And finally had come affection. In public, they observed all military form. In private, they relaxed as friends. From time to time Lothar suspected there was a strong homosexual element in the other man. It didn't bother him. In his own social and cultural circles, anything was admissible except indiscretion, which was why he had always embraced the indiscreet. But he guessed that to Loewenhardt self-knowledge might be destructive, so he held his peace.

'What's the time?'

'Half past five.'

'Broad daylight outside. There'll be no attack today.'

The two gunners had been summoned from the artillery lines four days earlier to help interpret the intense bombardment which the enemy had been mounting for almost two weeks.

The first three days they had risen an hour before dawn and made their way from Brigade HQ to the reserve defence lines where they'd sat with a squad of machine-gunners, recording and analysing the bombardment.

Yesterday their report had gone to the Brigade Commander, probably simply confirming what he already knew. It was almost certainly pre-offensive, though it might be a feint. And it was causing far less damage than the enemy probably believed. A large proportion of their shells were duds, they had very little wire-cutting capability, and in the two years of stalemate on this section of the front, the Germans had dug their defences too deep to be troubled by anything less than a direct hit from a thousand-pounder.

They had also used the time to make themselves incredibly comfortable. Lothar still smiled every time he looked at the chamber in which they were housed. Two sergeant-majors had been shifted out of here to make room for them. The walls were lined with planks, smoothed and stained almost to the order of panelling and hung with a sentimental painting of a mother and child. The light bulb had a fringed shade and on the boarded floor was a small square of carpet.

He could feel the floor trembling gently as he shaved. The bombardment was still going on half a mile away.

They breakfasted separately. After breakfast they expected to be summoned before the Brigade Commander to discuss their report, but at nine o'clock word came that he would not be available till midday. There were reports of a massive British offensive to the north and the General had more important things to do than waste time on artillery men.

'I think I'll go and say goodbye to Klipstein,' said Loewenhardt. 'You coming?'

Klipstein was the young machine-gun Oberleutnant who had been their host for the past three days. He tried rather too hard to cut a dashing figure and his studied nonchalance was a little too studied, but beneath it all he was a likeable and perhaps a slightly frightened young man.

'Why not?' said Lothar.

Outside, they stood in silence for a moment, letting the douche of golden sunlight slough off the foetid atmosphere of the command post. Here they were on the eastern side of a long chalk down so far only lightly pocked by the French bombardment and with green vegetation fringing the white scars. As long as you kept a keen ear cocked for the odd shell which overflew its front line target, you could walk up to the crest in comparative security.

'I heard from my sister yesterday,' said Loewenhardt as they walked up the slope. 'All's well in Württemberg. You get any mail?'

Lothar knew this was a discreet way of asking directly or indirectly if he'd heard anything about the conduct of the war from the top. His father, the Graf von Seeberg, was an official of the War Ministry in Berlin.

'No,' he said shortly. 'I heard from my mother at Schloss

Seeberg, but she hears as little from my father as I do, which is hardly at all. But I'll write to my father if you like and invite him to spend a couple of days with Klipstein and give us all the benefit of his wide experience of modern warfare!'

Loewenhardt said, 'You don't give a damn, do you? I'm not certain if you're really a serious person at all.'

'Who wants to be serious on a day like this? Lovely weather. And nothing to do but stroll in the sunshine and listen to those 75s sing!'

He was referring to the French 75 mm field-gun which could rattle off rounds at the rate of thirty a minute. The din of the bombardment had risen to a new pitch of ferocity in the last few minutes and Loewenhardt said, 'You know, if it wasn't the middle of the morning, I'd say that was pre-attack. But that would be lunacy!'

'Since when did that make it impossible to the military mind?' said Lothar. 'Besides, what better time to attack than when no one's expecting it?'

As he spoke, the guns suddenly fell still. The two artillery men looked at each other in uneasy surmise.

'Raising their sights?' said Lothar.

'Not wanting to hit their own advancing men?'

Then with one accord they were running towards the crest. As the first shell exploded close by they dropped into a trench, exchanging speed for protection. This trench intersected at right-angles with another which snaked along the whole length of the ridge, with more short deep trenches branching off into tunnels leading into the dug-outs of the defenders.

By the time they reached the ridge, earth and air were shaking and the sun's brightness was dimmed with smoke and dust. In a landslide of rock and dirt they slid into the depths of the machine-gun emplacement, which was their goal.

''Morning, Loewenhardt,' said Lieutenant Klipstein breezily. 'Wondered if you'd come. Care for some coffee?'

'Thank you,' said Loewenhardt. 'It sounds as if you may get that action you're hoping for today.'

'You think so? I hope you're right! Not that it'll bother us,' he went on gloomily. 'Those lucky bastards down below will get all the fun.'

Down below meant the forward line, five hundreds yards before and two hundred feet below them.

'You don't think the Frankers can get through the line then, sir?' said Lothar.

Loewenhardt shot him a warning glance, but Klipstein replied, 'Alas, no, Sergeant. It's quite impenetrable, I fear. But don't be afraid. In the remote contingency that they did, we're ready for them.'

That at least seemed to be the truth. Not quite as domesticated as Brigade HQ, the dug-out still gave an impression of order and solidarity. The roof was high; it and the walls were firmly boarded. All around the gunners were sitting by the light of hurricane lamps oiling and checking their Maxim 08s. Nothing short of a direct hit from a thousand-pounder could reach them here.

They sat and drank coffee and ate black bread and cold sausage. After half an hour the bombardment perceptibly slackened, and without needing instruction the machine-gunners began to prepare to manhandle their guns out of the dug-out. From time to time the Oberleutnant was summoned to the field telephone. After one of these calls, he spoke to Loewenhardt at some length. A little while later the captain joined Lothar.

'What's happened,' the sergeant asked.

'The French have broken through in a couple of places,' murmured Loewenhardt.

Lothar began to laugh.

'Our impenetrable line and they're through it in forty minutes! At that rate, they'll be in Berlin by the weekend!'

Loewenhardt did not reply but went up the dug-out steps and Lothar followed him.

In the trench the platoon sergeant, a weatherbeaten veteran, was supervising the deployment of the Maxims with stolid efficiency. The Oberleutnant's face was flushed with an excitement that did not fall far short of hysteria. It was a state which could produce acts of great courage, or great madness, thought Lothar. Probably the alternatives would appear indistinguishable.

He joined Loewenhardt at an observation post.

'Are they holding?' he asked.

'See for yourself.'

The captain passed him the small telescope which was so much less dangerous for distant observation than binoculars. He peered through it at the forward line, at first only able to penetrate momentarily the swirl of smoke. Then gradually the fragments began to form pictures. The ground before the forward line, ploughed and harrowed by shellfire, was now sown with bodies. They lay on the broken earth and hung in the tangled wire; in some places they had fallen so close that they seemed to huddle together for a comfort they would never find. It seemed impossible that the attack had not been repulsed.

But still they were coming, these *poilus* with their dusty blue uniforms and long rifles made even longer by their ferocious bayonets, no longer advancing upright and confident in the trail blazed by their artillery, but progressing in short dashes, scrambling from one shell-hole to another, slipping forward behind drifting smoke and under the cover of tossed grenades.

'Incredible,' said Lothar almost to himself. 'The Frankers are going to do it.'

'Rubbish! *Rubbish!*' said Klipstein, shouldering him aside. 'I know those men down there. They'll not give an inch of ground, not while there's German blood running through their veins!'

Lothar caught Loewenhardt's eye and made a worried face. The captain didn't respond, but the platoon sergeant noticed and when Lothar moved past him, he whispered, 'Never worry, I'll take care of the lad. He just gets a bit excited, that's all.'

'He seemed so light-hearted about it all before,' murmured Lothar.

'Light-hearted? You'd be light-hearted too, mate, if you had his breakfast.'

The sergeant made a schlurping sound with his lips.

Lothar looked at him in surprise. Naïvely, he'd never associated Klipstein's gaiety with drink.

'The bastards! The cowardly stinking bastards!' screamed the Oberleutnant behind him.

Lothar sprang back to the observation post. The cause of the officer's indignation was not far to seek. Some of the men in the forward line had decided enough was enough. In all directions grey-coated figures were scrambling back along the communication trenches, with one or two actually sacrificing cover for speed by running across the open ground.

It was, thought Lothar, either a rout or a tactical withdrawal, depending on who was making out the report. There was no question which viewpoint the Oberleutnant was taking.

'Sergeant Muller!' he cried. 'Give those yellow-backed swine a long burst from Number One gun! Hurry!'

'Sir?' said the sergeant. 'But they're our lads, sir.'

'Ours? They're bringing shame on us, Sergeant, do you understand that? Shoot them!'

The sergeant glanced at Loewenhardt, who said tactfully, 'It could be they've just been ordered to fall back on stronger positions.'

And indeed it was true at least that most of the retreating infantrymen had stopped running, not because they were keen to set up a new line of resistance, Lothar guessed, but because they had the wit to recognize the dangers of further flight in broad daylight up the slope to the next ridge.

But three of them were in such a state of panic that they kept on coming. Two of them were shot down still some distance from the ridge, but one seemed to have a charmed life. Less than a hundred yards away now, his features were clearly visible, distorted with terror and physical effort, but still the features of a mere boy.

The Oberleutnant had drawn out his long-barrelled wooden-stocked Luger.

Lothar said, 'Dieter!' and Loewenhardt turned and saw the younger officer taking aim.

It is turning us into lunatics, this war, thought Lothar desperately.

Then the war, which was never afraid of taking its own

decisions, resolved the crisis by dropping a shell twenty yards behind the running boy. A red hot fragment of metal ripped open and cauterized his left thigh from hip to knee in a single operation. He screamed and fell, rose again, pushing his chest off the ground with his extended arms and lay there staring with fixed mad eyes at the parapet of the trench on the ridge and let his scream tail down to a dog-like whimper of pain and despair.

Klipstein was still pointing his gun, but the barrel was wavering now and on his face was a dawning horror which Lothar guessed had as much to do with what he was beginning to see in himself as the scene before him. But this was no time for clinical observation. Lothar sighed like a weary lover, then went over the edge of the trench in a single athletic rolling movement.

Now the choice was speed or stealth. He opted for the former, running forward over the uneven ground towards the wounded boy who watched his approach with wide unblinking eyes. The fifty or sixty yards between them did not seem to diminish. Shells were still bursting, rocking him with their blast and sending drifts of black smoke which momentarily hid the boy from view, and when the smoke passed, he never seemed any nearer, till all at once a single stride seemed to cover a dozen yards and Lothar was almost stumbling over the recumbent body.

He did not look at the torn leg. Nor did he speak. He just stooped and in a single movement lifted the boy over his shoulder, compounding the long high scream in his ear with the scream of French shells through the bright sunlit air. How blue the sky was over the ridge! Downhill it had seemed such a shallow slope, now the return seemed steep as a flight of stairs. The boy's scream was louder now. No, not louder, just unique. For a moment the French guns had paused, like an opera chorus falling silent to leave the soloist in complete command of the melody. They must be raising their sights again. Fresh targets! Fresh attacks! There was no standing still if you wanted to win!

The ridge was here. Loewenhardt and the Oberleutnant were reaching up to take the wounded boy from his arms. His screaming had stopped; dead or unconscious, it hardly seemed to matter which. Instead of dropping into the trench he paused, standing upright on the edge looking up to the rising sun, stretching his arms and closing his eyes as his face blushed in its warm caress.

Loewenhardt looked up at him and screamed, 'Lothar! For Christ's sake!'

Lothar looked down at him and smiled.

A Frenchman, wishing he was a hundred miles and ten years away, strolling through the golden fields of his home village near Arles, taking potshots at any wildlife he disturbed, was able to put two bullets into the crucified silhouette before it fell forward out of sight.

3

The grass was lush and long, waist high in places, and the hot sun reaching down to the still damp lower stalks was releasing rich odours.

Suddenly almost beneath Josh's feet something moved. He caught a glimpse of grey, German field-grey it seemed for a moment, and his heart squeezed tight and small as a walnut kernel.

Then the flash of colour was gone and only a ripple of fronds on the grass-sea ahead showed him its passage.

'Here!' he said excitedly to Wilf. 'I nigh on trod on a bloody hare!'

Wilf laughed. He looked flushed and excited. He'll probably be disappointed if there's no fighting to do, thought Josh. Himself, he'd be content if they saw nothing more frightening to shoot at than a few hares! Their steady progress had taken them over the first slight undulation, and now the line was breasting the second. They were in the smoke now. So far there had not been the slightest sign of resistance. Overhead the shells were still screaming, and seemingly still all in the same direction, to fall more distantly now on the enemy reserve and supply areas. They reached the top of the second undulation and stepped out of the drifting smokescreen at almost the same moment.

There, less than two hundred yards away, was their first objective, the German front line trench. All seemed quiet.

'It's a walkover!' someone cried enthusiastically.

Hope crescendoed in the new boys, but now the sensitized ears of the old hands picked out a terrifyingly familiar pizzicato amid the soprano screams and bass rumble of the artillery bombardment. Josh hardly heard it, but looking to his left he saw a strange phenomenon. The feathery tips of the tall grass were rising in the air like chaff beneath a flail. And in strict tempo to the flailing grass, the figures of his comrades in the line were spinning and twisting in curious grotesque movements, reminding him of an old woodcut in one of his father's few books, before they fell out of sight beneath the turbulent green.

It must have been a matter simply of seconds, three or four at most, but he had time to observe, to grow curious, to recall the woodcut, to identify it as representing something he'd never really understood called *the dance of death*; but he felt no fear till in the final second he saw Ed Birkett's tunic shred bloodily; Jimmy Todhunter's steely muscles twist like rubber as he was spun round, his mouth screaming, towards his brother; and laal Jockey unable to return the scream as the machine-gun bullets tore his throat to tatters so that the unsupported head almost fell away from the little body.

This last Josh can hardly have seen, for he was already throwing himself sideways against Wilf, bringing them both crashing to the ground as the traversing machine-guns, brought up by the Germans from their untouched deep dug-outs, scythed grass and men all around in one undiscriminating harvest. But he saw it immediately afterwards, for his first reaction after a moment of still terror in the grass was to pull himself out of Wilf's restraining grip and wriggle towards where Jockey lay.

He couldn't believe it. He lay with eyes no more than a foot from the bloody ruin, his nose full of the fume of fresh blood, and he couldn't believe it.

He moved on, not driven by courage or friendship but by this sense of unreality. Jimmy by contrast looked scarcely touched, but the bullet which had pierced his heart had left him equally dead. Ed Birkett looked much more of a mess with his chest laid open to the ribcage, but curiously he was still alive. One word, Josh thought it was 'Outerdale', rose in a red bubble to his lips, then the bubble burst and he too was gone.

He crawled back to Wilf in search of comfort, of protection.

'Wilf,' he said. 'They're dead, all dead. Jockey and Jimmy and Ed.'

Wilf's arms folded around him and the brothers held each other tight. It took a few moments for Josh to realize that Wilf's clasp was not the comforting hug of an elder brother offering reassurance, but more like the despairing grip a drowning man takes on any would-be rescuer that comes near him.

'Wilf!' he said, 'Wilf! What shall we do?'

But no answer came unless a gasping dry sob were meant for an answer.

Josh raised his head slightly and peered backwards. The second wave was approaching at the same steady pace at which the first had strolled into the hail of bullets. It wouldn't do to be found lying here unwounded, while the attack was still in progress. Josh had some vague idea that they could be killed instantly for cowardice.

'Come on, Wilf!' he urged. 'We've got to get up. We'll go on with the second wave. Come on! Get up.'

It wasn't courage. It was partly a naïve young man's fear of military law. And it was also a desire to get as far away as possible from those dreadful shards of human flesh lying just a few feet away.

But Wilf refused to move. He had sufficient control of himself now to gasp. 'No. Rest here. Safer. *Josh!*'

The scream of his brother's name was his reaction to a huge explosion about thirty yards away. Even before the fountaining earth had ceased to spatter down on them, there was another explosion almost as close, then another, and another.

The German artillery, receiving the message that the attack was in progress, had commenced a bombardment of no-man's land.

Now the soft grass was no protection. Shells do not care if a man hides or not. Their sport is with the solid earth and the air itself, and they send shards of white-hot metal in search of cowering men, or toss them casually skywards with their hurricane blasts. Not even the dead can lie at peace. If one explosion throws a decent pall of earth over their shattered bodies, the next can disinter them. A huge blast to the left of Josh and Wilf set them hugging ground as though by sheer pressure they would bury themselves in it. The showering earth was full of heavy stones. One thudded into the ground just a few yards from their heads. Wilf opened his eyes and started screaming. And when Josh looked, he saw that the shell-burst had ripped Ed Birkett's head and torso from his lower body and dumped it alongside his friends.

And still the assault troops were coming. The second wave, what was left of it, arrived where the first had broken. The machine-gun was traversing again, but there was no long line of targets ready to be knocked down. The shells had seen to that. Instead, little groups of survivors moved forward, some at the old steady pace as instructed, others staggering even more slowly, and others again running now, as men hasten to confront that which they fear in the belief that it can be no more horrible than the agony of delay.

Josh rose, dragging Wilf up alongside him. He felt obscurely that this was what he'd come here to do and that not to do it to the best of his endeavour would be somehow to put himself in the wrong. Whether this was how the other advancing soldiers saw it, he couldn't tell. Whatever their reasons, they kept going forward and that was his direction too.

Wilf did not resist. Perhaps he had no sense of direction left and the feeling of movement was stimulus enough for him. They advanced slowly and whenever the scream of a shell seemed threateningly close, Josh pulled them both earthwards again. In this way, somewhere between the second and the third waves, they reached the crest of the final undulation. One more dip and they'd be up at the German front line.

But there could be no such dip. Along the hollow before the line, stretching as far as they could see in both directions, was an unbroken barrier of barbed wire.

Some men of either the first or second waves had reached it. They had attempted to pierce it with their wire-cutters. And now their bodies lay or hung there, reminding Josh of a keeper's vermin wire.

Josh and Wilf tumbled into a shallow depression left by one of the inefficient shells which had failed to cut the wire.

'What'll we do, Wilf? What'll we do?' whispered Josh, the old habit of dependency dying hard.

'Lie here till it's dark,' gasped his brother. 'Oh Jesus fucking Christ, why didn't I stay at home and shear sheep!'

They tried to burrow deeper into the ground, but it was hard and stony. At least, thought Josh, no one can say we didn't advance as far as we could, not with that lot hanging on the wire.

A few minutes later, he realized how wrong he was.

A familiar voice came drifting out of the eddying smoke to their left.

'Keep going forward, men! Forward! We've got them beaten!'

A moment later, Lieutenant Maiden came strolling along parallel to the wire. At least 'strolling' looked like the effect he was trying to achieve, but from time to time either the broken terrain or his own condition made him stagger so he looked like a drunken reveller making his slow way home. Perhaps the oddity of his behaviour made the enemy hold their fire, perhaps he simply bore a charmed life, but he seemed unhurt.

'Jesus, that's all we fucking need!' groaned Wilf.

Maiden paused now. A shell had tossed up a small platform of debris and he scrambled on top of this vantage-point and with his back to the enemy lines scanned the ground before him. His cap was at a raffish angle and his revolver dangled from its lanyard.

Josh raised his head a little to get a better view.

'Keep down,' hissed Wilf, but it was too late.

'You two there! I see you! Routledge, isn't it? Both Routledges! I know you! Skulking there! Come on, men, on your feet! Advance! Follow me!'

Reluctantly Josh began to scramble up the ramp of crumbling earth.

'You fucking idiot!' screamed Wilf, seizing his leg.

'Good man, come on!' cried Maiden, advancing towards them and leaning down to offer a helping hand.

Wilf rose and took it. Then with a short sharp jerk, he brought the officer slithering on his belly down the side of the shell-hole.

When Maiden rolled over on to his back, his face mottled with dust and rage, he found himself looking into the muzzle of Wilf's rifle.

'Listen to me, you mad bastard,' hissed Wilf, who looked if anything rather more deranged than the man he threatened. 'You want to get yourself killed, I'll oblige here and now. But you're not taking me and my brother with you.'

'Wilf!' pleaded Josh. 'Stop it! Please, sir, he doesn't mean it.'

'Shut your gob, Josh!' said his brother. 'This bastard knows I mean it, don't you, Maiden.'

Slowly the lieutenant pushed himself into a sitting position.

'Routledge,' he said in a trembling voice. 'I'm giving you one last chance. On your feet now and advance with me. That's an order.'

Slowly he rose. The rifle's muzzle rose with him.

'Go fuck yourself,' said Wilf.

'You, boy, are you coming with me?'

Shaking now with uncertainty, Josh who was still clinging to the side of the hole said, 'Wilf, we've got to go . . . we've got to . . .'

'No!' commanded Wilf. 'If you try to go, Josh, I'll shoot this silly fucker myself.'

Maiden was now standing upright.

'Right,' he said almost gaily. 'Here we go. One two three!'

He scrambled out of the hole, turned to offer Josh his hand once more.

The youngster instinctively reached for it. Wilf said softly, 'I mean it!' Josh said, 'Please, Wilf, no . . .'

Then Maiden straightened up, shrieked at a pitch almost above hearing, and fell back into the hole. For a second he looked undamaged, then blood came bubbling up along the right side of his tunic like water from an underground spring.

With apparent calm Wilf sat back against the broken earth, laid down his rifle and began to roll himself a cigarette. Josh stared in horror at the bleeding figure on the ground.

'Wilf, we've got to do something!'

'Bastard's a goner,' said his brother. 'Saved me the bother.'

'But Wilf . . .'

'You do what you like, Josh. It's a waste of fucking time.'

Josh knelt beside the wounded man. Normally blood didn't bother him, animal's blood, that was. Killing pigs, gutting rabbits, all this was normal in country life. But this was different. This was the very essence of a man, pulsing out before him.

Taking his bayonet, he ripped open the tunic to reveal the gaping hole. His hands reddened with Maiden's blood, he pressed a dressing to the hole. It did not seem possible a man could live with that damage done to him.

Above them the roar of the guns continued, but now it seemed distanced, remote. A pall of smoke drifted across the battlefield, darkening the sky, and with it came a figure, rolling over into the shell-hole with swift efficiency.

Wilf started up and grabbed for his rifle.

'Easy, lad,' said Sergeant Renton. 'What's this?'

'It's Mr Maiden, Sarge,' said Josh. 'He's hurt bad.'

Renton looked at the wound and whistled silently.

'You're right. That's bad.'

'In the back, you notice,' sneered Wilf.

Renton looked at him coldly.

'Let's see what we can do for him. We'll have to turn him over, see if it came out the front.'

Slowly he and Josh turned the officer over. As they did so, his eyes opened and fixed on Wilf.

'That man refused my order, Sergeant,' he whispered. Then they closed and his face went very still.

'Is he dead, Sarge?' said Josh.

'Not yet,' said Renton. 'Though from the sound of it, for your sake, Routledge, he better had be.'

He was looking at Wilf, who yawned with affected indifference.

Josh spoke in a whisper, not because he was afraid of drawing attention but because he knew if he tried to talk normally, his voice would crack with sheer terror.

'What shall we do, Sergeant?' he asked.

'Do?' said the sergeant, glancing up towards the great vault of blue sky. 'There's fuck-all to do till either night comes, or reinforcements. And my money's on night. So let's make ourselves and this poor sod as comfortable as possible, shall we?'

The long day dragged on. Renton was right. No one came. The sergeant attempted to wriggle out of the hole at one point, but a warning chatter of machine-gun bullets brought him slithering back down. All this time Maiden lay, white and scarcely breathing. Josh did not want the man to die but he kept remembering Renton's grim words to Wilf.

At last evening began to creep across the ripped and bloodied battlefield. It was a long time even then before it was full dark and the stretcher-bearers dared to penetrate so far forward of their own lines in their hunt for the wounded. The Germans seemed content to let this humane activity continue with only the occasional rifle round or mortar bomb to remind Tommy who was master of the field.

Renton made contact with a party of bearers and brought them back for Lieutenant Maiden.

'Blimey, hardly worth bothering about,' opined one of them as he inspected the corpse-like figure, and Josh was ashamed at the pang of relief that shot through his heart.

'You'd better bother,' said Renton. 'All right, you Routledges. Fuck off back ahead of us. We don't want to sound like a platoon on the move.'

Wilf who had spoken hardly a word all day led the way, his confidence increasing as they got a ridge between them and the German line. Even so, they made slow, nervous progress.

Finally, after a hoarse challenge from a nervous sentry, they fell into the assault trench which they had left so hopefully more than twelve hours before.

They were given hot tea laced with rum and hard biscuit which tasted like fresh-baked bread. Then, a little recovered, they were ordered to report themselves to the main command dug-out. Here the colonel, whose orders had forbidden him to participate actively in the attack, was sitting at a rickety card table with a bottle of Scotch at his elbow. A warrant officer with a face as grey as a winter sky wrote their names on a sheet of paper.

'You two brothers?' he asked in a dead voice.

'Yessir,' mumbled Josh.

'Pair of brothers here, sir,' the WO called, in a falsely animated voice. 'Routledge W. and Routledge J.'

'Brothers?' said the colonel. 'Something, I suppose. What's the count now?'

The WO examined his paper.

'That takes us over fifty, sir. Fifty-one.'

The colonel passed his hand over his face.

'Well done, lads,' he said, addressing the brothers directly. 'Well done.'

They stood there, uncertain of what it was they had done well. The WO ushered them out into the trench and said, 'All right, lads. Off you go. Get some rest while you can.'

'Sir,' said Josh. 'That fifty-one you mentioned. Were that fifty-one come back from our company, sir?'

He knew it was a stupid thing to ask even as he asked it. The WO was regarding him with weary pity.

He said softly, 'Fifty-one from the whole battalion, son. From the whole fucking battalion!'

PART TWO

◆

CASUALTIES

The Battle of the Somme lasted till November 18th, 1916

Total casualties were:	*German* – 437,500
	British – 420,000
	French – 203,000
	Grand Total 1,060,500

Allied gains ranged from one mile at the flanks to six miles at the centre of a twenty-mile front.

1

Lieutenant Neville Maiden had been taken to the Red Cross Hospital at Barnecourt to die.

The hospital was situated in what had been a medium-sized hotel, the irony of whose name – Hôtel de la Paix – had long ceased to cause amusement. Barnecourt itself was a small town, or rather a large village, straddling a tributary of the Ancre. During the two years of comparative quiet on the Somme Front, which at its nearest point was only three miles to the east, the hospital had been adequate for the demands made on it.

Since July the first, all that had changed.

Someone had had the foresight to order the erection of wooden huts in the grounds before the battle, but these had filled so rapidly that the fields behind the hotel had sprouted with bell-tents till the hospital's capacity had re-doubled four times.

And still it was not enough. In the warm summer weather,

men sat or lay on blankets in the open air and the crowded ambulances streamed in and out, day and night.

But Neville Maiden knew nothing of this. A surgeon patched him up, certain he was wasting his valuable time. At least once an hour a nurse checked to see if his bed could be vacated yet, but Maiden would not die, though his appearance mocked death. While green July matured into golden August he lay quite still, his eyes fixed on a stain in the ceiling plaster which his mind turned into a man's face. It seemed to him he made bargains with God in which he offered that man's life for his own.

At last one morning the surgeon allowed that he would probably live and could now be prepared for the journey home.

Maiden spoke.

'I'm sorry?' said the ward sister who was changing his dressing. She was not used to Maiden speaking. Her name was Sally Thornton, a slender, grey-eyed, auburn-haired woman who until she came to Barnecourt had thought of herself as a girl. Before, she had been working in a large, well-equipped hospital in Rouen. The work had been hard, but not too far removed from the world of civilian medicine. Then had come July the first, and she had come forward with many others to be close to the source of this unprecedented carnage.

'I want to speak to someone from the Provost Marshal's office,' said Maiden. 'At once! Before I'm moved.'

Sally Thornton recognized the voice of hysterical obsession. She knew better than to argue.

'I'll see what I can do,' she said.

She told the doctor in charge.

'Why not?' he said. 'It makes a change from sodding padres!'

There was a Military Police corporal directing traffic through the hotel grounds.

'It'll be the Assistant Provost Marshal you want, miss,' he said. 'Captain Denial. I'll pass a message on, if you like. Here, what time do you come off duty then?'

Sally Thornton smiled wearily at his lustful jocularity, but her mind was elsewhere. *Captain Denial*, the man had said. It wasn't a common name. To her it meant not a captain, but a young lieutenant in a hospital bed at Rouen, his left leg shattered to the point where amputation seemed certain. But the young man had refused with a quiet strength which had cowed even the godlike surgeons. That had been a year ago. But even if he'd kept up his resolution back in England, he must surely be still in a wheelchair or on crutches at least. It couldn't be the same man.

But somehow she knew it was.

She saw him next day before he saw her, standing at the reception desk, a slight, pale figure leaning unobtrusively on a walking stick. Then he looked up and saw her and his eyes widened in a shock of recognition.

It's just because I've changed, she told herself. He's wondering if it's really me! Her mirror told her every day that the colour and freshness of youth which had been hers a few months ago had faded, leaving her haggard and pale except where a smudge of dark shadows highlighted her eyes.

Captain John Denial wiped all emotion from his face as he watched her approach. A year ago he had been disconcerted to find himself in love with this girl. No thought of speaking his feelings had entered his head. A curious sexual diffidence made him doubt that he could be attractive to a woman, and the coldly analytical mind which had brought him success in his civilian job as a Metropolitan Police detective had told him that patients always imagined themselves in love with their nurses. He assessed it as a false emotion which would quickly fade. He had gone back to England and persuaded himself that it had.

Now here they were again, the false emotion and the real woman.

'Captain Denial?'

'Sister Thornton, isn't it? We met at Rouen.'

'Yes, I remember. Left leg, multiple fracture. You want to see Lieutenant Maiden? At least, he seems very keen to see you. This way.'

She set off up the stairs at a pace to match both her brisk manner and the rapid beating of her heart. By the time she got up the second flight, she was well ahead. Becoming aware of the gap, she halted.

'I'm sorry,' she said.

The usual coldly dismissive response to references to his leg rose on his lips. Instead, he tried a smile and said, 'I'm better going down.'

'How did you get back to France? I should have thought . . .'

Denial smiled again, grimly, remembering the categorical refusal of the medical board to certify him fit for active service, the eagerness of his Scotland Yard superiors to have him back in the Force, and then the suggested compromise from a staff officer who spotted his background.

At the outbreak of war, the Corps of Military Police had consisted of barely 500 men. It had expanded rapidly to meet the needs, not always obvious, of the biggest army ever known.

'It's an active job, you'll probably return to France, possibly quite near the front,' he'd been assured. 'It's not a popular job, but it's essential. It needs officers with the right experience who are in it for the right reasons, not shirkers or bullies, but men with a sense of service and duty.'

The right words had been spoken. Early in the summer, Denial had returned to France as an APM, responsible for the maintenance of military discipline, the flow of traffic, and the maintenance of good relations between the civil and military authorities in an area the size of Middlesex.

'I persuaded them that even a cripple has his uses.'

He meant to speak lightly, but he was not used to lightness, and it must have come out bitter. She glanced at him indifferently, and said, 'In here. Mr Maiden's third bed on the left. Good day, Captain Denial.'

As she moved away, his voice came after her, impersonal and authoritative.

'Nurse.'

'Yes?' she said, pausing.

'Will you go out with me one evening?'

She looked back at him and said in a puzzled voice, 'Are you challenging me to a duel, perhaps, Captain?'

Confused, he said, 'No. I'm sorry. All I meant was . . .'

She laughed at him, bringing back the fresh young girl of Rouen.

'Yes, I'll go out with you, if you like. I should get off at eight this evening. Would you like to meet me at the hospital gate?'

Then she was gone without waiting for an answer.

Amazed at himself, and at life, Denial went into the ward. The pale young man in the third bed looked about eighteen, but the eyes that opened at his approach were old and haunted.

'Denial, APM,' he said. 'You wanted to see me.'

'Yes,' said Maiden in a low voice. 'I want to make a report. About a man in my platoon. Refused an order. In the face of the enemy. That's a court martial offence, isn't it?'

'Yes, of course. And very serious,' said Denial, surprised. 'But surely you should be reporting this to your commanding officer? And discussing it with him first, perhaps?'

'He's not here, is he?' said Maiden. 'In any case, I don't want it smoothed over. Regimental punishment. That kind of thing. I want . . . justice!'

The word was spat out. For the second time in a few minutes the normally imperturbable Denial felt himself amazed, and this time it was not a pleasant emotion. Hate too could crop up as unexpectedly as love.

He said, 'If you're quite sure.'

'Yes! I'm sure.'

'Then I'll take your statement,' said Denial.

Thirty miles to the south-east, another young man lay in a hospital bed, his face as pale as the vestments of the priest who knelt by him to help him die.

'He's getting a good send-off for a sergeant,' whispered an orderly to a nurse.

'His dad's Minister of War or something,' said the nurse.

'Bastard. I wish we had *him* here for half a day.'

In the bed the young man stirred and opened his mouth. A sound which was scarcely more than a breath came out.

The priest leaned forward so their faces were only six inches apart.

'Yes, my son,' he said urgently.

Now the young man's eyes opened momentarily.

'Fuck off,' whispered Lothar von Seeberg.

And fell asleep.

2

Josh Routledge screwed up his letter and threw it away. It fell into a puddle, perhaps of rainwater, perhaps of urine, where it floated, a patch of white against the filthy brown liquid.

He started on another sheet.

Dear mam and dad, he wrote. And paused.

He had made the fatal error of allowing a note of truth into his first attempt. Jolly, generalized reassurance, that was the keynote. Nothing of the muddy gash in the earth in which he was living; nothing of the filth and the stink, of the lice which colonized his body, of the rats he could hear in the trench wall behind him gnawing and scraping at God knows what relic of the Boche withdrawal.

This was the furthermost point of the Cumbrians' advance. Four miles, less than a mile a month, and how many men a mile no one dared guess. The ruined battalion had been reconstituted, gone into reserve, returned to the front, suffered heavily again and gone through the same process, though with fewer casualties, another three times in the eternity since that hot July day. None of this had gone into any of his letters, though the deaths of Ed Birkett and the Todhunter brothers could not be ignored. But even here, the brief reference had been couched in terms of high heroism far removed from the reality of those obscenely twisting bodies.

He resumed writing.

We are back at the Front now and doing well. The weather is turning a bit colder here as I expect it is with you. Wilf is fine and sends his love and says he will write next time.

Another lie. Wilf showed no interest in letters home, either writing or receiving them. Since that first day in battle he was a changed man. There was a wariness in his eye, a restlessness in his manner, a readiness to take offence, a reluctance to do his share of any task however small and undemanding. All the old, lively good humour seemed to have been squeezed out of him and each day in the line exacerbated the change.

There are a lot of new lads with us, some of them from Keswick and they know one or two Outerdale folk so we can have a good crack in the evenings.

Lies, all lies. Josh had no time for new relationships, all his energies were concentrated on keeping some kind of restraining hold on Wilf, who was readier to strike a blow against, than an acquaintance with, a newcomer. Also, to tell the truth, Josh did not want the risk of new friendships. Such losses as he had borne in the sweet-smelling grass of no-man's land were not to be countenanced again.

A hand descended on his shoulder.

He looked up. It was Sergeant Renton.

'Your brother around, Routledge?'

'He's back there, Sarge. Asleep.'

'Wake him up, will you? He's wanted in the CO's dug-out, jildi.'

'What for, Sarge? What's it about?'

'I don't know, do I, son?' said Renton. 'CO doesn't take me into his confidence.'

He was lying. Even before he himself was summoned to the command dug-out an hour ago, he had known what it was about. All calls passing through the regimental exchange were monitored by the signals sergeant who made sure his fellow NCOs were kept up to date with all relevant matter.

'Sergeant Renton,' said the CO. 'This is Captain Denial, the Assistant Provost Marshal. He has some questions he wishes to ask you.'

The CO's voice was neutral, but Renton knew of his anger when he'd first heard of Maiden's hospital statement over the phone.

'There was a time before this blasted war when a man could

choose subalterns who knew the form,' he had barked out. Later he had checked Denial's own background.

'Man was a bloody bobby!' he had been overheard to confide in disgust to the Adjutant.

Now Renton stood in blank-faced silence before the 'bloody bobby' and waited for this slender, pale man, whose lack of expression matched his own, to start speaking.

For five minutes he warded off questions with *couldn't say, sir!* and *don't rightly know, sir!* till Denial said, 'The man's brother was present, I believe.'

'Sir.'

'Colonel Harrington does not seem to feel he ought to be questioned. What do you think, Sergeant?'

'He's a good lad, sir. But he's just a lad. He may even have been under age when he signed on, I reckon.'

'That would be a complication,' frowned Denial.

'And he worships his brother. It'd break him to be made to give evidence, sir.'

'And that wouldn't help either,' said Denial.

The colonel snorted in distaste.

'But unless your memory improves, Sergeant, we shall have no alternative,' said Denial with a mildness which did not disguise the threat.

So Renton with great reluctance had reported what he had seen and heard.

'Thank you, Sergeant. Can you wait outside for a moment?'

'Sir. Just one thing more, sir.'

'Yes.'

'This man, Routledge. Wilf Routledge, I mean. He was a good lad to start with. But since this lot began, he's changed. He's – not to put too fine a point upon it – a right bastard now, moody, uncooperative. What I'm trying to say, sir, is – he won't help himself much.'

'No, I take your point. Thank you, Sergeant.'

Outside the dug-out, Renton had unashamedly eavesdropped.

'What's the verdict, Denial?'

'No verdict, sir. But the next step is inevitable and has

been ever since I got this statement from Mr Maiden. These details must be passed up for a decision to be made on further action. Meanwhile Private Routledge must be placed under arrest.'

'Arrest? There's a bloody war to be fought out there less than a quarter mile away, don't you base-wallahs know that?'

'The act may be formal rather than physical, sir, but the man must know that he is under arrest,' said Denial coldly.

'Very well. Sergeant Renton!'

Five minutes later Wilf Routledge stood at attention before the two officers, resentment and suspicion in every angle and strain of his body.

'Private Routledge,' said Denial quietly. 'You stand accused of conduct prejudicial to good order and military discipline in that on July the first of this year you refused to obey the legitimate order of a superior officer. Until further action is decided upon, you will be held within your regimental lines under close arrest. Do you understand?'

Wilf made a noise which might have been assent, indignation or a simple jeer.

And outside the dug-out Josh, who had assumed Renton's eavesdropping stance, ceased to hear the ever-present skitter of bullets and crump of shells and began to shiver like a man benighted on a wintry fell, far from home.

3

Autumn splashed the summer leaves with red and still the battle rumbled on. At last drenching rains brought hope that the seasons might do what the generals couldn't and bring the fighting to a halt. Young men in bed in England with wounds taken in the heat of summer were already being urged by a grateful nation not to overstay their welcome. Among these was Lieutenant Neville Maiden whose face as he bumped towards the front in a Crossley supply truck was almost as pale as when he had left it all those months before.

He passed through Barnecourt without recognizing the tiny town. To him it had been little more than a discoloured hospital ceiling and some glimpses of sky.

But Barnecourt had more to offer the weary British soldier than just medical care. In the main square were two establishments devoted entirely to his comfort and well-being.

One was an estaminet known as the Doze-Off, a corruption of *deux œufs*, the most popular because the only item on the menu there.

The other was Holy Mary's, where nothing was corrupted except innocence and, from time to time, genitalia.

On this autumn night which had the first frost of winter in its breath Arthur Aloysius Viney of the Australian Army sat in the Doze-Off with an almost empty wine bottle at his elbow and chewed the end of a stub of pencil. He was a huge man, clad in shirtsleeve order despite the cold of the night. His shirt was bleached a light beige, evidence to the expert eye of long exposure to the sun of Gallipoli. The three stripes on his sleeve were strained almost to horizontal lines by the swell of his biceps muscles, and on his right arm the wings of the tattooed butterfly stretched like a bat's under the strain of composition.

Before him lay an envelope. It was postmarked Melbourne, December 1915, and addressed to his regiment in Lemnos. It had followed him by fits and starts and numerous misdirections all around the Mediterranean and finally caught up with him, nearly nine months after its despatch, on the Western Front.

He had had it now for six weeks and still had not found a way to reply. Now once again he gave up the attempt. He crushed the letter in his hand, then smoothed it out and put it back into his tunic pocket. His own small beginnings he ripped viciously into a hundred pieces and forced the shreds down the long green neck of the almost empty wine bottle at his elbow.

'Going to toss a message over the side, mate?' enquired a tipsy English private staggering by. Viney looked up at him.

He had dark green eyes which at moments of high emotion shone with terrifying intensity. Suddenly much soberer, the private moved hurriedly on.

Viney rose and went out in the chilly night, stooping to get his six feet eight inches under the low lintel. Holy Mary's was on the far side of the sloping square, a useful arrangement, making it as natural in geography as it was in physiology for men who had devoured their fill of fried eggs and plinkety-plonk to pass from one establishment to the other.

There was a long queue tonight, despite the fact that the five ladies on duty had reduced the average running time from six to four-and-a-half minutes.

Viney crossed over and made his way along the queue, not pausing till he was nearly at its head. Here two men with corporal's stripes were passing a long bottle of cheap brandy between themselves to keep out the cold. One was almost Viney's equal in size, but unlike the hard and muscular sergeant, much of his bulk was fat. This was 'Blackie' Coleport, a garrulous and superficially jolly man whose recitals of the ballads of 'Banjo' Paterson were a highlight of impromptu concerts, but whose mean, hard eyes belied his easygoing exterior. The other, 'Patsy' Delaney, was a spare, rangy man with a lantern jaw and an expression of cold violence which belied nothing.

'Viney, you've changed your mind! All that thinking about us enjoying ourselves got you horny, did it?' called Coleport.

'No, I just got fed up of waiting for you bastards,' said Viney. 'Thought mebbe you'd got inside and couldn't raise it.'

'No problem,' said Coleport. 'Just stick it out the window in this frost and it'll soon get stiff enough. What say, Patsy?'

Delaney showed big yellow teeth in a humourless smile.

'Hot or cold, I'll manage,' he said.

The door of the brothel opened, three men emerged and three more entered, taking the Australians almost to the head of the queue.

'You jokers took your time,' observed Coleport good-humouredly. 'Forgotten what to do, had you?'

'Mine was about ninety,' replied one of the men in a Welsh lilt. 'Takes a bit of working up to, that does.'

But one of his companions, less inclined to good humour, answered, 'It'll be all right for this lot. They're used to making do with bloody kangaroos, aren't they?'

Coleport laughed and said, 'Fair do's, cobber. Last time I had a Welshwoman, it was like excavating a coal-mine, so an old kanger didn't seem so bad after that.'

The Welshman made a move towards him, but the first man to speak grasped his arm and muttered in his ear, nodding down the street. Three military policemen had appeared, one a sergeant-major, and were making their way slowly towards the brothel. The Welshmen set off rapidly in the opposite direction.

When the MPs reached the queue, they stopped and regarded the men in it with cold disapproval. Viney, who had been about to return to wait for his friends in the warmth of the estaminet, now plucked the bottle from Coleport's hands, took a long pull and said insolently, 'You'll have to wait your turn, cobbers.'

The sergeant-major stepped towards him and said, 'Who the hell you think you're talking to, Sergeant?'

'Couldn't say, seeing as we ain't been introduced,' said Viney.

The sergeant-major stuck his face close to Viney's and wrinkled his nose.

'What's that you're drinking, Sergeant?'

'Cough mixture. Like some, mate?'

The sergeant-major drew in a deep breath and his cheeks went almost as red as his hat. But before he could speak, there was a cry of upraised voices from inside the house, English and French confused in anger, then the sound of breaking glass and a woman's shriek.

The MPs rushed into the brothel.

'I knew the bastards'd find a way to jump the queue,' observed Delaney.

A couple of minutes later the two MP privates emerged, dragging between them a soldier who was screaming loudly and trying to pull up his trousers which were round his knees. The queue broke into spontaneous applause.

'Good on yer, bare arse!' yelled Coleport.

The sergeant-major appeared in the doorway and glared at the waiting men till the noise subsided.

'That's it, you men!' he yelled. 'I'm closing this place down as of now. Get back to your units. Go on, get a bloody move on!'

Grumbling, the men began to disperse, but Viney stepped forward and said, 'Hold hard, friend, you can't do that.'

'Can't do what?'

'Can't just close this knocking-shop down. Why? What's your reason?'

Drawn against his judgement into a debate, the sergeant-major said, 'It's a disorderly house. There's been fighting in there, so that's why it's closed down, and you'd better get yourself out of here before I lose my temper!'

'Fighting!' echoed Viney, mocking the sergeant-major's intonation. 'What the hell do you cherry nobs know about fighting? You should have been at Jolly Polly, cobber. Or back there' – pointing his bottle dramatically eastwards – 'that's where the fighting is. Why don't you close *that* little lot down too!'

The dispersing men had paused and now a ripple of laughter passed among them.

'Right!' screamed the sergeant-major. 'That's it! I've had just about all I'm going to take from you fucking Aussies! Let's have your name and number. You're under arrest!'

Viney's reaction filled the non-Australians in the queue with horrified admiration.

He threw back his head in a full-throated bellow of laughter and said, 'You'll need more than a fancy hat to arrest me, sport! Better stick to old tarts and little lads with their pants down.'

'Atten-shun!' screamed the sergeant-major. Then when this produced no effect whatsoever, he seized Viney's right arm close to the dangerously swelling butterfly and cried, 'You two! Over here quick!'

The two men dragging away the de-breeched soldier looked round just in time to see Viney smashing the brandy bottle over the sergeant-major's head. For a second they were paralysed in disbelief, then, releasing their prisoner, who gathered up his trousers over his thighs and set off at a staggering run into the protective darkness, the MPs dashed back towards the brothel, truncheons at the ready. Another surprise awaited them. While the generality of the soldiers turned to flee, Coleport and Delaney came walking to meet them. One of the MPs was intimidated enough to slow down and take out his whistle to summon help. He managed one long blast while Coleport got his companion in a vicious half-nelson and Delaney drove his boot into the helpless man's crutch. And he'd started on a second blast when Coleport's fist crashed the whistle into his mouth, breaking several teeth and piercing his palate to the depth of half-an-inch.

But help was close. Another five or six MPs appeared at the furthermost end of the street. The soldiers who'd made up the queue, awed spectators till now, began to scatter. But when Coleport called, 'Come on, you Dinkums, if you can't have a fuck, next best thing's a fight!' some of the Australians paused, and some of the British troops too. And when the MP reinforcements arrived, they suddenly found themselves confronted not by a trio of men resisting arrest, but by an incipient riot.

For a moment things hung in the balance. Then as the new group of MPs led by a full corporal advanced purposefully across the square, Blackie Coleport declaimed,

'Now this is the law of the Overland that all in the West obey,
A man must cover with travelling sheep a six-mile stage a
   day!'

Then advancing to meet the corporal, he added mockingly, 'You know what that means, mate? We've got fucking sheep in Aussie can move a bloody sight faster than you red-caps!'

The corporal said, 'You're under arrest, soldier,' and swung his wooden truncheon at Coleport's head.

The Australian ducked, evading the blow easily, and came up with a gleaming bayonet in his hand.

'You're all wind, Corp,' he said. 'Let me ventilate you.'

And he drove the blade in the MP's stomach.

After that the riot broke out in earnest. The MPs were driven out of the square by the fury of the mob of soldiers who then turned their attention to the Doze-Off and Holy Mary's and when these had been wrecked, went on the rampage through the rest of the tiny town.

It was an hour before reinforcements sufficient to quell the riot appeared. By this time Viney, clutching a full brandy bottle, was half a mile up the track which led to his unit's encampment. There was a keen frost now and as he stood drinking, his breath rose above him like white smoke from a barrage and he could feel the night air tracking the runnels of sweat down his slab-muscled body.

After a while there was a movement in the darkness and shortly afterwards Delaney, also armed with a bottle, joined him.

'You seen Blackie?' asked Viney after a while.

'Yeah. Silly cunt tried to lug one of the girls from Holy Mary's back with him. Last thing I saw, he was on the deck with ten redcaps kicking shit out of him.'

'He'll survive,' said Viney with the certainty of one to whom all merely military troubles were survivable. 'Let's get back then.'

They walked a little way in silence.

'It'll be ninety degrees up a penguin's arsehole in Melbourne,' said Delaney suddenly.

'Shoot your foot off, they'll ship you home,' said Viney.

'Mebbe I'll do that,' said Delaney. 'I'm ready for a change.'

It was a question, despite its form. Viney did not reply straight away, and the normally taciturn Delaney persisted with a more direct approach.

'You never think of home, Viney?'

'What's home?' said the huge Australian savagely.

He took a long pull at his bottle, then hurled it into the night.

'Let's get Christmas over and a bit of a thaw, then we'll see,' he said more evenly.

'You're the boss, Viney,' said Delaney.

Far ahead in the east a star shell arced skywards and hung still for a moment of incandescent beauty. In companionable silence the two men headed towards the distant light.

4

At first Wilf Routledge's arrest semed to mean very little. He remained with the regiment and did all his duties. Once the idea of being under arrest had become familiar to him, he even began to complain with savage humour that by rights he shouldn't be in the line at all but down at base somewhere, in a nice comfy cell, with three hot meals a day and no shelling.

His comrades laughed and Josh, desperate for reassurance, joined in. But Renton didn't laugh.

Not even the return of Lieutenant Maiden to the regiment seemed to change things. He looked a little paler, a little thinner. Rumour had it that the colonel received him very coldly, but as the regiment's junior officers had changed almost entirely since July, the replacements accorded him the respect due to a survivor and there was no general boycott.

Josh avoided him. Wilf openly sneered every time he saw him. Renton kept the two apart as much as he could. And after a couple of weeks, everyone began to assume that the whole business had somehow been kicked under the carpet.

As usual in the Army, when things happened, they happened quickly. Josh returned from a raid, bone-weary and with his nerves in tatters, to discover that Wilf, Renton and Maiden had gone. He rushed to see the colonel and managed in his desperation to get as far as the RSM who said not unsympathetically, 'Listen, lad, the only way you could be with them was as a witness, and the colonel wouldn't wear that. So think yourself lucky.'

Five days later Maiden and Renton returned alone.

'They found him guilty, son,' said the sergeant, who came

straight to Josh. 'They had to. No choice. You know the Army. Fucking rule-book all the time!'

'What'll they do to him, Sarge?' asked Josh fearfully.

'No sentence yet,' said the sergeant. 'They usually take a bit longer to decide the sentence after the verdict. That's good; make sure they get it right, that's the idea. Chin up, lad. It'll be all right, likely.'

Renton could not be drawn on a forecast, but other prognostications ranged from a jail sentence (suspended till after the war) to a couple of weeks of Field Punishment No. 1, followed by six months of all the nastiest jobs in the battalion.

Frequent repetition of assurances that whatever happened Wilf would be back with them by Christmas finally restored Josh to something like his old self. The worst seemed over, and this seemed confirmed a few days later when the battalion was ordered out of the line into rest.

It was always a strange journey to make, like moving from one world to another. After what seemed hours in a tangle of communication trenches which seemed to lead nowhere, at last they filed out on to a real road which ran straight as an arrow through the support lines. At first as they marched they still adopted the stooping, almost crouching stance which became second nature in the trenches. But slowly as they passed by supply dumps, MT depots, horse lines, all full of bustle and chatter, the men began to straighten up and eventually they began to talk, often with an exaggerated even slightly hysterical loudness as if to demonstrate to themselves that the front line need for wary whispering was over. And when at last the support area was left behind and they entered a landscape where houses were not ruins but stood intact with light in the windows and smoke curling from the chimneys, and trees were not stumps but rose tall and erect, leafless only because of the season and edged with snow like Christmas card illustrations, several of the men collapsed by the roadside as their minds acknowledged safety for the first time.

The first morning of the rest period, Josh awoke feeling better than he had done for months. Suddenly survival for himself, and for Wilf, seemed a real possibility. Wherever his brother was at this moment, he was at least safe from German shells and bullets. Christmas was close. His heart swooned to think of what Christmas had meant in Outerdale. He let his thoughts drift into a fantasy of being given Blighty leave for Christmas. He wouldn't tell anybody, but just appear at the farmhouse door. He could hear the screams of delight from his sisters, see brother Bert's broken-toothed grin, feel baby Agnes's hands gripped tight in his hair as he ducked round the room bearing her high on his shoulders. His father coming in from the fields would show little emotion, but would press his hand in a grip of steel. And his mother would put her arms around him and press her cheek to his. And then ask, where was Wilf?

The fantasy faded.

There was no way he could pretend they were going to let Wilf go home this Christmas. But oh, for himself, in almost his first purely selfish thought since July, he would give anything for himself to be sent home for Christmas!

But vain longings must not be allowed to spoil present pleasure.

He rose to see that it was the best kind of winter's day, bright and crisp. There would be a decent hot breakfast to eat, then a pay parade. This was no seaside holiday, there would still be military duties to do, but if a man kept his head down, they needn't be too troublesome. Later perhaps there'd be a football match. But what he was really looking forward to was getting right away from everybody to take a look at this countryside. There'd be livestock to examine, and he might with a bit of luck meet up with a dog and make friends with it. He thought of Sailor, his personal favourite among the half-dozen Border collies who lived and worked on Beck Farm. Did Sailor miss him? he wondered, but turned quickly away from the thought as he saw the road it was leading him back to.

He went out into the crisp cold morning and surprised several of the men of his section with the energetic good humour of his greeting.

But to his own surprise as he stood in the queue for breakfast, Sergeant Renton stopped as he walked by, looked close into his face, and said, 'Here, Routledge, you look a bit peaky. You'd better go sick.'

'But I feel fine, Sarge, never better,' protested Josh.

'Well, you don't look it, son. Better safe than sorry. You get yourself on sick parade and that's an order.'

It was an order which filled Josh with irritation and some concern. If the sergeant's eye, so expert at spotting attempts at malingering, had in his case diagnosed illness, there must be something to it. His irritation subsided a little as the news went round that the word had just gone up on company orders that the whole battalion was to parade in half an hour. *For an inspection*, went one rumour, *They're asking for volunteers to go behind Jerry's lines dressed as peasants*, went another. But accompanying all the rumours was a resentment at this bit of bullshit being allowed to muck up the rest period so soon.

Renton's order would at least spare him *that*, thought Josh.

But when he reported to the MO he found no one in attendance except an orderly who told him brusquely that everyone who could walk had been commanded to the parade and he'd better get along there sharpish if he didn't want trouble.

He found the battalion already forming up in a hollow square on a huge ploughed field whose furrows had been set solid by the hard frost. The irregularity of the ground and the slipperiness of the surface made proper marching impossible and Josh was able to slip easily among the slow shuffling men till he found his platoon. Sergeant Renton spotted him and glared angrily at him, but was unable to speak as at that moment the sergeant-major's scream called them to attention.

At the open end of the square, a trestle table had been erected. Three chairs were placed behind it and in the centre one of these sat the colonel, flanked by his second-in-command and adjutant. The other officers of the battalion stood behind and to the sides.

For a moment everything was perfectly still and silent, the only movement being the drift of mist as the men's breath rose visible in the cold air. Distantly they heard the strained whine of an engine in low gear and eventually a Crossley General Service truck with ambulance markings came bumping over the set furrows towards the table.

'They got a dozen Fannys in there. They're going to raffle 'em off,' whispered a wag, bringing a scream of 'Quiet in the ranks!' from at least three NCOs.

The truck stopped. The man sitting next to the driver got out. He was wearing the red cap of a military policeman. He saluted the assembled officers and went round to the back of the truck. The tailboard was lowered. Another redcap got out and then the two of them assisted a third man to the ground. He needed assistance because his hands were manacled.

He turned his head slowly from side to side as if bewildered by the location and dazzled by the light, and his profile confirmed to Josh's eyes what his heart had told him the moment he saw the bright red of the escort's hat.

It was Wilf.

The sergeant-major now took over, screaming orders addressed to *prisoner 'n' escort*, bringing them to attention, sending them in a wheeling march to the front of the table; but his hectoring tone lacked something of its usual conviction and when Wilf stumbled on a frozen clod and nearly fell, his arm shot out to steady him and in the clear air it was possible for everyone to hear his quietly spoken, 'Easy, lad.'

Finally they came to a halt before the table, each man standing rigid at attention, Wilf hatless, his golden hair thick and tousled above his shaven neck, the two redcaps glowing spots of colour in that wintry-hued scene; the sergeant-major slightly to one side, staring fixedly at some point about forty degrees over the horizon.

The adjutant stood up. He probably intended his voice to sound firm and strong, but his first attempt came out as a high-pitched yap and he had to try again. This time it came out as a rapid but still audible gabble.

'Private Routledge having been found guilty by field general court martial of the charge of refusing to obey a legitimate order given by a superior officer in the face of the enemy, the sentence of the court was that the accused should suffer death by being shot, which sentence has been confirmed by the Commander in Chief and will be carried out at dawn tomorrow.'

The sergeant-major started screaming his orders again almost before the adjutant had stopped speaking, the little procession marched and wheeled. Wilf disappeared into the back of the truck which immediately bumped its way out of sight and then out of earshot across the field.

The proceedings had taken less than five minutes.

The cold air then was filled with the bellow and scream of commands as warrant officers and NCOs went about the task of unpiecing the human square. But Sergeant Renton's voice was silent as he pushed through the ranks of his platoon to catch Josh as he fell.

5

Jack Denial lay in his narrow bed and stared at the ceiling. He had been lying thus so long that his eyes had grown accustomed to the darkness and he could pick out the wooden crossbeam quite clearly.

Or perhaps it was nearly dawn.

He pushed the idea out of his mind.

'Jack. You awake?'

He had been lying stiff and straight so that his sleeplessness would not disturb the girl next to him, but now he relaxed his body against hers and that was answer enough.

He still could not fully comprehend his good fortune. That first night he had felt so wretchedly awkward. He had made no plans. There was nowhere to go, nothing to do; at least, nothing that would not involve putting Sally Thornton under the prurient gaze of his brother officers. So he had brought her back to this room and offered her a drink and told her in the blunt controlled terms of a man who expects nothing, and so has nothing to fear, that he loved her.

The outrageous absurdity of the declaration still made him sweat at its memory.

But she hadn't struck him, or laughed at him, or called for help. Instead she had kissed him gently. And he had kissed her not so gently. And after that nothing was clear, till at last they had moved far enough apart on this same narrow bed for him to get a proper view of her and he had seen once more the pretty young girl of their first encounter, her pale, pinched features flushed and softened in the aftermath of pleasure.

Since then she had come to him every night their duties left them both free. Where it might lead, he could not guess, dared not speculate. Instead he treasured every meeting as if it were the last.

Yet tonight he wished she had not come. His body had not responded to her caresses. Her warm nakedness at his side seemed only to accentuate the chill and darkness around.

'Jack, it's not your fault. It's your job,' she whispered in an attempt at comfort.

'What *is* my job?' he replied harshly. 'Back in the Force, I thought I knew. Protecting a society that on the whole deserved protection. Here, what do I protect? An inhuman system engaged in a monstrous war!'

'If that's how you feel, Jack, you have to tell someone.'

'I've tried,' he said grimly. 'Not quite so directly, maybe. I added a few comments of my own to the report on that riot we had. All it got me was a rocket!'

What he had suggested was that such incidents, though not common, were no longer rare. The Somme had changed the common soldier's attitude to the war. There was a new cynicism about the generals, a new despair about the future. He had gone as far as suggesting that a large-scale mutiny could not be discounted.

The report had come straight back from his immediate superior with the laconic comment, 'The British Army does not have mutinies. It does have junior officers who are invalided out with shell-shock. Please emend.'

He had emended the report, but not his opinion. Whatever the traditions of the British Army might be, there were other armies in the field alongside them. The French were vociferous in their outrage at the conduct of the war, and as for the Australians who were appearing in increasingly large numbers . . . ! He thought of the one they had in confinement at the moment, charged with sticking a bayonet into Corporal Parker during the riot. There were plenty more like him, not as criminal as he judged this one to be, but with the same basic indifference to authority that was marked only by rank or red caps. By all accounts, the Australians fought as bravely as any men in the line, but they were never afraid to ask the reason why. Such independent notions were contagious.

These ideas and arguments marched through his head, but if Sally's warm body could not bring oblivion, there was no hope that military debate could take his mind off what lay ahead.

He couldn't deceive himself any longer. The dark winter night was turning to grey.

He shifted slightly and immediately her arm tightened round him.

'It's not time yet, surely?' she murmured.

'Soon,' he said. 'Morning. First light. I'd better start getting ready.'

He slipped from her grasp and out of the bed. His head was throbbing.

She said urgently, 'Jack, darling, it's not your fault. It's a duty, nothing more.'

'Nothing more,' he agreed. 'Except that . . .'

'Yes?'

'Duty should be more. More than . . . mere necessity, that is.'

He laughed harshly.

'Mere necessity is what that poor bastard is feeling. Duty should be more than that, wouldn't you say?'

She ignored the question.

'What will he have been doing? Will they wake him early? Or . . . just before?'

'I doubt he'll need much waking,' said Denial. 'Or if he does, it'll be from drunken insensibility. They'll have given him a whole bottle of whisky to himself. Such generosity from the officers' mess! Such luxury for a poor Tommy!'

'Jack, it's not your fault!' she repeated. 'It shouldn't happen, but it isn't your fault.'

'No,' he said. 'It shouldn't happen.'

He went and stooped over her. She pushed the blankets down and pulled him close against the soft curves of her tiny breasts, kissing him passionately.

'I love you, Jack,' she said desperately. 'I love you.'

He held her in a strong grip for a long moment, then broke away.

At the door he said, 'You'll be gone when I get back?'

'Yes.'

'I'm glad,' he said. 'I wouldn't want to return to you from . . . that.'

She heard the door open.

Out of the darkness his voice drifted.

'Sally.'

'Yes?'

'I love you.'

The door closed behind him.

6

Morning.

First light.

Wilf Routledge shut his eyes against it, but there was no keeping it out. He tried to stand up, but failed. His legs were like bits of four-by-two, limp and strengthless.

Not fear. Fear had gone with the whisky, washed away with the whisky. But his strength too, that had been washed away.

He felt sick. He struggled against it, didn't want to be sick, didn't want to stain his uniform.

Then he thought of the stains it would bear shortly and he let out a wild laugh. Curiously, that seemed to do the trick. His nausea retreated.

A man looking down at him, an officer, with a red-cap thug at either shoulder.

'I'm Captain Denial, Assistant Provost Marshal. Can you stand?'

He returned blank for blank.

'Do you hear me? Can you stand?'

Blank.

The captain said, 'Pick him up. Carry him out.'

'What'll we do with him out there, sir?'

'See if the air helps. If not, take a chair.'

The air cold on his face, cold as if it had come blowing down off Bleak Fell into Outerdale. He drank deep. It gave him strength, strength where it was needed, in the spirit. But not to his legs.

'Bring a chair.'

They brought a rickety upright kitchen chair. They sat him on it and bound him to it. His head was lolling. He tried to keep it upright.

'Captain,' he croaked.

'Yes?'

'Captain.'

'Yes?'

'Tell Josh . . .'

'Yes . . .'

'Tell Josh . . . I'm sorry . . . I'm sorry . . .'

The head slumped to hide the fast-flowing tears.

It rose once more just before the fusillade. The weeping eyes looked over the blurred line of men to the rising sun beyond. Looked into it without blinking.

Surely . . . yes! Surely that was Bleak Fell he could see! The familiar ridge, the peaks and hollows picked out by the rising sun . . .

Morning.

First light.

Last light.

PART THREE
━━━◆━━━
RETURNS, DEPARTURES

In Germany the potato crop failed in 1916 and in the hard winter that followed, the turnips used for cattle fodder became a staple of the common man's diet.

At the Front, the German Army straightened its line between the Scarpe and the Aisne by withdrawing to the strongly fortified Siegfried-Stellung, *known to the Allies as the Hindenburg Line. The area of withdrawal was devastated. This, with the ground fought over during the Somme campaign, left a tract of ravaged land up to twenty miles deep behind the Allied line – the Desolation.*

The British Army meanwhile had bolstered its claim to being a democratic institution by executing two of its officers, one in December 1916, the other in January 1917.

1

'*Awake! Awake! and greet the happy morn!*'

Lothar von Seeberg lay in his hospital bed and dreamt that he was a boy again, wrapped in an eiderdown in his own room at Schloss Seeberg in the Harz mountains, waking up on Christmas Day to the joyous carolling of his brother, Willi.

Suddenly the eiderdown was dragged off him. He opened his eyes to see and his mouth to curse the unmannerly nurse who had so shattered his dream. And in the twinkling of an eye, almost swooned with delight as he reversed dream and reality.

Yesterday, only yesterday, it *had* been a hospital bed, in

the Military Infirmary in Frankfurt to which he'd been transferred from France after many months of relapse and recovery. But yesterday, only yesterday, his mother and sister-in-law had come to collect him and take him to Schloss Seeberg for Christmas.

But nothing had been said about Willi coming home too. Now here he was, incredibly dashing in his elegant Rittmeister's uniform, a smile on his lips as he sang the rousing old carol.

'Oh Willi, Willi!' said Lothar, putting his arms around his brother's neck and embracing him so tightly his wounds began to sting. 'I can't tell you how glad . . . oh Willi!'

Now followed a time of pure bliss, marred hardly at all by the absence of their father on state business, and only ending with the return of Willi to his regiment in Galicia early in January. By now Lothar was beginning to heal quickly and thoughts of his own return to active service bothered him. But with his recovered health came other, nearer problems.

In his invalid state it had been easy to think of Sylvie as a sister. She was in fact a distant cousin and had been a frequent visitor to the castle since childhood. To the Seeberg twins she had indeed seemed a kind of sister, though adult tongues had early forecast that she might one day make one of them a suitable bride. But Lothar was astounded when he came back from Heidelberg one summer to hear the news that Willi and Sylvie were betrothed. Only then did the truth of his own feelings burst in upon him. He returned almost instantly to Heidelberg, where he added to the list of his wild excesses fornication on a scale which brought forecasts of his early demise from exhaustion or disease.

The marriage took place just before the war broke out. Lothar had seen little of Sylvie since, but now, daily in her company, he found it each day more difficult to hide his feelings.

There was a spot by the lake, a deep pool in a narrow bay formed by the arms of two rocky crags. Here they had picnicked and swum as children, slipping into the water from a narrow ledge at the foot of the crags. Here they came one

February morning, shivering in the cold air and recalling those hot summer days.

'Oh, those were happy times, Lott,' sighed Sylvie.

'They'll come again,' Lothar assured her.

She turned to him, smiling. Her foot slipped on the damp rock and she cried out in alarm and fell forward against his body. Instinctively he put his arms around her and drew her tight. It might have passed for a fraternal embrace if he had kept it brief, but his muscles seemed to contract like springs. She turned her face up to his and for one swooning moment he thought she was offering her lips to be kissed. Then he saw the bewilderment of her expression and realized that his groin was pressed so hard against her abdomen that his excitement was unmissable.

He broke away instantly.

'I'm sorry,' he said. 'I haven't . . . I'm not . . .'

'Poor Lott,' she said gently, putting her hand on his arm. And then with a little smile she added, 'We must find you a . . . friend.'

She thought it was just a generalized lust! A soldier's healing body, too long deprived of female company! Or perhaps, Lothar saw with sudden insight, this was what she knew it was best for them both to pretend. Thereafter he took great care to avoid over-closeness, and so did she, and this was a little barrier between them.

A short time later, he travelled to Blankenburg for an Army medical board. There he was certified fit to return to duty, but not directly. First he was to go to Berlin to be invested with his Iron Cross. Such a presentation was normally done by some handy general at Headquarters behind the front line, but Lothar could see his father's hand in this. It was a heavensent opportunity for the Count to show the Berlin establishment that though his younger son might be a madcap eccentric, with anti-social ideas, when it came down to it he was also a true German hero.

Both amused and annoyed, Lothar returned to Seeberg eager to share the joke and his indignation with Sylvie.

But as he entered the Schloss he felt a heaviness, drank in a silence which filled him with foreboding and blanked every other thought out of his mind. The old head steward was not in his place to receive his hat and cape. A maid came out of his mother's drawing-room, saw him, cried out and fled, sobbing. He pushed the door open and went in. His mother was seated on a tall, uncomfortable ornately-carved chair in dark oak against which her white face lay like a paper mask.

She looked at him. Something like hope flickered in her eyes, but died almost before it showed. She closed them and tried to speak. No words came, only a piping noise like a distant bird.

But no words were needed. Lothar knew without benefit of words what was being said.

Willi was dead.

It had been an almost spent bullet, fired probably aimlessly, probably a mile away, dropping out of the sky to where Willi was standing at his ease, chatting with friends, and with just sufficient force to pierce the skull to the brain.

Lothar raged against the futility of it, the absurdity of it, the bloody outrageous criminality of it; he raged against generals, against politicians, against God; he went to see Sylvie and found her being tended by a doctor. She was in a state of whitefaced, dumb shock. He had seen men like this at the front. She had no powers of response either to his words or his touch or the sight of him. He left her and went out and rode one of the poor nags in the stable to the point of exhaustion.

When he returned to the castle, he went straight to his bedroom with a bottle of brandy and set about drinking himself insensible.

It proved impossible. At midnight he was still awake, still sober. None the less he stripped his clothes off and got into bed, trying to close his brain like his eyes to the images which beset it.

Eventually he fell into a light, troubled sleep and when his bedroom door opened an hour or so later, he awoke instantly.

It was too dark to make out anything but a pale shape moving slowly across the room towards him, but he knew it was Sylvie. His mind and body were too full of the sense of her to be mistaken. She reached the bedside and he said, 'Sylvie', as she pulled back the eiderdown. He could see her face now in the thin moonlight filtering through the ill-drawn curtains. It was pale and set and trance-like. And he could see more than her face. Her nightgown was unbuttoned to the waist and the soft curves of her small round breasts gleamed white around the deep-flushed aureoles of her hard and pointed nipples.

'Sylvie, what do you want?' he asked in a voice edgy with panic.

She lay beside him, her hand reaching out to clasp his naked body. He felt her breasts soft against his chest.

'Please,' she said. 'Oh please . . .'

Her voice was unreal, strained with the desperation of a phantom striving to be heard at a seance. Her lips touched his neck, his cheek, his mouth. They were cold but urgent. And still they murmured.

'Oh please . . . oh please . . . Willi! Willi!'

His body was aflame. His mind screamed out in protest, but his flesh shuddered and ached and reached out. His mind made one last effort, flashed one last picture of where he was and who he was and what he was doing. There was a moment of complete and utter horror, an eyelid flicker in the blind face of desire, but what he saw in that flash was enough to set him leaping from the bed. Gathering the woman in his arms, he strode rapidly from his room and down the long corridor to her bedroom door. She hung limply, offering no resistance, no inducement. He laid her gently on her bed.

She looked up at him with the mildness of despair and said, 'Lott.'

He left. The brandy was finished. He dressed and went downstairs and fetched another bottle from the cellar. This one worked, and he was still half drunk when an excited servant burst in at ten the next morning to tell him that Frau Sylvie had disappeared.

The search did not take long. He led them without thought down to the lake, to the deep pool beneath the crags, and stood in a silence as still as the very rocks as they lifted the slight pale figure from the water.

He felt nothing. He could neither give nor take comfort. He carried out the duties his place in the home demanded, made arrangements, received the coffin with Willi's body in it at the railway station, saw it laid in its place alongside Sylvie's in the castle chapel, and felt nothing.

But there were feelings deep inside which he dared not trust.

When the Count arrived from Berlin for what was going to be the most magnificent funeral the district had ever seen, he was met at the station by his son in his artillery sergeant's uniform with a small travelling bag at his feet.

'What are you doing?' demanded the Count.

'I'm rejoining my unit,' said Lothar.

'But the funeral . . . your medal . . . the Kaiser himself is going to . . .'

'Tell the Kaiser his soldiers don't want medals, they want peace,' said Lothar contemptuously. 'As for funerals, Father, out there at the front, every second of every minute of every day there are funerals to attend. If they dug a grave big enough to bury all the dead, they'd have to excavate half of Europe. Don't you realize that? Don't you and those idiots back in Berlin even begin to realize that?'

He realized he was shouting and that he had become the focus of all attention on the platform.

He picked up his bag.

'Goodbye, Father,' he said quietly.

An hour later he was on a train puffing westwards across the chill dead face of Germany towards the dying agonies of France.

2

'Well, well, well!' said Sergeant Renton with an excess of cheerfulness. 'Look who's here!'

The truck had moved off, leaving its load of supplies, and standing by them a figure so still that he almost passed unnoticed in the evening gloom.

It was Josh Routledge.

His mental collapse had been so bad that even the chronically sceptical army doctors had been persuaded that here was a case beyond either m. and d. or the threat of a charge of malingering.

So Josh had got his wish and gone back home to England for Christmas.

They didn't keep him in hospital very long. Demand for beds was great; treatment for psychiatric disturbance was basic; while they would have been reluctant to send a gibbering idiot out into the world, a physically fit young man who could dress himself, feed himself and respond to simple commands and requests was clearly just occupying valuable space needed by the really sick.

The fact that he could not laugh, could not smile, never initiated or developed a conversation and spent much of the night awake and silently weeping did not occupy their concern even when it caught their attention. Home was the best place. A few weeks at home would be the perfect restorative.

Home had been a hell worse than that first day on the Somme.

His young brother and his sisters had greeted him with delight but even that had been muted by the atmosphere of grief and shock which lay around the old farmhouse.

His father had shown him a letter from the Infantry Records Office.

I am directed to inform you that a report has been received from the War Office that your son, Private Routledge, Wilfred, was sentenced by court martial to suffer death by being shot, and this sentence was duly executed on 12th December 1916.
I am sir,
Your obedient servant.

It was signed illegibly.
His father wanted to know what had happened.
Telling him was agony.
His mother only wanted to know if it was true. And when he told her it was, she had turned away with a pitiful cry of 'Why? Why Wilf?' which had torn his barely mended sense of self apart.

The trouble was there was nothing in him to resist the notion that it would have been better if he had died and Wilf had lived to bring home the news. Wilf; handsome, tall, merry, strong, laughing, lovable Wilf; the leader, the man of ideas, the man for all seasons and all sexes; Wilf the golden lad.

He forgot the morose, self-centred, backbiting, aggressive Wilf of the last half-year and thought only of his dearly beloved brother as he had always seemed, and he could not quarrel with those who reacted as if he had a case to answer in still being alive when all that talent and promise lay dead. At home, no such reproaches were spoken, but his mother's endless, unremitting grief and his father's long brooding silences sent him early to his bed or late out into the frosty darkness where the drystone walls ran like a sandbagged parapet down to the eerie light of the distant lake. And finally one day, his little sister Ruth, six years old, drawing her conclusions from what she had overheard and what she could see, asked him innocently why he had left Wilf all by himself in France to die.

The next day, he asked to go before a medical board and because they were very busy, and very patriotic, and concerned only with muscle and limbs and lungs, they avoided looking into those light blue eyes which were a window on to a landscape of despair and passed him fit to return to duty, to France and to the arms of his comrades.

'How're you doing, Josh?' went up the welcoming cry.
'How's all them lovely little pushers back home? I hope you gave 'em one for me!'
'Dump your gear here, Josh. Plenty of room. This isn't a bad billet, quite cushy in fact. Fancy a fag?'
There had been no formal agreement to make a fuss of Josh, but now spontaneous expressions of welcome and goodwill showered on him from all sides.
It was Renton's belief that the boy hardly noticed. He took his place in the platoon and did his work, and replied if spoken to; but there was something not there, some part of him was either dead or absent, and this slight figure who lived among them was as spectral as those ghosts which must crowd the air above no-man's land.
He expressed no interest in or curiosity about the Cumbrians' new situation. They had moved recently to the town of Arras which had been wrecked by shelling and deserted by nearly all its inhabitants long ago. But its cellars had been interlinked and joined up by tunnels to a huge honeycomb of caves just outside the town.
This electrically-lit underground complex was still being developed and Josh and his companions found themselves put to work unloading crates which seemed for the most part to contain either ammunition or tinned soup. These were then stored in newly prepared chambers under the eagle eye of a Supply Officer who carried a grand blueprint of the whole and was often heard expounding the dire consequences of putting a tin of Maconachies into the hands of a soldier expecting rifle bullets.
'I know which I'd bloody prefer,' muttered someone. The same wag, when an old regular complained that this wasn't fit work for a fighting man, said, 'You do my fighting, dad, and I'll stack your boxes.'
But Sergeant Renton put the damper on this momentary good humour by saying, 'You'll all get your share of fighting soon enough. What the fuck do you think this is all in aid of?'

Silence fell. The men all knew. There was no need to bring it closer by saying it.

Meanwhile, at much the same time about four miles to the south-west, another wounded soldier was being welcomed back.

'Sergeant, you're back! We thought you'd given us up!'

'Have a coffee, Sarge! Real German acorns, none of your French muck!'

The unaffected pleasure of the greetings brought Lothar close to those tears which he had thought were dried up for ever. He had found his unit in a wood near Neuville in support of the Siegfried Stellung. This was yet another allegedly impenetrable line of defence to which the Germans had recently withdrawn, leaving a wide tract of devastated land for the enemy to occupy.

'Sergeant Seeberg! Welcome back. Come to my quarters and I'll bring you up to date.'

He followed Dieter Loewenhardt into his billet. As soon as they were out of sight of the men, the captain turned and embraced him warmly.

'Welcome back, Lott. It's good to see you. I heard about your brother . . .'

'Yes,' said Lothar sharply. 'War's a great leveller.'

Loewenhardt said, 'I'm sorry. I was pleased about your medal, though. Congratulations.'

'Oh. They haven't told you, then? No, I suppose they would prefer to keep it quiet. I turned it down!'

'You're joking,' said Loewenhardt in amazement and alarm. 'For God's sake, why? You earned it.'

'By my courage, you mean? The only true act of courage left to any of us, Dieter, is to stop this carnage,' said Lothar with sudden passion.

'If you feel like that, why the devil have you come back here?'

Lothar said, 'I'm a soldier, I obey orders. Isn't that what we're paid to do? Besides, it would be easy back home to make a fuss and cause a scandal, but the Government would soon have found a way to silence me. Hospitalized with shell-shock. Or, at the best, taken into "protective custody" like so many others.'

'And you imagine the Army will let you speak subversion?' said Loewenhardt incredulously.

'The Army is more than the idiots who direct it,' said Lothar. 'The Army is these fine fellows in our section, and thousands like them. This war could be over sooner than you think and then the real battle begins.'

'The war over? What do you mean?'

Lothar laughed. 'Don't they let you hear news from abroad out here?'

'You mean America coming in? For God's sake, Lothar, that's defeatism!'

And Lother laughed again.

'Not America!' he said. '*Russia!* They're having a revolution there, Dieter, a *people's* revolution. That's where the hope lies, in *their* success and *their* example!'

Loewenhardt grasped his arm and looked closely into his face.

'Lott,' he said urgently. 'Why are you talking like this? I can respect a man's beliefs, even when I don't agree with them. But this sounds like mere wildness! Listen to yourself! Are you sure it's not just a bit of playacting, another scene in the famous Lothar von Seeberg melodrama?'

Lothar dragged his arm free. He felt a terrible and irresistible rage surge up inside him.

'Is that what you think of me?' he said bitterly. 'In that case perhaps you'd better start examining your own motives for being so familiar with an NCO. What do you like best? Me? Or my money and family influence? Perhaps we'd both better stick to our appointed roles in future. *Sir!*'

He gave a formal, very correct salute, turned and marched away.

It was curious. Later as he sat alone on his bed and analysed his anger it did not diminish one jot with the recognition that much of it was caused by the large amount of truth in what Loewenhardt had said.

Was there anything, had there *ever* been anything, deeper inside him than his own pain, his own uncertainties, his own sorrow? Hadn't everything he had ever said or done been simply an act of self-indulgent exhibitionism?

He thought, not for the first time, of suicide. Perhaps the thing to have done was to have followed Sylvie into that deep, drowning pool in the wintry lake.

Immediately the thought disgusted him. Again the posturing, again the self-dramatization! No, if he was after death, let it be some simple uncomplicated, anonymous kind of death. Stick his head up over a parapet in the front line! Become a statistic, one of hundreds of thousands!

That was the reality, the only reality to grasp at. Those deaths had happened, and they need not have happened. He knew no way to stop them, but there was in him, as there must surely be in any man, a sense of outrage at them, a genuine pain which linked itself surely and unbrokenly with the pain he felt at Willi's death. Sylvie was something else. Sylvie was a secret to be locked in the innermost chamber of his feelings. But this other deserved to be spoken out loud.

So he started to talk. He talked reasonably, dispassionately, about the war, its causes, its course, its casualties. He talked about the situation at home, the effects of the harsh winter, the Allied blockade, the failure of the potato crop. He talked about what was happening elsewhere in the world, about America's entry into the war and Russia's exit from it, and the reasons for these things. He spoke openly to all ranks of all units, to anyone who would listen.

Loewenhardt made no effort to remedy the coldness between them. Indeed he seemed deliberately to distance himself, leaving more and more of the everyday running of the battery to his under-officer, Lieutenant Bermann, a cold, unyielding young man with dead eyes and a grey complexion.

Bermann listened to what Lothar had to say, openly taking notes on occasion.

'You realize, Sergeant,' he said in his flat, grating voice, 'that at home there are men and women kept under protective custody orders for saying less than you. And others jailed for subversion for saying as much.'

'Then perhaps you should order me to be jailed or put in protective custody, sir,' said Lothar calmly.

'We are all in protective custody here, Sergeant,' said Bermann, his lips twitching in his version of a smile. 'And it will take a stronger man than I think you are to break out of it. I doubt if such a man exists on either side.'

'We must live in hope, sir,' said Lothar.

'Or die in it,' said Bermann. 'Carry on.'

It was as if he were addressing the war. And obediently the war carried on.

3

Sergeant-Major Maggs hated Australians. He was used to the common soldier bending in the gale of his command, and to the common officer cooperating in whatever he cared to do to the common soldier.

But Australians were different.

As soon as his cracked head was bound up after the riot in Barnecourt, he had headed out to the unit of the fat Australian corporal they had in custody.

To his amazement, the captain in charge, indistinguishable from his men by any badge of rank, had flatly refused to organize an identification parade. Further, when Maggs had by chance spotted the huge bulk of the sergeant who had assaulted him, the captain had asserted without turning a hair that he must be mistaken, for Sergeant Viney had been on duty all that night to his personal knowledge.

But at least Corporal Coleport was his.

It had given him great satisfaction to hear Denial refuse the Australian captain's request to have Coleport transferred into unit custody.

'The matter is too serious,' Denial had said. 'This man is charged with maliciously wounding one of my NCOs.'

'Charge him then.'

'As soon as the wounded man is recovered sufficiently to give evidence.'

Denial was a cold fish in a lot of ways, but at least he didn't stand any shit, thought Maggs. Meanwhile they had Coleport nice and handy for a dose of 'Mr Dunlop' whenever things got dull in the Cage. The Cage was the name given to the Military Police restraint compound in the buildings of a worked-out mine some three miles out of Barnecourt, while 'Mr Dunlop' was an unofficial rubber truncheon made out of tyre rubber, much favoured because it left fewer body marks.

Coleport seemed almost impervious to pain, however, but sat in his cell in the Cage, driving everyone mad with *The Man From Snowy River* and other similar recitations. He'd been there for nearly two months now. The trial had been set up once, but Corporal Parker had had a relapse on the day he was due to be discharged from hospital and was once more on the serious list.

It occurred to Maggs as he stood at the huge metal gate which blocked the entrance to the pit yard that Parker's death might be every bit as handy as Parker's evidence. These fucking Aussies had their own rules, in which serious matters like cowardice and detention weren't capital offences. But even in their ragbag of an army, they still topped murderers!

'Mounties coming, Sar'nt-Major,' called one of his men.

This was what Maggs was waiting for on this cold February day. 'Mounties' was the semi-mocking term given to cavalry troop sections who from time to time combed the Desolation in search of deserters and stragglers. It was not a duty many of the horsemen cared for. But it was a necessary task and offered a substitute for the real cavalry action this war seemed unlikely to require.

They came clattering out of the early twilight now. It was a poor bag, just two gaunt men, each sitting in front of a reluctant trooper. Maggs sneered in disapproval. If he was in charge, the miserable bastards would have been walking, or dragged along at the horses' tails if they couldn't walk. But that silly fucking cowboy in charge was soft as shit.

He snapped up a salute.

'' 'Evening, sir!' he roared.

'Good evening, Mr Maggs,' said Lieutenant Augustus Cowper, who liked this detail even less than Maggs imagined. He was a young man in his twenties and his delight at getting a commission in an albeit unfashionable cavalry regiment had long since been eroded by an awareness of the kind of war this really was.

'Is Mr Denial about?' he asked.

'At his quarters, I think, sir. Or you can catch him later up at the Château, sir. He's dining there tonight.'

'The Château' was the so-called Château d'Amblay, two miles to the north, where divisional HQ was housed.

'Thank you. Take care of these two, will you? Give them some food straightaway. I think the poor devils are starving. Corporal Ackerman, take over.'

'Sir,' said the corporal.

He watched the officer's horse out of earshot and added scornfully, 'Here's another poor devil who's starving! If we did it my way and just shot this scum on sight, I'd have been eating my supper by now.'

Maggs regarded him approvingly. He had a singsong Welsh accent, but he spoke some sense for all that.

'Open the gate!' he called.

As the new prisoners were being thrust inside a redcap came running across the yard.

'For fuck's sake!' cried Maggs. 'Don't run unless I tell you to run or there's a minnie on your tail.'

'Sorry,' gasped the MP. 'But I've just heard on the blower, sir. It's Corporal Parker. He's dead.'

'Dead? Are you sure?'

'Yes, sir. It was the hospital. They wanted us to get word to Mr Denial. A hæmorrhage or something, they said. That Aussie bastard's murdered him!'

Maggs absorbed the news. They could probably catch Denial before he went off to dinner. Or later at the Château. On the other hand, it was a pity to spoil his evening and the morning would do just as well.

He smiled in the sweetness of anticipation.

'You're right there, lad,' he said. 'That Aussie bastard has murdered him, right enough. Let's go and tell him, shall we? Let's go and tell him!'

At just about the same time as Maggs was deciding that Jack Denial could be left in ignorance till the morning, the APM, was hearing the news of Parker's death.

He'd called at the hospital to tell Sally Thornton that he doubted if he'd get back from HQ in time to see her that night.

She was sitting at the reception counter, bringing records up to date, and assured him that she had enough to do to keep her there till midnight. Then she told him about Parker, and all other thought went out of his mind.

'I'd better get out to the Compound,' he said.

'Why? We've rung through.'

'That's why,' he said laconically, and left.

She watched him limp away and wondered what kind of future there could be with this rigid, duty-bound man. They never talked of the future. Fear was her inhibition, but what was Denial's she could only guess.

Meanwhile Denial was bumping along Barnecourt's cobbles on the motorbike with sidecar which was his preferred method of transport. An APM had to be able to circumvent traffic jams if he was to clear them. Eventually he left the cobbles for the frozen ruts of the country road which were even worse, and it was a good ten minutes before he finally arrived at the Cage. Maggs, forewarned by the engine's rattle, came forward to meet him, a little out of breath.

''Evening, sir. Didn't expect you back this evening. Thought you were dining at HQ.'

'I was till the hospital contacted me. Corporal Parker's dead, I'm sorry to say. Had you heard?'

'Yes, sir,' said Maggs. 'Just now. Dreadful business.'

'It means we'll need to take special care of that Australian. Wheel him out, will you? I've got to apprise him of the new situation.'

'Sir,' said Maggs, 'I'll do it if you like, sir, if you want to get on to your dinner.'

'Thank you, Mr Maggs, but no. Just fetch him.'

'Sir,' said Maggs.

He went into the Cage and returned a short while after, looking concerned.

'Coleport's ill, I'm afraid, sir. Looks genuine too. Must have been something he ate.'

'Let me see,' said Denial impatiently.

He strode into the Cage followed by Maggs. Right at the back in the darkest corner, covered by a blanket, he found Corporal Coleport.

'Get out of the way, Mr Maggs!' Denial ordered the sergeant-major, who was stooping over the recumbent man, making cooing noises as though to a sick baby.

Maggs stood aside and Denial bent forward. The Australian's right eye flickered up. The left was too swollen to join in.

Through bleeding gums and crushed lips he whispered,

'So he spread them out when they left camp wherever they liked to go,
Till he grew aware of a Jackaroo with a station-hand in tow.'

'Stow that!' commanded Maggs. 'Officer present.'

Slowly Denial drew back the blanket. Gently he placed a hand on the man's chest. He gave a little groan of agony and went slack.

'Mr Maggs!' said Denial in a quietly terrible voice. 'Go and send for an ambulance!'

4

'Hey, Delaney!'

Patsy Delaney paused and turned towards the speaker, who was the same scruffy captain from Brisbane who had so offended Sergeant-Major Maggs.

'Yeah, Cap?'

'How's Viney? He snapped out of it yet?'

Delaney didn't ask what the captain meant. For more than a fortnight Viney had been plunged in a black and gloomy mood which had men walking on tiptoe and speaking in whispers around him.

Delaney said to the captain, 'Why don't you ask him?'

'Not me,' said the captain. 'I'm not going to *tell* him either.'

'Tell him what?'

'That redcap Coleport gutted. He's died. I've just heard. That's curtains for Blackie.'

'What do you mean?'

'Wake up, Patsy! He was a redcap and Blackie Coleport stuck a bayonet into him in front of half a dozen other redcaps. That's murder, and it doesn't matter which man's army you're in, they string you up for that. Even if he just got jailed, them redcaps would kick him to death. I hear he's been taken to hospital already. I can guess how that happened.'

'Bastards,' said Delaney.

'Yeah, it's a bastard. So you tell him, Patsy. He won't like it. You three jokers have been together a long time, haven't you? Some Holy fucking Trinity!'

It was a trinity which had formed at base during training.

There'd been no obvious reason for the trio to come together. Their backgrounds were very different. Viney's father was an itinerant stockman, Delaney was an Irish dockworker's son, while Coleport had revealed in his cups that his father was a Baptist minister in Alice Springs. It had seemed an act of nature that Delaney and Coleport, both notorious tough nuts, should gravitate towards Viney, who in an assembly of tough and violent men was accorded the ultimate compliment of never being called upon to prove his place in the pecking order. They had got through Gallipoli together, come to France together and so far survived the Western Front unscathed. Around Viney, you got a sense of immortality.

Now he had to be told that for Blackie, the immortal days were over.

Delaney found Viney lying on his bed, staring blankly at the ceiling. He told him what he had heard but there was no response. Delaney shut up and sat down in a rickety old chair and lit a cigarette. Thirty minutes later his feet were surrounded by dog-ends and his head wreathed in smoke. Viney had not once spoken or moved in this time; his eyes had scarcely blinked.

Delaney put another fag in his mouth and struck a match. It was nearly dark now and the flame picked out the big sergeant's profile like the marble effigy of some mailclad knight on a tomb.

The eyes closed, then opened again. Then he swung himself off the bed and on to his feet in one easy athletic movement. Delaney held the match an inch away from the unlit cigarette and waited.

'Let's go get him,' said Viney.

At the Red Cross Hospital in Barnecourt, Sally Thornton was sitting behind the reception desk, trying to bring some records up to date.

She'd been there earlier when the ambulance carrying Coleport had arrived. Jack Denial and Sergeant-Major Maggs had arrived simultaneously on the motorcycle combination. She had smiled in welcome, but Denial had been businesslike to the point of brusqueness. He wanted the man treated immediately and he didn't want him in a ward with anyone else.

'Then he'll have to go into one of the huts,' she told him sharply. 'We can't clear out a room in the main building, not even for the Assistant Provost Marshal!'

This at last had won a smile from his reluctant lips.

'I'm sorry,' he said. 'It's just that he's a prisoner. There'll be a constant guard. You wouldn't want MPs hanging around a general ward, would you?'

'A guard? He doesn't look as if he's going anywhere in that state. What happened anyway?'

The smile went.

'That's what I'm going to find out as soon as he can talk.'

It was easy to provide an empty hut. After the bloody excesses of the summer, winter had stifled the killing impulse to an occasional outburst, and the hospital's population had shrunk to numbers containable under the hotel roof.

Coleport had been found to be suffering from severe bruising, cracked ribs and torn muscles, and he had lost several teeth. When the doctors were finished with him, Denial had set two men to guard him and then gone into the hotel with the chief MO for a discussion which lasted about twenty minutes. During all this time, Sergeant-Major Maggs had stood stoically in the doorway.

Finally Denial had reappeared.

'We'll go and see our man now, I think, Mr Maggs,' he said loudly. Then, stooping to Sally, he had added in a lower, gentler voice, 'Our woman we hope to see later.'

'Later?' she said, not looking up. 'You mean tonight? Tomorrow? Next week? Next year?'

Her ill-tempered response was not due simply to his earlier brusqueness. They never talked of a future beyond their next meeting, and suddenly fatigue and irritation combined to make her long for plans outside the uncertainties of this war.

He was silent a moment, then said unemotionally, 'All of those. And half a century more, I hope.'

His lips brushed her cheek and then he was gone, heading towards the hut, limp hardly noticeable, with Sergeant-Major Maggs in tow.

For the next couple of minutes Sally sat in a delicious turmoil, staring at the records but seeing nothing except a future she had hitherto refused to contemplate.

She had just got herself in hand again and begun to work properly when the door opened once more to admit a pair of soldiers in Australian uniform. One was a lantern-jawed man who looked as if he'd seen everything and found it wanting. His eyes ran assessingly over her body as if her trim uniform didn't exist and he was noting every physical manifestation of her recent excitement. The other was a huge rock of a man with slabs of solid muscle moving beneath his tunic. There was nothing sexual in his gaze, but those eyes, green as emerald and just as hard, conveyed an impression of still menace which was far more disturbing.

But when he spoke his voice was low and respectful.

'Excuse me, ma'am,' he said, coming to attention before the desk. 'Looking for an Aussie corporal, Coleport's his name. He's under arrest. We're his escort.'

'Escort?' she said, puzzled. 'But he's not going anywhere tonight.'

'No, ma'am. Probably tomorrow, ma'am. We've been detailed to escort him back to his unit for trial when he's fit, so if you could point us where he is.'

It all sounded very reasonable, but Jack had said nothing of this, and there was something about this huge man which made her feel very uneasy.

She stood up and said, 'If you'd just wait here, Sergeant, I'll fetch someone to deal with this.'

As she moved from behind the desk, the big man stood aside politely. She gave him an instinctive smile which froze on her lips as his fingers gripped her forearm like a vice.

'Sister,' he said softly. 'Just take us to where Corporal Coleport is.'

There was no threat, no attempt at persuasion, but she felt a wave of pure terror sweep up through her body and she led them out of the main door as if her will had been invaded by a foreign power. That piece of herself which remained intact deep down was praying that Denial and Maggs would have completed their visit and left, but her heart sank at the sight of the motorcycle combination outside the hut.

Viney said, 'How many in there with him, Sister?'

'Four,' she said.

Viney frowned. For a second she felt hope. Perhaps these were worse odds than they were willing to accept. But the hope quickly died. This man did not deal in calculating odds. He dealt in removing obstacles.

'Right, Sister,' he said. 'In you go.'

He rapped on the hut door with a German officer's Luger pistol, opened it, and thrust her inside.

Jack Denial was studying the patient's temperature chart. In the furthermost corner of the room, Sergeant-Major Maggs was dressing down the two redcap guards whom he had caught playing brag.

Denial's eyes opened wide as he saw Sally. She gasped. 'Oh Jack!' and flew into his arms, effectively immobilizing him. Maggs, whose back was to the door, continued his scolding and so strong was the power of discipline that for a second the two redcaps hesitated to interrupt him. Then they made a sudden start for their weapons and Maggs spun round in alarm.

'Steady, boys,' said Viney, holding the Luger like a toy in his huge hand. The redcaps went quite still, like children playing statues.

In the bed, the plastered and bandaged figure of Blackie Coleport said, 'G'day to you, mates. Patsy, you mean Irish bastard, have you brought me no grapes?'

Denial held the trembling nurse tight in his arms and said quietly, 'What's the meaning of this, Sergeant?'

'It's that bastard who scarred me!' said Maggs urgently.

'Shut up,' said Viney and silence fell.

Delaney, moving with great care, now removed Denial's revolver and put the other MPs' arms out of reach.

'You fit, Blackie?' he asked, giving his friend his full attention for the first time.

The fat corporal rolled out of bed and stood upright, stripping the linen nightgown from his frame with exaggerated but not ungenuine groans. His ribs were tightly strapped and dressings were taped at regular intervals on his arms and legs, and the visible skin was blue and mustard yellow with bruising.

'Jeeze!' said Delaney.

'Resisting arrest,' said Coleport. 'Twice. Here, give's hold of that.'

He took Denial's revolver from Delaney's hand and limped towards Maggs.

'Fancy a bit of resistance now, sport?' he said.

Maggs from his great height regarded him with hate and contempt, but the gun was poking at his belt buckle and he did not speak.

'Get dressed, Blackie,' said Viney. 'Time to go.'

'OK,' said Coleport.

He turned slowly away, then swung back in a movement of tremendous violence, bringing the revolver up hard between Maggs's legs. The sergeant-major let out a high-pitched, almost soundless shriek and doubled up. Coleport wrapped his arms around his ribs and grimaced, saying, 'Christ, that hurt! But you saw him resist, didn't you, friends? You saw him resist?'

The two redcaps, whitefaced now, nodded their agreement. Viney gestured at them with his pistol.

'Take his tunic and webbing off and tie him up.'

Delaney had brought a small pack with him in which were several lengths of rope. As the two redcaps bound the sergeant-major, Denial said quietly, 'Sergeant, I'm sorry, I don't know your name.'

'Just call me Viney,' said the big man. 'Forget the sergeant. I'm a civilian now.'

'Just like that?'

'Why not? I volunteered in. Now I'm volunteering out.'

'You realize what you are doing could cost you your life?'

'Not in my army, sport,' said Viney. 'We ain't so liberal with death penalties as you Poms. Man in your job should know his law.'

To his mild surprise, Viney saw the gibe actually stung. He also saw that there was something more than mere terror in the way the girl clung to this officer.

'You two acquainted already?' he said.

There was no reply and the sergeant laughed and said, 'Patsy, toss us some rope.'

He proffered a couple of lengths to Denial and said, 'Tie her up.'

Denial ignored the rope and spoke softly to the girl.

Viney said impatiently, 'Patsy'll do it if you don't.'

Denial looked at Delaney who, having supervised the binding of Maggs and one of the redcaps, was tying up the third man himself, pulling on the rope till it cut cruelly deep into the flesh. Coleport was meanwhile slowly and painfully getting into his clothes.

Denial spoke urgently to the girl.

'Lie on the bed,' he said. 'It's better this way. You'll be all right, trust me.'

Speechless with fright, the girl obeyed. Denial took the rope and began to tie her.

'Now you,' said Viney. 'Get on the bed with her if you like.'

It was a generously intended suggestion. Denial looked at

the narrow bed and shook his head. Instead he sat on the floor and leaned back against the bed-frame.

While Delaney tied him up, Viney tested the girl's bonds.

'Nice job,' he said.

'Yes, ain't she?' said Coleport, lecherous still despite his condition.

Denial, though he had decided that there was little sexual danger to Sally in this expedition, spoke quickly to snuff out this flicker of eroticism.

'Where will you go?' he asked.

Delaney jerked his arms behind his back with a violent pull on the rope and laughed, showing large yellow teeth.

'He wants a map, Viney. Real joker!'

Viney said, 'Marseilles, mebbe. Catch a boat home.'

Denial's eyes flickered to Coleport and then he said rapidly, *'Et votr' ami, croyez-vous qu'il peut voyager si loin?'*

Viney smiled. The cunning bastard was checking how much Frog he spoke! The answer was practically none. He guessed the trend of the officer's thoughts. Not speaking the lingo and with a sick man on their hands, they wouldn't go any further than they needed to. Well, he was right, but much good would it do him!

He said, 'Finished, Patsy? Now gag them.'

Delaney set to work with lengths of bandage and balls of cotton wool taken from a store-cupboard in the room while Viney assisted Coleport in getting into his boots.

Denial said, 'Do you have to gag the girl? She'll say nothing.'

A wad of cotton wool was thrust into his mouth.

Viney replied, 'Sorry, Sister. But you'll find your voice quick enough when we've gone.'

As if to prove him right, Sally found it now.

'Oh Jack,' she said. 'Jack. Please. I love you . . .'

'Touching,' said Delaney, cutting off the words with a lump of cotton wool and winding a bandage round the pale cheeks.

Viney meanwhile helped Coleport into one of the MP's tunics and set the red-covered cap on his head.

'Suits you,' he said as he struggled into Maggs's tunic. The sergeant-major was a big man, but the seams were stretched to tearing point by the time Viney got into it. Delaney followed suit with the third uniform.

'There we are,' said Viney. 'It'll take a brave cunt to stop us now. You OK, Blackie? How're the legs?'

For answer the injured corporal, clearly exhausted by the effort of getting dressed, gasped,

'So it's shift, boys, shift, for there isn't the slightest doubt
We've got to make a shift to the stations further out!'

'You'll do,' said Viney, putting his huge arm round the man's corpulent body. 'Come on, Patsy.'

The three men left. A few seconds later Denial heard the roar of his own motorbike-combination and then the putter of the engine faded into the night.

Above him on the bed he also heard Sally struggling against her bonds and he willed her to lie still. In these circumstances the danger lay in overmuch activity and in panic. After a while to his relief she stopped moving and thereafter, apart from the occasional shift of position as one or other of the captives tried to obtain relief, they all lay still and waited to be discovered.

It was two hours before they heard the door open to admit the redcaps' relief guard.

'Jesus Christ,' exclaimed a voice, and next moment one of the two newcomers was trying to undo Denial's bonds while the other loosened his gag.

He spat out the cotton wool and croaked, 'On the bed. The nurse! See to the nurse!'

The man who'd removed his gag stood up and turned his attention to the bed.

Denial heard him gasp, 'Oh sweet Jesus!'

His hands came loose. He forced himself upright, ignoring the fiery agony as blood tried to re-enter constricted veins.

On the bed Sally, still bound but with her gag removed, lay with her eyes wide open, her mouth agape. He touched her head. The skin was cold, the head lolled to one side and he saw into the open mouth.

She had swallowed her tongue. She was quite, quite dead.

PART FOUR

◆

ARRAS

The Battle of Arras began on Easter Monday, April 9th, 1917. It continued till the end of May, solely to distract German attention and forces from the French Front on the Aisne, where a disastrous offensive had resulted in such huge losses that many French units mutinied and refused to return to the front line. Pétain took over from Nivelle and restored order and morale. Over 400 death sentences were passed and 55 executions carried out.

Allied casualties at Arras were 158,660. It was on the whole adjudged to be a successful operation.

1

On Easter Monday Josh Routledge came out of the caves of Arras. The April air was white, not with the seasonal blossom which the unkind winds too quickly tore off the orchard trees in Outerdale, but with snow.

Snow, blossom, it mattered not which. He was as indifferent to its presence as he was to the whining bullets and glowing shrapnel which ripped through the white pall the following day when the attack began.

He rose with his fellows on command, went over the top, advanced till told to halt, then crouched in whatever hole they had reached with as much indifference to its safety as he had shown to the peril above. Up and forward, down and consolidate, up and forward . . .

After a week they came out into reserve. Most of the men were excited by confirmation of the news, already rumoured at Easter, that America had at last entered the war. The stories grew in number and absurdity. A thousand ships were on their way packed with guns, plans and a million men. Battalions of Red Indians had been formed, warriors of incredible stealth and cunning! Western heroes like Buffalo Bill were training crackshots capable of putting a bullet in a Hun's eye at five hundred yards from the back of a galloping horse!

The futility of galloping horses was brought home to the men as they returned to the front line. Some fool had been hopeful that this push would lead to something like the rapid mobile kind of warfare which all the manuals described. The reserve lines were full of cavalry and, desperate for glory, someone had even brought a regiment within range of the German guns.

The soldiers returning to the front passed their ghastly memorial. Mounds of dead horses, unburiable because of the hard chalky ground, lay rotting and stinking on every side. The weather had changed, sunshine pushing aside the snow, and the dead beasts, ripped and broken by shell and shrapnel, were given new life by a pulsating skin of feeding flies.

Josh, untouched for a long time by human suffering and degradation, felt the sight of these unthinking, unquestioning animals burn into his soul almost to the point where pain might once more begin.

Us too, he thought. Dumb beasts for slaughter. Us too.

But then the sight was behind him and he was once more in the familiar hell of the battle line where trenches, shell-holes, explosions, screams and the contorted bodies of the maimed, the dead and the dying were to his senses what fell and lake and the screeching of ravens and the bleating of sheep had once been.

Sergeant Renton noticed Josh's condition and was sufficiently concerned to remark on it to Lieutenant Maiden. But the officer who needed all his slender resources to hold his own tattered nerves together answered, 'He's still functioning, isn't he? Still taking orders? For God's sake, Sergeant, we're all in a bad way! Kick him up the backside if he gives any bother. I'd have thought his bloody brother had given us all the trouble we needed from one family!'

Silently, Renton turned away and went down the trench. Maiden was clearly no fucking use to man or beast, he thought. Trouble was, when he cracked he could be responsible for the deaths of many others, and all that'd probably happen to him was being shipped off home to rest.

He saw Josh lying in a shallow dug-out in the side of the trench. From the earth above his head protruded a German boot. Renton stopped to speak to the boy but saw that he was asleep. In sleep, the distant, haunted look had faded from his face to be replaced not by peace but the inconsolable sadness of an unhappy child and from each closed eye ran a round clear tear, crystalline in perfection; last fragile symbols of purity and innocence in a world of the broken, the befouled and the betrayed.

Sergeant Renton took a deep breath and passed on.

2

'Sergeant von Seeberg,' said Lieutenant Bermann. 'A job for you.'

'If it's in my power, I am yours to command, sir,' said Lothar.

'If, or not,' said Bermann softly. 'Our front line is being firmed up along its present position. The English will be trying to press on with their advance, so we have to give our men every ounce of artillery support. Pin-point accuracy is what we need. You are to set up a forward observation post *here*.'

He stabbed his finger at the map he held.

The enemy's expected spring attack had been going on for almost a month now. To the north, Canadian units had taken the heights of Vimy, a victory balanced in staff eyes by the tremendous damage being inflicted on the French above the Aisne away to the south. To the man in the line, however, north and south didn't matter, not above a mile anyway. What mattered was the enemy in front of you which in this case was the English. They'd been advancing steadily and seemed determined to keep on advancing even though it made more sense for them to consolidate. At command level, this was recognized as a device to take pressure off the French on the *Chemin des Dames*. At front line level it was recognized as a bloody nuisance.

At the moment along the half-mile of front covered by Lothar's regiment, no-man's land consisted of a long mound, like a basking whale. Wooded and with a concrete blockhouse, it had been a defence strongpoint till a ferocious British bombardment had razed the trees and shattered the concrete. The Germans had retreated and put down an equally fierce barrage, forcing the enemy advance to halt at the foot of the ridge.

Those on either side were not unhappy to have the rare luxury of a sheltering land mass between them and the enemy. But such a key feature could not be allowed to remain unoccupied for long. And this was where Bermann's finger was stabbing.

Soldiers give their landscape familiar nicknames. Some wag had called this long ridge the Brocken after the famous haunted peak in the Harz mountains in whose foothills Lothar had grown up. Suddenly he laughed.

'Someone must wish me ill, I think.'

'What do you mean?' said Bermann, oddly irritated.

'Only that the witches gather on the Brocken to worship Lucifer on Walpurgisnacht. That's April 30th. Tonight! Just a joke.'

'People can get tired of your kind of jokes, Sergeant,' said Bermann sourly. 'You'll go at midnight. There'll be two signallers with you to lay the telephone wire.'

'One will be enough,' said Lothar. 'Less chance of making a noise and disturbing those witches.'

Outside the lieutenant's dug-out, it occurred to Lothar that perhaps Bermann's reaction to his joke had been because there was too much truth in it. It *was* a dangerous job and there were plenty of precedents for getting rid of troublesome soldiers by putting them in exposed positions.

On the other hand, when he reached the crest, there was no doubt it was a militarily justified risk. The battery hurled over a few big ones to keep the enemy's eyes and ears busy while he made himself comfortable in the ruins of the blockhouse, and the flash of their explosions plus the light from a fitful moon revealed a panoramic view of the British lines.

The signaller checked the telephone, then beat a grateful retreat. Lothar took a last glance westward through the spyhole he had created for his narrow telescope. The battery had fallen quiet again, not wanting to draw attention to the possibility of activity in no-man's land by a too enthusiastic bombardment. The moon suddenly swam brilliantly into a large, open patch of sky. It was far too light for there to be much danger of raiding parties advancing from either side and Lothar allowed himself to relax, rolling over on his back and staring up at the stars and the riding moon with unblinking eyes.

He let his thoughts drift, first eastwards across the chalk plain of the Champagne and the long German miles beyond to Schloss Seeberg. Nothing but pain lay there and he changed direction, heading to the west, crossing over the long line of British troops perhaps already massing for the next attack, over their artillery emplacements, over the desolation of a countryside which the war had already left pocked with shell-holes and wormeaten with trenches, then on to the spring-fresh, untouched landscape of Brittany and Normandy, across the Channel to Hun-hating England, across the Atlantic to America, preparing now to fling new supplies of eager young men into the boiling pot.

There was little for comfort in that direction either. He felt himself totally isolated, caught here at a point of balance between forces whose slightest shift could destroy him.

He dared not move. He lay quite still, staring at the moon and the stars as though he might at last absorb, or be absorbed into, their cold and cleansing light. It was weeks since he had felt so much at peace.

Amazingly, and if he were caught, criminally, he drifted into sleep.

In his sleep he dreamt he dived deep into the lake-pool near the castle. Deep, too deep. He reached the bottom and the drifting weeds caught at his limbs and coiled around them and held him there to drown. But he knew they were not really weeds but the tangles of Sylvie's hair.

With an effort of will he broke loose and forced himself to the surface of waking.

It was dawn. There was sunlight already on the eastern slope of the Brocken. He rolled over on his belly, stiff with cold and immobility, and peered through his narrow telescope. The face of a soldier, pale and strained, leapt up the hill at him and he cried out in shock. Then he looked again traversing from left to right.

There was no mistaking. This was no isolated patrol he'd picked up. As he lay asleep the British forces opposite had come out of their trenches without the usual preliminary bombardment and were advancing steadily up the protecting slope towards him.

3

'All right lads, not long now,' said Sergeant Renton's voice, confident, comforting. 'Take your time, go at it steady. Remember, they can't see us till we get to the crest, so no hurry. You know what day it is! May the first. May Day! Let's make the buggers do a right morris dance, shall we? What do you say, Josh? All right, lad?'

Josh Routledge did not reply.

He stood quietly by the fire-step looking up at the distant ridge, already rimmed red with the advance fire of the rising sun.

But though he said nothing, the sergeant's words set up resonances in his mind.

May Day . . .

Red sky in the morning, shepherd's warning . . .

May Day, holiday. Not holy day. Dancing, a queen, drinking, merriment. In Outerdale a special day . . . walk a mile out of the valley and men still working . . . but in Outerdale . . .

Red sky in the morning . . .

Mist. But above the mist, red sky . . . beyond the moon, red sky . . . at the height of noon, red sky . . . blood red, flame red, shame red . . . always warning never delight . . .

Tea sweet with rum . . . everything clear beneath the red sky . . . the pool of urine at his feet where he'd pee'd a minute or a year ago . . . a rusting tin of Maconachies . . . an insect licking it clean . . . his silent companions looking inwards . . .

Red sky inside . . . the shepherd's died . . .

Words . . .

Renton's . . . confident . . . reassuring . . .

Maiden's . . . high-pitched . . . blustering . . .

Sounds . . .

No guns yet . . . no warning . . . speed, surprise, up the hill and catch them sleeping . . .

But whistles now . . . distant, mist-muted, drifting up to the fells from the village football field on a still winter's day . . .

Over the plonk . . . the hardest part . . . the burden heavy . . . a broken-legged ewe carried down from the Crags . . . his father's knife swift . . . red sky . . .

Uphill . . . always uphill . . . never a summit . . . never a downward plunge . . . riding the scree as the storm breaks around . . .

Now the guns . . . ours, theirs . . . indifferent shells . . . no distinction . . . the earth's May Day dance . . . red sky at our feet . . .

Voices beside . . . cheerful, reassuring . . . Wilf in the fell race . . . grasping my arm, clapping my back, then going on to win . . .

Wilf!

No. Renton. Immortal . . . enduring . . . indestructible . . .

Renton!
Gone in a puff . . . gone in a flash . . . part of the red sky.
Almost the top . . . just once more the summit . . . look down from a peak . . . below, field and fell . . . below, Outerdale . . .
Never to reach it!
A ring of volcanoes . . . earth, stone and fire spiralling upwards . . . the summit crumbling . . . sliding towards him . . . not reach the summit but be reached by the summit!
Noise, blast, heat, silence.
A tarn of silence scooped just for him . . . lie in it, float in it, drown in it . . .
But not alone . . . down earthy sides a body slithering . . . Lieutenant Maiden . . .
Mouth working, tongue wagging . . . but still the silence.
Hand pointing . . . mouth gaping . . . but still the silence.
Gun pointing . . . shaking but pointing . . . commanding . . .
Leave the silence.
In the pistol is silence forever.
In the pistol . . . Wilf? In the pistol . . . Outerdale?
Perhaps not.
But in the pistol, silence. For ever.
Sit in the earth and hope for the pistol.
Hope for the pistol. Be part of the red sky.

4

His conditioning must have gone deeper than he believed possible. For a few minutes after his initial shock, Lothar von Seeberg became the perfect, efficient military machine.

Cranking his telephone, he passed on news of the attack in terse sharp phrases, giving distances, dispositions and strengths. The response was rapid. Within the space of a minute the shells of his own guns were being hurled over his head and dropping on to the slope below. They would slow but not stop the advance and it wouldn't be long before his own position was overrun. Mechanically he continued to give reports on the enemy's progress and the accuracy of the artillery range.

The expected response from the British was not slow in coming and as their belated bombardment began, it struck Lothar that for the moment the crest of the Brocken was the only area within a quarter of a mile in either direction which was not being shelled. It was a very temporary safety. The advancing soldiers and his own shells would soon be arriving here and he could expect no mercy from either. It looked as if the authorities would very soon have their von Seeberg problem solved for them. He only wished that they could have known how indifferent he was to the imminence of death.

A voice was speaking in his ear. It came from the telephone and it was calling his name.

'Lothar? Hello, hello! Are you there, Lothar?'

It was Loewenhardt's voice.

'How pleasant to hear from you again, Dieter! I thought Lieutenant Bermann was doing all your talking nowadays!'

'Lothar, what's the situation up there?'

'Well, we're giving them a nasty time,' said Lothar. 'But they're still coming in steady bursts. They've learned a lot since the Somme, it seems, I estimate they should be up here in ten to twenty minutes.'

'Lothar, you've got to get down from there.'

'That may be difficult,' he answered almost gaily. 'Unless you can persuade the British to stop shelling. In any case, I quite like it here. You could see the peak of the Brocken from the old tower at Schloss Seeberg, you know. I always wanted to spend Walpurgisnacht up there to see if the witches really danced with the Devil. Well, they do, Dieter, believe me. They do.'

He was sounding light-headed, almost hysterical, he knew. But a man about to die, *eager* to die, was entitled to a little hysteria.

There was a long silence at the other end and he began to suspect the cable had at last been severed.

'Hello? Hello?' he said, realizing self-satirically that he was disappointed at losing a sympathetic audience for his last words. 'Dieter? Are you there?'

The reply came in a rush, as though the words had forced themselves out.

'Lothar, the ridge is mined. It'll be blown in five minutes' time. Get down off there now!'

The line went dead. Lothar let the receiver drop. It was useless. No one was going to talk to him on that any more.

He was astounded. How very curious. Despite all he knew of them, despite the warnings he had received from Dieter, despite Bermann's note-taking, despite his own crazy desire to push them to the breaking-point, despite *everything*, it had never occurred to him that they would actually conspire to kill him. Expose him to danger perhaps. But not stake him out to die!

But it was such an obvious stroke! Instead of a living nuisance, a class traitor, a possible focus within the army of growing discontent with the war, they would have a dead hero. Lothar von Seeberg, true to the great traditions of his class and family, remained at his post in the face of the advancing enemy!

A few moments earlier, he'd been ready to die. But that had been his choice. He felt in no mood to oblige them by accepting *their* choice.

He rose to his knees and slung his rifle over his back, his mind wrestling with the problem of what to do next. An enemy on both sides and a mine beneath his feet – there was nothing in Marxist dialectic to deal with this situation! But movement was of the essence, and the quicker the better. He pushed himself to a crouching position, but that was as far as he'd got when a huge explosion flung him to the ground again.

So much for Dieter's timing! he thought madly as the earth rippled beneath him and the ruined blockhouse rocked and started to make its own way down the western slope of the Brocken independently. The previous shell-blast must already have half-deafened him and now the process was completed so that he found himself sliding slowly towards the enemy on a toboggan of concrete in total silence.

The slide seemed to last a long time. Then it was over as if someone had switched off the sun.

He awoke, perhaps seconds, perhaps hours later. He was in what seemed like a cave with a small triangle of light at his feet. Carefully he felt himself all over. There were many areas of pain, but none of agony. His nose was bleeding. The blood was still hot and sticky, so probably that meant it had been seconds rather than hours. He began to slide towards the light and immediately found himself stuck. It took several moments of fighting an old terror at the thought of being trapped in an enclosed space to realize that it was his rifle which had become wedged in the opening. He managed to draw his bayonet from its scabbard and sawed through the rifle-sling across his chest. Then he resumed his wriggling once more and a little later was out in the open air, if you could call this dust and smoke-filled atmosphere air.

He had exchanged a small hole for a big one, he realized. The smaller one was formed by a delicate interdependence of bits of concrete which in any less subtle arrangement would probably have crushed him to death. The larger, he surmised, had been excavated by the mine. The sappers' plans for total destruction of the Brocken seemed to have been thwarted by a long fault in the rock which had channelled the blast out sideways rather than upwards. It was in a long, narrow and deep trench running along the probable line of the fault that he found himself. And he wasn't alone.

About fifty yards away through a haze of dust-motes he could see a curious tableau which struck him instantly as having something Biblical in it, Abraham about to sacrifice Isaac perhaps. Two British soldiers were involved. One of them, an officer, was standing over the other, a private, who sat perfectly still in a rather childish position with one leg crooked under the other and his rifle resting across his lap. It struck Lothar that he couldn't hear what the officer was saying despite the visual evidence that he was shouting.

He put his little fingers into their respective ears and excavated a vast amount of dust. Slowly sound returned, albeit through a dull ringing which was the aftermath of the explosion.

Now he could hear the officer.

'You cowardly little bastard!' he was screaming. 'They should have shot you with your brother! I'm giving you an order. Are you going to obey it?'

The seated man looked up at him indifferently.

The officer drew his pistol and flourished it wildly. He was, Lothar guessed, completely hysterical, and this crazy scene had as much to do with his own reluctance to get out of the safety of this hole as with the boy's.

Still, it wasn't himself he was about to shoot.

And he *was* about to shoot, there seemed no doubt of it.

'For the last time, Routledge,' he said in a voice breaking like a pubertal boy's. 'I order you to get up and move forward. Now!'

The pistol was held six inches from the sitting man's head. The muzzle trembled violently, but at that range the acceptable margin of error was huge. The sitting man looked up still unconcerned. Then slowly his dust-caked face split open in a gentle smile.

'You're a traitor!' cried the officer, now reverted to full alto. 'A traitor and a coward. In the name of the King . . .'

Lothar slid his hand back into the hole, grasped his rifle butt, pulled it out, brought the weapon up to his shoulder and aimed it in a single smooth movement learned not from any army weapons instructor but from a childhood of handling hunting pieces in the forests of the von Seeberg estate. The pistol muzzle was touching the sitting soldier's head as Lothar squeezed the trigger. The officer jerked sideways and fell.

It was, Lothar realized, the first shot he had fired at a human being since the war began.

And then he mocked himself with the memory of all those thousands of high explosive shells he had been in part responsible for hurling into the enemy line during the past three years.

Slowly he advanced. The sitting soldier did not move. As Lothar got closer he saw that he was nothing more than a boy.

He greeted Lothar's appearance with as little interest as he had greeted the threat to his life.

Lothar examined the officer. He was dead. A young man also, not as young as the other, but young for ever now.

Lothar sighed, laid his rifle against the corpse and crouched beside the boy.

'*Hey, Knab*', *wie heisst du?*' he asked. 'What are you called, young man?'

At first he thought there was going to be no response, but at last the wide grey eyes, bloodshot with lack of sleep and excess of dust but still childish in their unblinking curiosity, turned to look at him. Another long moment passed. Then the boy nodded as if acknowledging that in whatever twilight land his mind inhabited, there was no incongruity in being addressed in English by a German soldier who had just saved his life.

'Josh,' he said.

'Well, Josh, I am Lothar. My friends call me Lott.'

There was no place for social niceties in the twilight land. The boy looked away indifferently.

'Why was this officer going to shoot you, Josh?' asked Lothar. 'Why?'

Again the long pause, as if the words had much more than a physical distance to carry.

The answer when it came was incomprehensible.

'So I could be with Wilf in the red sky.'

'And what will you do now, Josh? I think your friends are winning the battle.'

His hearing was getting better all the time and it seemed that the main noise of conflict had moved away a little. The comparative failure of the Brocken mine probably meant the British were over the ruined crest and advancing behind their barrage towards the German lines.

The boy stood up. All this time he had been clutching his rifle across his lap. Now he hurled it away with all his young strength.

'No more battles!' he cried.

He began to scramble out of the hole on the downhill side.

'Hey, you! Soldier! Josh!' Lothar called after him. 'They

will kill you, your own people, *nicht wahr*? Be careful, *Kind*! How can you hope to get home?'

The boy stopped and spun round. His eyes were blank no longer, but blazing with emotion.

'Home?' he screamed. 'Home? I can't go home, can't go home, can't go home . . .'

The words tailed off into sobs, then he turned once more and resumed his scrambling progress.

And what of me? thought Lothar. Can I go home? What awaits me there? Willi dead, Sylvie dead; the authorities preferring me dead! Well, why not? Let them have their dead hero . . . For a time.

Perhaps another time would come when at last the ordinary people drew back from this madness and began to count the cost. *Then* would be the time for the survivors to emerge, with their rage burning high, joining with their comrades in Russia and France and Britain to cleanse and purify this stinking sty of a civilization!

He found himself laughing at his own fervour. Playacting again! he mocked himself. Dieter was right. Uncertain of a role, his natural inclination was to turn to melodrama.

On the other hand, even forgetting the fine speeches and toning down the histrionics, it might be amusing to survive a little longer, and perhaps even help this youngster who was turning his back on war to survive also.

'Hey, Josh!' he called after the retreating figure. 'Wait a little! Wait for me!'

BOOK THE SECOND

Desolation

PART ONE
———◆———
THE WARREN

In June 1917 the offensive of the British Second Army at Messines was preceded by the explosion of nineteen mines placed in tunnels beneath the German lines. The noise of the explosion was heard in England.

During 1917, over a hundred military executions also took place. Little noise was heard of this in England, though in November, Bonar Law, the Leader of the House, told Parliament that in future dependants of men executed would be told they had died on service.

1

'That was how I met Josh, five days ago I think,' concluded Lothar. 'I had no plan. I threw away my Mauser, found the boy another Lee Enfield. We pretended I was his prisoner. I was content it should end so for me, but I knew there was no such good end for the boy.'

'What's that mean, sport?' growled Viney.

'For me, a prison camp till the war was over. But for him, whatever it is you do to your soldiers when they will not fight. Shoot them, isn't it? From what he has said, I think they have shot his brother already.'

There was a silence, the first which could be called sympathetic.

'So you decided to help the lad? Why, Fritz?'

'Why not? He had given up the war. I too wished to give up the war. It was not planned. Perhaps that is why we succeeded. No one challenged us that first day. There was much confusion. We hid in an old dug-out that night, then travelled on. After a while I realized we had gone so far, my own uniform would attract attention. So I took this tunic off a dead man. By now we were on the edge of this desolate land. We travelled on till today we were chased by the horsemen. Then we were rescued by you. For which I thank you.'

'Save your thanks, Fritz, till we decide what to do with you,' said Viney. 'All right, you bastards. There's no rank here, everyone gets a say, you know that. What should we do with 'em?'

The answer came swiftly from the scab-faced man called Foxy.

'Kill the fucking Hun for a start!' he yelled.

'I second that,' said the gingery Taylor who had previously drawn Viney's wrath. A mutter of agreement, not unanimous but not audibly opposed, ran round the chamber.

Viney turned towards Foxy. In that moment, watching that burly, hard-muscled figure standing there with the arrogant confidence of a pride-leader in every line and lineament of his body, Lothar knew they were safe. The invitation to democratic debate had not been given with any thought of bowing to majority opinion but simply to demonstrate yet again that what *he* thought *was* majority opinion.

'That's what you think, is it? Like I said before, Foxy, anyone want to kill Huns, that's where they're killing Huns – out there! Out there, you get paid to kill Huns, they give you medals for killing Huns. Did someone speak?'

No one spoke till the big paunchy man with the same twangy accent said, 'What say you, Viney?'

'You asking my opinion, Blackie?' said Viney with mock surprise. 'All right. Here it is. These lads here are on the run from the Army, it doesn't matter which Army, there's only one sodding huge Army in the whole world. And they look hungry. Christ, don't they look hungry! Let's feed them, that's the first move. That's what I say, Blackie.'

There was no visible sign of opposition, and Lothar and Josh were thrust to the floor in a corner and a little while later handed chunks of cheese and stale biscuits. The avidity with which they devoured them won some small applause and also a little sympathy from men who knew the feeling well.

'Poor buggers is starving,' said a big red-faced man.

'We've all been starving, Heppy,' said another, less sympathetic voice. 'What use are they going to be to us, Viney? A fucking Hun and a kid who looks like he's doolally tap. What use are they?'

Viney looked mildly at the questioner.

'You want us to start asking what use everyone here is, is that it?' he asked softly. 'I'll tell you what, Strother, there'd be precious few would pass *my* test!'

Strother, a squat ugly man, forced a smile in an effort to seem to be sharing a joke rather than folding under a threat, but the smile didn't reach his eyes.

There was no further discussion.

The food finished, the two newcomers whose status, despite Viney's pronouncement of welcome, was clearly still ambiguous, were taken out of the large chamber and thrust into a smaller dug-out which was sealed with a corrugated iron sheet. A foul-smelling blanket had been thrown in after them. Lothar wrapped it round Josh and when the boy put his arms around him, draped the loose end over his own body. Sleep came quickly for the youngster, but Lothar lay awake, bothered by the enclosed space despite his fatigue, and trying to guess what the future might hold for them in this place.

After perhaps half an hour he heard the corrugated sheet being drawn aside and suddenly a lantern was shining in his eyes and the blanket was pulled off his body.

It was Viney standing colossus-like astride them. He said in a low menacing voice, 'Fritz, I hope you've not been fucking this lad.'

Shielding his eyes against the light, Lothar said, 'He is unhappy, lost, demented even. They shot his brother, his own side. He has suffered more than he can bear and needs peace and comfort and someone to trust. I am his friend. He stays close to me waking and clings tight to me sleeping. I am his friend. That is all I am.'

Viney stood in thought for a while, then let the blanket fall.

'All right, Fritz,' he said. 'So you're his friend. Be a good friend. But I'll be watching you, *friend*, never doubt that. I'll be watching you.'

The light retreated, the corrugated sheet was replaced, the darkness was once more complete. Lothar slept.

2

There was a thunderous knocking at his door, a voice calling his name.

Jack Denial awoke.

It was half past midnight. He noted the hour without resentment or relief. To be awoken from nightmares to memories was merely to change one kind of pain for another.

He opened his door. Sergeant-Major Maggs said, 'Sorry to disturb you, sir, but I thought you might like to hear this straightaway. There's been some bother in the Desolation, some of them mounties killed. Mr Cowper missing. That corporal, Ackerman, came back alone, I thought you'd want to talk to him.'

'Give me two minutes, Mr Maggs,' said Denial.

Quickly he dressed. In appearance and action, he gave no sign of having changed at all since Sally Thornton's death. His reputation as a cold fish had been considerably enhanced by his refusal to accept an offer of leave or a transfer. Only Maggs from time to time had glimpsed the depth and violence of his grief, and that only in reflected hints, in an involuntary tightening of muscle, hooding of eye, lowering of voice, whenever Viney and the escaped Australians were mentioned.

They had found the motorbike combination abandoned on the edge of the Desolation where the terrain became impossible for anything but a tracked vehicle. And there the trail had ended. Denial knew better than most that it was likely to be a permanent end, for he guessed what his superiors and the General Staff refused to acknowledge, just how vast a population of deserters had taken refuge in the Desolation and elsewhere.

But a greater force even than duty now energized his whole being. Hate.

Corporal Ackerman's mad flight had carried him a good way towards the edge of the Desolation before the thickening light and his horse's fatigue had combined to negate his riding skill and good luck. A foreleg plunged through the spokes of a half-buried gun-carriage wheel had brought man and beast crashing down. The man rose, the beast couldn't. Pausing only to shoot the crippled animal, he had continued on foot, unmanned more than he would later care to admit by what had happened; and he had not stopped for rest till he found himself once more in countryside where the trees had leaves on them and the earth bore crops.

Eventually in the early hours of the morning he found himself in the presence of Jack Denial.

'These men who attacked you,' prompted Denial.

'Yes, a dozen of them, maybe more,' exaggerated Ackerman. 'They came swarming out of nowhere, out of the bowels of the earth, that's how it looked. Well, it did, really it did, sir.'

He defended his fanciful phrase with the indignation of one who had seen it and had been amazed.

'Describe them,' said Denial.

'Well, they was scruffy, dead scruffy,' said Ackerman with all an NCO's distaste for this condition. 'Sort of rugged and wild-looking, more like a bunch of pirates than anything. There was one big brute, nigh on seven foot tall he looked. Well, six and half anyway, and broad to match. He looked like he was in charge.'

'Did they say anything?'

'They was dead quiet at the start, but there was a bit of shouting at the finish. It was just noise, sir. *Get down!* and *Shoot!* that kind of thing, I couldn't make it out proper.'

'But the accent, you could make that out, perhaps?'

Ackerman looked puzzled.

'I mean, did they sound Scottish, say? Or Welsh?'

'Oh, I get you, sir. No, not Scottish. But come to think of

it, there was a little bit of a twang. More like them New Zealanders. Or them Aussies, perhaps.'

Denial nodded. Sergeant-Major Maggs opened his mouth to speak, but the captain closed it for him with a glance.

'And your comrades, they were all killed?'

Ackerman hesitated. Dawes he was sure of. The troopers on the far side of the ridge he hadn't been able to see, but the screams of that horse had been enough. They'd been down; there'd been rifle fire. They must have been dead. Certainly he'd be mad to give even a hint that he might have abandoned his fellow soldiers in dire straits.

But his hesitation had already done some damage.

'You don't seem certain, Corporal,' said Denial gently.

'Yes, sir. I mean, no, sir. What I mean is, the lads were definitely dead, sir,' said Ackerman trying to retrieve his position. 'But Mr Cowper, sir, they shot his horse from under him and I saw him go arse over tip, begging your pardon, whether he was actually hit himself, I couldn't say, but he seemed in a fair way to breaking his neck when he fell, and in any case, he was right in the middle of them bastards and they didn't look like they was going to let anyone get up and walk away.'

He spoke defiantly. Denial's unblinking gaze filled him with the unpleasant sensation that every scrap of truth and evasion and supposition was being separated inexorably into its proper pile. But all the APM said was, 'Thank you, Corporal Ackerman. You'd better go and get some rest.'

After the corporal had gone, Maggs said, 'I'll see if I can rustle up a mug of char, sir.'

He returned with two steaming mugs ten minutes later.

Denial sipped the scalding liquid carefully.

'I don't like the sound of this, sergeant-major,' he said after a while. 'This smacks of organization. Pirates, the corporal called them. He may be right. Take a scattering of fugitives, organize them, and suddenly what you have is a band of brigands, a focal point for disaffection.'

'Yes, sir,' said Sergeant-Major Maggs politely. His interest in the present case was much more personal, as he suspected was Denial's also.

'Sir,' he said impatiently. 'Could it have been him, do you think? Description fitted.'

There was no need to specify who was being referred to, and Denial did not pretend puzzlement.

'Australian, possibly. Big, possibly. Corporal Ackerman's ear is not testable and he was in a mood to exaggerate, I think,' mused Denial. 'No. It's all too vague, Mr Maggs. Put it out of your mind, right out of your mind.'

But Maggs had seen the tensing of the fingers and the rapid flicker of a pulse under the jaw-line, and he smiled as he saluted and took his leave.

3

Lothar awoke. Josh had turned in the night and pulled the blanket off him and he was cold. In addition his bruised and exhausted body had stiffened as he slept and his muscles ached as he tried to flex them.

But it was neither cold nor pain that had awoken him. It was a sense of something changing permanently in his life, an altered state deeper and more distant far than the lunatic events of the past few days. He rose and pushed aside the corrugated sheet shutting off the dug-out. He made as little noise as possible in order not to wake Josh, but he was less successful with the men posted to keep an eye on them.

As he crawled through the gap he'd created, a hand seized his hair and forced back his head and a voice borne on a breath like a sewer said hoarsely, 'Where'd you think you're going, Fritz? Going to slit a few throats and win yourself a medal?'

Though he'd heard it only once before, Lothar recognized it as belonging to the man called Strother who'd asked what use the newcomers were.

'I wish only to relieve myself,' he said.

'Well, get back in there and piss in your pocket, chum!' said Strother. 'Next time you sticks your nut out here, I'll crack it in half. *Comprenny?*'

A match spluttered in the dense gloom and a lantern bloomed into light.

'What's up here, Strother?' asked another familiar voice.

'Nothing I can't deal with, Hepworth,' said Strother. 'He says he wants a piss, I've told him to piss off back in there.'

He laughed harshly at his witticism, his ugly face twisted grotesquely in the lamplight.

The man with the lantern proved to be the red-faced man who'd remarked sympathetically how starved they were.

He now said, 'If the poor bugger wants a piss, let him go. It's bad enough round here without encouraging more stink.'

'It's your funeral, mate,' said Strother.

He released Lothar's hair and he crawled all the way out.

'Come on, Fritz,' said Hepworth. 'Follow me.'

He was a big thickset man, and Lothar noted once more how his complexion had retained its ruddiness despite this subterranean life which seemed to have turned most of the rest pale grey.

'Here we are,' said Hepworth. 'Latrines. Your lot did a right job when they dug this place out, I'll give 'em that.'

The remark confirmed what Lothar had already guessed. This was an old German HQ complex. No English or French would have delved so deeply or supported so strongly.

But it was badly dilapidated now. The planks which had lined the walls had in places disappeared either through decay or, more likely, by being ripped off to provide firewood. And the once efficient systems of lighting, ventilation and, by the vile stench which greeted him, sewage disposal had obviously broken down.

Hepworth caught his grimace of distaste and laughed sourly.

'Couple of days in the Warren and you'll not notice the stink, Fritz,' he said.

'The Warren?' said Lothar, his excellent English for once failing him.

'Aye. Place where rabbits live,' explained Hepworth. 'Though it's not fucking rabbits we share it with. Get out of it, you bugger!'

He swung his booted foot at a shadow which had detached itself from the wall. It was a rat, one of the biggest Lothar had ever seen. It skipped nimbly aside, paused a few feet away, its eyes gleaming in the dim lamplight as it seemed to debate whether to retreat or attack, then slipped away into the darkness.

Lothar swore in German and Hepworth said grimly, 'Not what you're used to, Fritz? Best have your piss quick afore he brings his mates back for a look.'

Hastily Lothar obeyed. He had just finished when there was a sudden uproar in the central area of the Warren, a babble of voices cut through by a sharp cry of pain.

Josh! This was Lothar's instant thought.

Ignoring Hepworth's injunction to go slow, he set off at a stumbling run back down the narrow corridor. Ahead he saw a strong pool of light and shadow which as he reached it settled into a circle of men with lamps and torches. At their centre were three figures. One was the English officer, waxen-faced and unconscious, a long wound at the side of his head bleeding copiously. Beside him, sitting upright and holding his jaw was a small dark man. Standing over them, wielding the billet of wood which had clearly been used to strike the officer down, was Viney.

'What the fuck are you playing at, Taff?' he snarled. 'For Christ's sake, do I have to do everything myself?'

'He came at me sudden-like,' said the man with the injured jaw. 'Took me by surprise. Tell the truth, I didn't think the poor bastard'd have the strength!'

He seemed to have satisfied himself that his own injury was superficial and now he turned to Cowper.

'No need to knock his head off, boyo,' he said reproachfully. 'I'd have held him, no bother, without this.'

'Mebbe,' said Viney.

Lothar was interested to note the man called Taff seemed completely undaunted by Viney and Viney for his part didn't find it necessary to confront him into subjugation.

'Fetch me some water, one of you lot,' said Taff, examining Cowper's wound.

'Let the fucker bleed to death,' said Foxy, the scab-faced

man with a high laugh. No one else joined in and someone went off and returned a moment later with a mess tin full of water.

Emboldened by this show of concern and also by Viney's relatively mild manner, Lothar said, 'Please, the water, before you bathe his wound, has it been boiled?'

All eyes turned on him. Viney exploded, 'Jesus! What's fucking well going on here? Who's supposed to be taking care of the Hun?'

Before anyone could answer Lothar said, 'Even the Hun must urinate, Viney. I see the state of things down here and I think perhaps this water is more dangerous than your club!'

'What're you on about, Fritz?' demanded the Australian.

'I don't think he liked the look of the ablutions, Viney,' said Hepworth, causing a general laugh.

'It is filthy,' said Lothar calmly. 'There is high danger of infection.'

'It's fucking Hun infection then!' snarled Foxy.

'There's nothing wrong with our water, but,' said Hepworth. 'We go out and collect it from a stream.'

'Think how many dead men lie beneath this ground,' said Lothar. 'Think how many bones this water may have washed through.'

'Cheerful cove, ain't he?' said someone. But another man went off and came back with a can full of steaming water.

'I was just brewing up,' he explained.

Lothar knelt beside the officer and, while Taff held his head, he bathed the wound, using a strip of the man's own shirt.

'You some kind of medic, Fritz?' enquired Viney.

'No. But I spent some time in a hospital,' said Lothar, looking up at him. 'I pay attention to what I see.'

'Yeah. I bet you do,' said Viney with a frown. 'Get that joker back under restraint when Flo Nightingale here's finished with him, and make sure he don't go wandering again! Tie him up if needs be.'

'And what about Fritz here?' demanded Strother who had joined the group. 'Do you want him tied, Viney?'

The Australian glowered at Strother.

'Yeah, he's a real bloody menace, ain't he?' he said. 'Just like them Huns in the papers, you can see he's just prowling around looking for some nuns to rape and cut their tits off. But as we're short of nuns and the nearest we've got to a tit is you, Strother, I reckon he's out of luck!'

He spat on the floor, turned and left.

'What does he mean?' enquired Lothar, who was not certain he had accurately caught the drift of the Australian's remarks.

'He means you can go and piss by yourself, I reckon,' said Hepworth. 'Let's get sonny boy here back to his cot and then mebbe we can all get some sleep!'

When Lothar returned to his dug-out a few minutes later, Josh turned as he crawled back beneath the blanket and said, 'Wilf?'

'It is I, Josh. Lott,' said Lothar gently.

'Lott,' echoed the boy. And after a few moments his breath resumed its even rhythm and once again he slept.

4

Settling into the Warren was so like settling into a new army billet that Lothar wondered if any of his fellow 'volunteers' had noted the amusing irony.

He was not going to antagonize them by asking, however, not until he knew them a lot better.

Even their resentment was a familiar part of the process. An aristo slumming it in the ranks of the Kaiser's army gets used to being resented. But Lothar knew it would fade almost, though never quite, to vanishing point if he ignored the baiting and did his job efficiently.

Having a job to do in the Warren was the problem. To his surprise, Viney himself solved it within forty-eight hours.

'Fritz,' he said, 'this drum you're giving us about the state of this shit-heap, any chance of you putting your money where your mouth is?'

Lothar's blank expression clearly delighted the big Australian.

'What's up, Fritz? Don't these relatives of yours speak the King's English?'

When the question was translated, Lothar didn't hesitate. His assumed expertise on the sewage and ventilation systems of an underground complex was based on a boring tour of Brigade HQ with a houseproud sergeant-major. But to have a function within the 'volunteers' was a step to safety.

'I think I can work things out,' he said. 'But as for repairing, I should need practical help.'

'That's easy. Taff here's a miner, or was. Knows all about tunnelling and such.'

The small dark man who had stood up to Viney when he knocked out the English officer held out his hand.

'Carwen Gwynn Evans,' he said. 'But these poor ignorant bastards call me Taff.'

'Any hard labour needs done, put these idle fuckers to work,' continued Viney, gesturing at the other men present.

'I didn't ship out of the line to be ordered about by no Hun and no Welshman,' muttered a bald-headed man called Nelson, one of Strother's circle.

'It's me who's doing the ordering,' said Viney. 'Fritz, what do you reckon we can catch in this shit-hole if we don't take care?'

'Cholera, typhoid, plague perhaps. I am no expert.'

'You'll have to do. That answer your objections, Nelson? Too right it does! Now, let's see some action!'

In the event it turned out the work was most easily done by two men alone, or rather three, for Josh never strayed far from Lothar. It was good for the boy to work, thought Lothar. And for himself too. And a real bonus was the company of Taff Evans whose independence of spirit made him indifferent to his workmate's nationality. Though he had little formal education, he had a sharp enquiring mind and was terrier-like in pursuit of an idea.

It was with Evans that Lothar first discussed the question of what they were doing there, why they had deserted. He approached the topic cautiously. It was a strange thing. All around him he could see the haunted eyes, the nervous tics, the explosive irritability, which were the surface signs of that inner breakdown of endurance which must have caused most of these men to desert. Yet it was mixed in many of them with an inhibiting guilt, a conditioning shame, which made most references to their military lives almost hysterical outbursts of self-justification. Evans, however, was clearly pleased to have a serious listener.

'I made a mistake,' he said. 'But it wasn't all my mistake, see, and I paid for it all right; I paid dear. But they wouldn't let me stop paying. I was a miner, see, that's a protected job. I could've stayed back home, doing my work, no bother. But I thought, here's this war, boy, best chance you'll have maybe to see a bit of life. There's a whole world out there you know nothing of. This old valley's still going to be here waiting when it's all over, so why not take the chance, sow a few wild oats? So I tossed away my pick and went and got myself a rifle. Well, two years I had of it, two years of mud and trenches and shells and bullets and, well, you know what it's like. Two years. I went to see my officer and I said, "Look, I'm a miner, it's a protected job, I'd like to get back to it," but all he said was, "You should have thought of that sooner, Evans. You shouldn't have been so hasty." We was in rest then, so when the order came to go back up the line, I just wandered off. No plan, like; just wandered off into the Desolation.'

'Where you found Viney.'

Evans grinned suddenly.

'He found me, boyo. Haven't you noticed? You don't find him, he finds you. That was a couple of months ago. Coleport, his oppo, was a bit sick then. He'd been beaten up by some redcaps or something. Bastards! So they were holing up here till he got back on his feet. It wasn't such a bad place then when there were only a few of us. But others came drifting along. You don't notice, see, when things happen gradual. It's like you said, get used to anything, we can! And we'd best get back to work before we get used to being idle, I think! Hey, Josh, boy! Don't lay around all day!'

Josh, who had been resting against a wall with his eyes

closed, opened them and smiled. Though he spoke little, he had rapidly become a favourite of the majority of the men. Perhaps it was his complete, almost childlike air of vulnerability which touched their hearts. With most of these fugitives, the bitter experience of war and their own collapse in the face of it had been an ageing experience. Taff Evans, for instance, was twenty-five, but he looked nearer forty. Only with Josh had the horrors eroded rather than added the years, sending him back into boyhood so that he could easily have passed for sixteen.

They were digging out a collapsed ventilation shaft. Josh was very useful here, his slim frame being able to wriggle through spaces which even the Welshman's small but chunky body threatened to get wedged into. Lying on his back using techniques quickly learned from Evans, he was shovelling the fallen earth back down his body with his bare hands.

Suddenly there was a slight creaking noise and then, with no further warning, earth came tumbling in upon the supine boy. His legs kicked wildly. Lothar screamed, 'Josh!' But Evans in one smooth movement seized the kicking legs and drew him back down the shaft, spinning him over and banging his back to help him retch out mouthfuls of earth.

To Lothar's relieved surprise, the boy seemed none the worse for his experience. Instead, as soon as he could speak he said excitedly, 'Look, Lott! The sky!'

They peered up the shaft. The fall of earth had just been the last thin layer barring the way to the surface. There, sure enough, was a blue circle of evening sky with a trace of white cloud drifting across it. Josh was regarding it as if it was his own personal discovery. In a way it was. They had been allowed up under careful supervision to taste fresh air, but only after dark. This was the first piece of daylit sky they had seen since their arrival, and Lothar suspected it was the first piece of daylit sky Josh had taken real notice of for many long days before that.

Now his eyes were wide with the pleasure of it, and the sight of that pleasure brought a respondent joy to Lothar's heart.

'Can we go up, Lott?' asked the boy excitedly.

Lothar raised his eyebrows interrogatively at the Welshman, who looked uneasy.

'Best not,' he said. 'You know how Viney is about going outside.'

The rules which governed life in the Warren were few and unwritten. Disputes were settled by simple reference to Viney. But the one great commandment which had been clearly spelt out was that no one went above ground without the Australian's knowledge and express permission. A sentry stayed on duty night and day in a hide close to the one entrance. His job was to check would-be exiters as well as to warn of intruders.

'Please, Lott,' begged Josh.

Lothar hesitated. He had no desire to antagonize Viney on whose favour everyone's survival depended.

On the other hand, he thought, with a sudden flash of that arrogant wilfulness which had been the spring of so many of his youthful outrages, why the hell should I, the son of an old and noble family, the possessor of a first class degree from Heidelberg, kowtow to an unlettered brute from a British penal colony?

His saving sense of irony immediately added: What an absurd socialist you make!

But already, ignoring Evans's protests, he was helping Josh to scramble back up the shaft and out into the open air.

'Be careful!' he urged. 'Do not let yourself be seen!'

In fact because the shaft opened on to the side of the tumulus, under which lay the Warren, opposite to the entrance, Josh was out of sight of the sentry. This was fortunate, for he had no thought of concealment, but stood there half in earth, half in air, like some newborn spirit in a classical allegory, his nostrils flared, his mouth and eyes wide, as if he would drink in all he saw at one draught.

It was a fine clear evening, the end of a golden summer day. As heat declined, clarity had grown, and though the foreground and middle distance showed only an uninterrupted view of the Desolation, looking westward almost into the setting sun, Josh could see beyond the churned and ravaged ground to a horizon fringed with silhouetted shapes which his keen eyes and keener longing told him were trees in full leaf.

He tried to pull himself fully out of the shaft, but Lothar had been prudent enough to reach up and take a hold of one of his feet.

'All right, Josh,' he said. 'Come back now, please.'

There was a moment's hesitation. It made Lothar anxious, but it also gave him pleasure. For the boy to be even thinking, no matter how briefly, of independent action was a sign of his recovery.

Then Josh slid back down, his eyes shining with remembered pleasure.

'I could see trees, Lott, away in the distance. Trees!' he said excitedly.

'That is splendid,' said Lothar gravely.

'Can we go where the trees are, Lott?'

'Soon, yes. Very soon.'

The boy appeared content with the answer.

Evans said, 'We'd better get that hole camouflaged. Some kind of grille made from deadwood should do the trick, I reckon. Let the air in, but keep bodies out, that's what we need.'

'Yes,' said Lothar thoughtfully. 'Though it could be useful perhaps to have an alternative door, could it not, friend Taff?'

'What the hell are you getting at, boyo?' said the Welshman, alarmed.

'All I mean is an emergency exit. For fires or other alarms,' said Lothar, all wide-eyed innocence. 'That is all I mean.'

'It's a sight too fucking much to mean,' said Evans savagely. 'And if there's any sense left in that Jerry head of yours, you'll understand that! Now, let's get on with the job.'

Grinning to himself, Lothar obeyed.

5

Corporal Arnold Tomkins was not happy. By rights he should have been drinking his evening cocoa prior to going to bed in the Château d'Amblay which was currently housing Division HQ. But disaster had crept up on him from behind. There had been some confidential orders for an RFC unit. The staff sergeant in charge of despatch had suggested to the general's ADC that they could go with the RFC supply truck, shortly due to leave. 'Must be delivered by hand,' said the ADC importantly. 'And that doesn't mean one of those disgusting drivers' hands.'

And the eye of the staff-sergeant, who hated him, had fallen on Tomkins.

So here he was crushed in the cab of a Crossley GS truck, bumping along what might once have been a road till shells and rough usage had flayed off its surface, autumn rains washed out its hardcore and the frosts of a violent winter split it down to the raw earth.

As for the surrounding countryside, the incidental devastation of the Somme battles merged into the deliberate devastation of the German withdrawal and to Tomkins, brought up in the crowds and bustle and endless buildings of London, it was like a landscape of hell.

The driver, a rough-tongued Scot called Shawcross, had listened to this balding, thick-spectacled, middling-aged moaner for thirty patient minutes, then advised him, 'Haud your whisht, Jimmy, and count your fuckin' blessings.' Shawcross didn't hesitate to complain himself whenever his truck encountered a more than usually deep pothole, but he was on the whole not too discontented with his lot. It was late evening and the sun was setting behind the trees in a pretty watercolour of gentle pinks and pastel yellows. But it was not from the west that he derived his comfort. In the eastern sky there were colours too and these would grow brighter as darkness fell. A hoarse grumbling of artillery rolled across the intervening miles without cease, and, to his mental ear, a deeper grumbling of men as they crouched in their trenches for the evening stand-to. To be spared that made the discomforts of this obstacle course of road seem an easy penance.

He negotiated a roughly filled shell-hole which had the old engine labouring like a bronchitis case. Tomkins touched his arm to draw his attention above the din and pointed ahead to where two figures, vague in the gloaming, had appeared by the road.

Shawcross, sensing his unease, shouted, 'Pair o' fuckin' Huns, Jimmy. Wi' a bit o' luck, they'll mebbe cut your tits off and only rape me.'

But his ghoulish humour faded too when he saw that the waiting men were wearing the red caps of the military police. Police meant trouble just the same on the Western Front as they did in Glasgow.

One of them, a big man wearing a sergeant-major's crown on his sleeve, waved him down.

'What's up, sir?' enquired Shawcross through the open window, as the truck stopped.

'Let's see your orders, laddie,' replied the sergeant-major. He had a voice and manner used to command. His accent sounded vaguely cockney to the Scot, but not to Tomkins.

He took the proffered papers and said to the corporal, 'Who're you?'

'Tomkins, Sergeant-Major. Orderly room corporal at Div. HQ. I'm on my way to the Squadron with some papers.'

The sergeant-major grunted and studied the orders Shawcross had handed over which included a load inventory.

'Do themselves well these fly-boys, don't they?' said the sergeant-major returning the papers. 'You've got trouble ahead, Jock. That bit of road along the marsh, you'll know it, I reckon? Well, it's finally gone under.'

'Oh shit,' said the driver. 'Shall I have to go back?'

'No. We'll get you round all right. Nice little diversion that some of the lads have been clearing all day. But it's a bit hard to spot till you're used to it. Move over.'

Tomkins found himself shoved hard towards the driver as the sergeant-major clambered in beside him. The truck swayed as the other MP, a private almost equal in bulk to his sergeant-major, clambered over the tailboard.

Shawcross started up the truck and, following the sergeant-major's instructions, turned off the road about fifty yards ahead. Five minutes later he began to grow worried as the new route got progressively bumpier and the expected swing back to their former direction never came.

Finally he said, 'Are you sure you've got this right? It feels like we're half way up the fuckin' Trossachs!'

'I think we're OK, Jock, but if you stop here I'll take a look-see.'

Relieved, the driver brought the truck to a halt. The sergeant-major got out and walked round the truck, banging on the tailboard in passing, and coming to a halt by the driver's door which he opened with a commissionaire's flourish.

'Are we OK, then? Can we make it?' enquired Shawcross.

'We've made it, Jock. Step down, cobber, and stretch your legs.'

Shawcross with growing alarm heard the tailboard crash down.

'Here, what the fuck's going on?' he cried.

For answer the sergeant-major seized his tunic front and dragged him with easy strength from the cab.

'I said, step down,' he murmured. Then, putting two fingers in his mouth, he whistled twice.

Slowly out of the gloom figures began to emerge, stepping forward awkwardly, even shyly, like animals lured from their lairs by curiosity or by hunger. Six or seven of them there were, with grey, unshaven faces, hollow cheeks, and the wide unblinking gaze of night-predators whose eyes are thirsty for every drop of light.

'Come on!' ordered the sergeant-major, 'let's be having you! Get this truck unloaded, sports! Blackie, when it's empty, run it along to that nice big hole we found and sink it.'

'Sure thing, Viney,' said the man in the MP private's uniform. 'What about laughing boy here? Too proud to join us, Corp?'

Trembling, Tomkins climbed out of the cab. But his legs would not hold him and he slumped to the earth where he lay, ignored by the bearded, ragged figures moving with rapid stealth despite their burdens on all sides.

Suddenly there was an upsurge of voices, loud among them Shawcross's shouting, 'You're no' making a fuckin' deserter out of me!' Footsteps running, a dark shadow merging rapidly with the surrounding gloom. Then a loud bang, a momentary stab of flame, a short cry.

Silence.

'Stupid bastard,' said Viney, ejecting the case from his rifle.

Some of the pack of men-predators had detached themselves and gone out after the fleeing Scot. After a while one of them returned.

'He's dead, Viney,' he said. 'What'll we do with his mate here? Sink 'im with the truck?'

Tomkins felt himself being seized by the collar and dragged upright. His bowels were opening but he didn't care.

'No need for that,' said Viney. 'We're not going to have any trouble with *you*, are we, sport? You'd much rather stay here with us than be parking your bum on a hard office chair, wouldn't you?'

Tomkins's mouth was too dry for him to answer, but the man who'd travelled in the back of the truck laughed and said, 'Sure he would, Viney. *Though we work and toil and bustle in our life of haste and hustle, All that makes our life worth living comes unstriven for and free.* It's your lucky day, your lucky fucking day!'

Viney listened to his henchman's cheerfully violent badinage with an emotion which would have surprised the fat man. Basically it was envy. He had no illusions about Coleport's true make-up. He knew more than any other the depths of sheer mindless viciousness the man could sink to. He was truly amoral. Nothing bothered him.

But it sometimes bothered Viney that he should have power over such a man. And it bothered him now that he felt a passionate upsurge of desire to be as uncaring as his lieutenant. There was a careless ease about Coleport's attitude to life and all its circumstances which Viney envied. Give him a drink and a warm berth and he'd be happy anywhere. There'd been a box of Scotch in the Crossley, and Blackie even now was pouring down the odd couple of ounces and making contented little schlurping noises as they made their way through the dark with their loot.

But at Viney's centre there was a blackness which no amount of drink had ever been able to wash away. Sometimes it seemed to swell up and fill his mind, like pustules of swamp-gas eructating on the smooth surface of a pool. There was melancholy there, and weakness and failure, and hate and shame. There were times when he felt like two people, sometimes more than two. There were times when his body acted without his mind. He had raised the rifle and put a bullet through the Scotch driver's head without a thought. He felt no guilt in the act, yet he felt bewildered that such an act was possible for him without thought.

In the beginning he had welcomed the war. Here at last was proof that his condition was not unique but part of humanity's common chaos! Gradually, however, he came to realize that war, far from being chaotic was in fact just a symptom of that same order which told him what he ought to be, and do, and think. So he had at last turned against war itself and taken refuge in a situation and landscape which echoed the desolation at his core. This was no pacifist declaration. The enemy was the War and all those who wished to continue in it. The Scottish driver had deserved death because he had fled towards it.

They were back at the Warren. Contact was made with the sentry who had strict orders that without an exchange of agreed signals, anyone approaching the Warren was to be shot – 'even if it's me and you can see that it's me!' added Viney. "Cos if you don't, I'll bloody well shoot you, believe me!'

The men had believed him.

Safely below, there was a general gathering in the largest of the chambers, the former general control chamber of the HQ and command post. A successful raid was more than just a way of replenishing supplies. It reconfirmed the Volunteers' belief in their capacity to survive. And when, as now, it brought in a fresh 'recruit', there was a chance of catching up on news of the world at large.

The chamber was jam-packed. Viney realized with a slight shock that his strange force had now almost outgrown its adopted quarters. There were about forty of them, men of almost all nationalities involved in the War except the French, whose natural inclination when ducking out was to head for home or disappear in the nearest large town. The only man from the French army with the Volunteers was a jet-black Moroccan who spoke next to no English. There were two other coloured men, light brown Indians from a Punjabi cavalry regiment which had fought as infantry and suffered hugely. To all the other horrors of war, distance of space and climate and culture had added an extra dimension for them. One was a hook-nosed middle-aged man with a single stripe, the other a youth, classically handsome with dark liquid eyes perpetually clouded with bewilderment. Their names had been elicited, discarded as too long and foreign, and they were addressed with rough familiarity as Dell and Nell.

As for the rest, the British predominated with a good admixture of colonials.

The Volunteers were in cheerful mood tonight. Viney had permitted three bottles of whisky to be opened, reserving the rest against future need, and also careful to ensure no one was going to get stupidly drunk. The stores removed from the truck were of excellent quality and made a welcome change from the diet of canned beef and Maconachies which recent lean pickings had reduced them to.

Tomkins found himself treated as father of the feast and reacted with nervous relief, gabbling out answers to the rapid-fire questions as quickly as he could.

Lothar, standing by the chamber entrance, observed with curiosity that the men still seemed interested in the progress of the war, as though its outcome would somehow alter their situation. Others spotted this irony too, and when one man expressed disappointment that Tomkins could report little more than 'no change', Taff Evans said, 'What's it to you anyway, boyo? Just because we win this fucking war don't stop you from being shot for a deserter, do it?'

'Aye,' added Hepworth, whom Lothar now knew to come from Yorkshire, a kind of English Saxony. 'Mebbe it's best for us if Jerry wins. Then we'd all be prisoners and get treated the same, would that be right, Fritz?'

Lothar shrugged and did not reply and Viney ended matters by saying, 'Right. Enough gabbing. Anyone got anything sensible to ask?'

There was silence, broken by Lothar who said, 'Among these stores, were there any medical supplies?'

Viney glanced at Coleport, who grinned and held up a whisky bottle.

'Only this, but you can't beat it!' he called to general laughter.

'Sorry, Fritz. Anything special in mind?' said Viney.

Lothar said, 'Everything. Dressings, disinfectant, drugs – we have nothing.'

'Who needs 'em except that fucking cavalry officer?' called Strother. 'Better just to let that bastard die, says I.'

'Mebbe you'd like the pleasure of killing him yourself,' said Hepworth.

'Wouldn't say no,' said Strother, baring his rotting teeth. 'And there's plenty of others here as feel the same.'

Foxy (this had turned out to be merely an extension of the man's name, Fox, but it fitted his pointed predatory face very well) agreed with vicious force. 'Kill the cunt straight off, no messing,' he said vehemently.

There were one or two grunts of agreement. Viney said, 'All right, stow it. Let Fritz have his say.'

'This is nothing to do with Lieutenant Cowper who is much recovered,' said Lothar. 'Every man here is at risk every day. Even cuts and scratches are dangerous. Break a bone and you will probably die of gangrene. You, Foxy, need a proper dressing for that sore on your face. Perhaps even our friend Strother might get an ache in his one good tooth.'

This brought a laugh except from the Cockney, who screamed, 'I'd let the fucker ache rather than let a dirty Hun come near it!'

'I said stow it!' boomed Viney. 'I'll not stand that kind of shit. We're all together in this, you'd better understand that, sport. No enemy, except them as don't agree with us. Fritz here's done more to make life tolerable round here in the short time since he joined us than most of you jokers could manage in fifty bloody years. Ain't that right?'

He glared at the men, defying contradiction. In fact there was no way of denying the improvements Lothar, with the technical expertise of Taff Evans, had made to ventilation and general hygiene.

Patsy Delaney defused the situation by saying mildly, 'He ain't got round to fixing the lights yet, Viney.'

Viney smiled and everyone relaxed.

'Yeah. What about that, Fritz?' asked the big Australian in an almost jocular tone.

'We need power supplies for electric lights,' said Lothar. 'Perhaps you brought the battery from the truck which you held up? No? A pity. It would have been a start. The toolbox then? Or some pieces from the engine? Certainly at least you brought some oil and some gasoline? No? Such a pity!'

What made him reply in this provocative fashion, Lothar did not know. It had been always thus. Since he was a child, whenever the sense of security and privilege accorded him by his birth and background had pressed too heavily upon him, he had reacted by some act of physical or verbal violence. Now it was Viney's public defence and approval which produced the response.

Viney's smile had faded and his more usual darkly brooding look had returned, many degrees intenser than before. Lothar regarded him with a light, superior smile on his lips and awaited his response. The atmosphere was one of surprise, rapidly, in the case of those like Strother who resented the German's favoured status, turning to delight.

Blackie Coleport drank some whisky and said slowly and apparently irrelevantly, 'What *are* you going to do with that fucking officer, Viney? We've got enough to do without wasting energy and grub taking care of a fucking prisoner.'

Viney did not appear to have heard, but after a long moment he said softly, 'What do you say, Fritz? What do you think we ought to do about the lieutenant?'

Lothar regarded Coleport with a new respect. There was a mind at work there after all.

He picked his words carefully.

'I think he should be dealt with as we would all hope for ourselves,' he said. 'With true justice, with true compassion, according to our true deserts.'

'Some fucking hope of that!' called out Harry Taylor, the

ginger man who'd been reproved by Viney for making jokes about Josh and Lothar being lovers.

Viney ignored the interrupter and said, 'Fritz, I don't know if you speak English better than any other cunt here, or just tie things up like you was speaking German, but I reckon what you mean is we'd all want a fair hearing, right? A fair hearing at a fair trial?'

Feeling himself being irresistibly manipulated, Lothar nodded.

'Well, that's what we'll give him,' announced Viney. 'You say he's on the mend? Right! What do you say, sports? Let's see that we give Lieutenant Cowper a fair trial, shall we?'

There was a general outburst of laughter and agreement. Only Lothar's voice was raised in open protest.

'You cannot do this!' he cried. 'You cannot drag this man in here and pretend to try him! It is a mockery.'

'Don't tell me anything I decide is a mockery, Fritz,' said Viney savagely. 'We ain't dragging him in here, like you put it. He'll be brought to trial tomorrow, right? That'll give him twenty-four hours to prepare his defence. Will that be long enough, Fritz?'

'I do not know,' protested Lothar. 'What is the charge?'

'Warmongering. Accessory to murder. That'll do for starters,' said Viney. 'And you ought to know if it's long enough, Fritz. Because I've just appointed you Lieutenant Cowper's defence counsel!'

6

Twenty-four hours proved more than enough for the preparation of Cowper's defence.

The young officer looked at Lothar with horrified disbelief as he explained what was to happen. Most of his cuts and abrasions had begun to heal, but the gash on his head where Viney had struck him after his escape attempt really needed stitches and remained a gaping wound beneath the makeshift dressing the German had applied. His young face was pale and haggard, good for sympathy in a civilian court perhaps, but not here in the Warren where such ghastly features were the norm.

'They have no right. No right!' he protested.

'They have the right of all men in uniform,' said Lothar. 'The right of strength.'

'I shall not answer them,' said Cowper.

'Please, you must consider seriously,' said Lothar. 'I fear that these men, some of them, would like an opportunity to have you killed.'

His plan to shock Cowper into cooperation failed. The lieutenant merely replied, 'Let them do what they will. I will not flatter such a travesty with speech.'

Perhaps he was right, thought Lothar. What was there to say which could possibly blunt the hatred in his accusers' hearts?

'One thing I should like,' said Cowper hesitantly. 'You seem a decent sort of chap . . .'

He hesitated once more.

Lothar laughed and said, 'For a Hun, you mean? How kind! Perhaps if I tell you that I am a graduate of Heidelberg, that one day I may be, perhaps already I am, the Graf von Seeberg, and that I have spent many happy holidays with our relations, the Drew-Emmersons of Suffolk, you will be even more willing to admit I may possibly be a decent sort of chap!'

'Good Lord!' said Cowper. Lothar's little speech had the desired effect of relaxing the young officer considerably. *He now sees me as a real ally, one of his own class, instead of both another rank and a foreigner!* thought Lothar as he listened to Cowper's suddenly animated voice as he sought for common acquaintance. But at the end of it he was still adamant that his response to the so-called trial would be silence.

His request turned out to be for a pencil and paper. Lothar provided him with these and the following day when he visited the man in his makeshift cell, Cowper diffidently put the paper folded into an envelope into his hand, and said, 'Look here, von Seeberg, it's silly I know, but better safe than sorry. I've scribbled a few words to my wife. If it became necessary, and always supposing you yourself ever got the chance, could you do me the kindness of sending it to her? I've printed her address on the back.'

'Of course,' said Lothar.

'Here, Fritz, you're wanted in the court-room,' said Strother behind him.

'Very well. Come, Lieutenant Cowper,' said Lothar.

'Nah. Prisoner comes in separate. Don't you Huns know nothing?' mocked Strother.

Cowper held out his hand and Lothar shook it.

'Jesus fucking wept,' said Strother. 'All pals together!'

Between the cell dug-out and the main chamber where the trial was to be held, Lothar paused and rapidly unfolded Cowper's letter. He felt a pang of distaste at the act but this was no time for the squeamishness of good breeding. He scanned the scribbled lines, then replaced the paper in his tunic.

In the trial chamber, the Volunteers were assembled and the atmosphere was almost festive. He saw Josh sitting near the entrance, looking relaxed and cheerful. The boy was responding daily to the almost universal kindness he received and for the last couple of days had shown himself willing to be separated from Lothar for longer and longer periods. He smiled broadly now as he saw his friend appear and Lothar smiled back.

The trial was being staged with a telling simplicity. Viney as President of the Court sat behind a trestle-table flanked not by his two Australian henchmen but by Hepworth and a round-faced, placid West Countryman called Quayle. What real function these two would serve was not clear, but they helped to give an impression of judicial impartiality which Lothar did not find reassuring. Seated at right-angles to the trio at the end of the table was the new arrival, Corporal Tomkins, the clerk. In his hand was a pencil, on the table before him was a wad of paper, and on his face a look of self-importance.

Two stools had been provided for Lothar and the prisoner, placed at the far side to Tomkins, leaving a clear well between the table and the spectators who for the most part squatted on the floor.

Viney rapped the butt of his Luger on the table and said, 'This court is now in session. Bring in the prisoner.'

His tone was sombre without the slightest hint of parody but when a moment later the slender form of Lieutenant Cowper appeared wedged tight between the grinning figures of Delaney and Coleport, there was a ripple of laughter, for the two men were wearing the red caps of military policemen. But the laughter was short-lived, for suddenly there was a scream of pure horror and Josh rose to his feet trembling so much that he staggered against the men seated around him and would have fallen without their support.

Lothar rushed towards him and enfolded the now sobbing boy in his arms.

'Take those stupid hats off!' he hissed over the blond head at the two Australians, who looked on in bewilderment.

At the table Viney made an imperative gesture, and they removed their red caps and tucked them out of sight.

'Look, Josh, it's all right, it's just Patsy and Blackie, that's all. Look.'

'Yeah, kiddo, it's only us. Don't take on. It's your old oppo, Blackie, see,' said Coleport. 'And this here long streak of misery's Patsy.'

Slowly Josh allowed himself to be turned towards them. After a while his sobbing subsided but his body was still shaking.

'Josh, son, you'd best go and lie down for a bit,' ordered Viney. 'Fritz, see him bedded down, will you? We'll take a five minute recess.'

Lothar led Josh from the chamber and took him to his bed-roll. A couple of minutes later as Josh relaxed under the German's constant flow of reassuring talk, Hepworth appeared.

'Viney says give him a shot of this if you reckon it'll help,' he said, producing a bottle of whisky. 'He also says we'll be starting shortly, so don't hang around too long.'

Lothar allowed Josh a single drink of the whisky. It seemed to have a calming effect, but he was still not happy to leave the boy by himself. But he also felt a strong need to be back in the chamber when the trial began.

'Josh, I will not be long,' he said. 'Rest now. Sleep if you can.'

'Right, Lott,' whispered the boy. 'I'll be all right.'

Back in the chamber, Viney asked, 'What was that all about, Fritz?'

Lothar said, 'His brother was tried by a court-martial and shot. Josh saw him brought before the regiment by two military policemen to hear sentence.'

Viney's face darkened.

'Bastards,' he said. 'That's what they call justice back there. Shooting their own lads! Bastards!'

There was a general murmur of indignant agreement, and Fox screamed out 'Fuck this for a lark. Let's shoot the cunt straight off and be done with it!'

Lothar said, 'Viney, couldn't we postpone this so-called trial? This atmosphere, it is not helpful to the lieutenant. It is unfair.'

'Unfair?' said Viney. 'I suppose it's not unfair to gallop around out there, rounding up men like cattle to take 'em back to be shot? I suppose that's not unfair!'

'It is not relevant,' asserted Lothar. 'The lieutenant here is accused of what? Of making war, I think you said? Of murder? These are absurd accusations . . .'

'You mean he's pleading not guilty?' said Viney. 'Make a note of that, Tommo.'

Corporal Tomkins scribbled away.

'No!' said Lothar. 'He makes no plea. He acknowledges no trial. He will refuse to speak.'

'Won't speak, eh? In that case, it don't matter much what the atmosphere here is, does it? But I'm a fair man, Fritz. I don't want Mr Cowper here feeling the deck's been stacked against him. So let's start off on a friendly footing, shall we?'

Cowper stared stonily at a point a couple of feet above Viney's head, but Lothar studied the big Australian's face anxiously. It wore a faint smile which matched his suddenly reasonable tone.

'Fair do's,' proceeded Viney. 'You're entitled to know who we are. Well, we're just what you see, a bunch of ordinary blokes who've tried the Army and done our bit and decided it's the civvy life for us from now on. It's a choice we reckon everyone should have. Wipe the slate clean, so to speak. Everything we've done in the past is the Army's fault. Now we're our own men. So here's what I'm saying to you, Mr Cowper. How about it? How would you like to join our happy little family?'

It was a masterly move, Lothar acknowledged. Cowper's resolve to meet accusation with silence was going to be hard to keep in the face of this absurd but courteous invitation. Sure enough, the young officer's gaze slowly dipped till it locked with Viney's. But when he spoke his words took everyone by surprise.

'It's Viney, isn't it? Sergeant Viney of the Australian infantry?'

'Used to be, sport. Have we met?'

'No,' said Cowper. 'But I've heard about you and read your description.'

'You hear that, friends?' grinned Viney. 'It's like the old days back in the bush. They've got wanted posters out on us!'

Most of the audience laughed, but Lothar was watching Cowper's face. It wore an expression of contempt which seemed to have its origins beyond his present situation.

He said, 'I knew I'd fallen in with a bunch of cowards and deserters, but I didn't realize till now just how low I'd fallen.'

An uproar of protest arose from the men. One or two of them scrambled to their feet. Viney too rose up, leaning across the table and bellowing, 'You chuck accusations around pretty easy, sport. Cowards you call us! There's men here done more real soldiering than you and all your fucking horses have ever managed! What've you lot been good for but galloping round the Desolation, chasing down poor bastards too weak to walk? What are you but a redcap's bum-boy, eh, sport?'

'At least I'm not a woman-killer!' flashed Cowper.

It was hard to think of a riposte which might have reduced the chamber to silence but this did the trick. Viney subsided on to his seat and said, 'You'd better explain that, sport.'

'It's very simple, Sergeant,' said Cowper. 'That nurse you tied up in the hospital when you removed your friend. She died. She's dead. At least that puts you on a footing with your British comrades, doesn't it? They don't execute men for desertion from your Army, do they? But they hang them for murder!'

7

Lothar watched Viney very closely during the next few moments. There was a buzz of speculation among the spectators. Slowly Viney's gaze moved away from Cowper and travelled round the room till it found Delaney. The long, thin Australian gave a minute shrug.

Viney, his face now expressionless, returned his attention to Cowper.

'It was an accident, sport,' he said. 'I didn't know. I wish it hadn't happened.'

'Accident?' sneered Cowper, enjoying his period of ascendancy. 'You don't look the kind of man to me, Sergeant, who has accidents.'

'The redcaps were OK, weren't they?' said Viney. 'If I'd wanted someone killed, don't you think I'd have started with them? There's millions of men all over France let out a cheer every time they hear a redcap's been killed!'

'You may wish you'd killed one of them before you're through,' said Cowper softly. 'I should not care to have Jack Denial on my tail.'

'Denial? He was the captain?'

'That's right.'

'And this nurse was his girl? Yes, I thought there was something,' said Viney softly.

He shook his head violently as though to clear away a pain.

'I'm sorry about the girl,' he said. 'Something else for you to answer for, Cowper.'

'Me?' cried the lieutenant in amazed indignation.

'Yeah. You. And all them like you who still order men to fight this sodding war.'

'Men don't fight because I order them,' protested Cowper, sucked into debate in spite of himself.

'Why then?'

'Because it's a soldier's duty. All soldiers of all ranks. That's what soldiering is about, fighting for your country. When you take the oath, that's what you're promising.'

'But most of these jokers here aren't soldiers,' said Viney. 'Not in the sense of choosing to be. Most of 'em were forced to it by conscription.'

It was a good argument, though probably inaccurate. Cowper shook his head stubbornly.

'It makes no difference.'

'You mean if I force you to make a promise, by putting this against your bollocks, say,' said Viney, producing his Luger, 'you'd feel obligated to keep that promise?'

'It's not the same thing,' said Cowper. 'War is different. A war like this threatens the whole country and everyone in it. *Everyone* has to take a hand in the defence. Individuals can't be allowed to pick and choose, any more than they can be allowed to pick and choose whether they pay taxes or wear clothes in public places. Everyone has to pull his weight in time of need.'

'Regardless of class, colour or creed?' said Viney.

'Yes.'

'All equal in the country's hour of need?'

'Yes.'

'All equal, eh?' mused Viney. 'When did you last go on leave, Mr Cowper?'

'Home, you mean? About two months ago.'

'And not home? Local. When did you last have some local leave?'

'Three or four weeks ago. What is the point of this?'

'Where'd you go, sport.'

'I went to Paris, as a matter of fact.'

'Paris, as a matter of fact! Enjoy it, did you? Had a bonza time, I bet! And before that when did you have local leave? And when was your home leave before last?'

'I can't recall precisely. This is absurd . . .'

'Can't recall precisely? Is that because the gaps are so long, Mr Cowper, sir? Or is it maybe because there don't seem to be any gaps at all?' Viney was half out of his seat now and shouting. 'You know how long it is since my men have had home leave? Not this bloody year, I tell you that, and with most of 'em, it wasn't last bloody year either! And local leave! Jesus Christ! What's that? These lads don't get local leave like your sort, Cowper. They get a few days in rest areas, that's what. And with a bit of luck they might find an estaminet where they can get pissed before their money runs out, or a knocking-shop where they get to the head of the queue before their bollocks freeze off! And that's their local leave, that's their bit of fun to keep them going another six months or more.

'And if they do get home, what's waiting there for 'em? No little place in the country, Cap. No dinners at the Café Royal and front row stalls at the theatre. No, there'll be a cold and draughty hovel; a wife who's so worn out capping bombs down at the munition factory, or skivvying for the likes of you, she's no strength left to comfort the returning bloody hero; kids in rags sitting in the gutter watching the landlords and the factory-owners and the gallant bloody officers driving past in their motor-cars.

'All equal when the country needs them, you say? I say *crap*! I say this isn't where men like these should be, fighting Jerry for the protection and profit of you and your sodding class! Home! That's where they should be, not fighting to protect what they've got which is mostly sod-all. But fighting to get what they ought to have, which is a fair bloody deal under a fair bloody law!'

He paused. There was a moment of almost stunned silence, then pandemonium broke out with men on their feet stamping and cheering so that the timbered walls of the underground chamber rang with the din.

Lothar himself remained seated only by an exercise of the will. Every cell in his body wanted to be upright, cheering and shouting. Here was the true spirit of revolutionary socialism made manifest! Here the soul of the people had been uttered forth in the accents of the people! Did it matter that it was Viney who'd done the uttering. Should the truth of the message be judged by the form of the oracle?

But as the applause died in the chamber, so the excitement died in Lothar's mind. Of course the form of the oracle mattered. Viney was no inanimate object, no mere cavern mouth or hollow tree, but a living man with plans and purposes which had nothing to do with the truth of the message. What they had just heard was no divine message, but mere demagoguery. Its aim was not the freedom of all men, but the death of one.

When there was silence again, the lieutenant tried to speak, but Viney silenced him with a simple gesture of one huge hand.

'I reckon we can see why you wanted to stay quiet, first off,' he said sympathetically. 'Every time you open your mouth, you start more trouble for yourself. We ain't heard much from the prisoner's friend, though. Come on, Fritz. Do your duty. You're an educated man just like the lieutenant here. Surely you can find something to say in his favour.'

Slowly Lothar stood up. There were a few derisive jeers but they faded quickly as Viney growled, 'All right. Enough of that.'

'Lieutenant Cowper,' said Lothar. 'How old are you?'

'Twenty-six.'

'What were you when this war started?'

'Twenty-three. Can't you count, Fritz?' some wag shouted, but Viney crashed the butt of his Luger on the table and roared. 'Next interruption gets the other end of this!'

Lothar resumed, 'What I meant to ask is, what did you do before the war?'

'I was just starting in the family firm. We are wine importers. I wasn't long down from Oxford, and I'd spent a year abroad partly learning the business, partly just getting a taste of . . . things.'

Cowper was answering with the frankness of a man who wants to hear his own answers.

'So already you were twenty-three, perhaps twenty-four, and you were starting work for the first time in your life in a wine-importing business owned by your family.'

There was a sigh of resentment from someone, but Viney's last threat prevented it from becoming the indignant protest it clearly longed to be.

'That's right.'

'I presume you had good hopes of doing well in this business?'

'One day I would own it.'

Again the sigh. This time Viney interpreted it.

'Come on, Fritz, you're supposed to be on this joker's side, not showing us what a useless fucking parasite he is!'

Lothar ignored him.

'Are you married, Mr Cowper?'

'Yes.'

'When?'

'Spring 1914. After I came back from the continent.'

'Do you have a family.'

'One boy.'

'Born when, please.'

'In April 1915.'

The officer's voice almost broke for a moment, but just held.

'So, Lieutenant Cowper, in August 1914 you were a young man, not long out of university, just starting a job you liked in a firm you would one day own, with a young wife, pregnant with your first child? A life of privilege, full of all the comforts, of pleasure, of love, with nothing but happiness either before or behind you.'

'Yes, you could put it like that,' said Cowper in a soft, melancholy voice.

'And what did you do in August 1914 when the war was declared?'

Peripherally, Lothar sensed Viney stiffening as he spotted the trend of the questioning.

'I joined the Army.'

'All this you gave up? Or at least put at risk?' asked Lothar in mock amazement. 'Why? Were you perhaps a military family?'

'No.'

'Were you perhaps persuaded by your parents or your wife?'

'No.'

'Then why? Please. The President of this Court has explained why so many men volunteered. They lost nothing but poverty and cold and physical and emotional starvation. But you . . . Why?'

'It seemed . . . the only thing to do. Most of the chaps I knew, university chums, boys I'd been to school with . . . everyone seemed to be . . .'

'So, it was not to protect your privileged life then that you joined?'

'No. I don't think so . . .'

'Yet you had very much to lose.'

'I had an idea, I think, that I would lose something else, I'm not sure what, by not joining.'

'Something else? What?'

'I don't know. Honour, perhaps. Or perhaps just an opportunity. It seemed perhaps this was a chance to, well, try oneself out, and one had to move quickly or it might go by, and then there'd be a lifetime of wondering . . .'

The audience was rapt. Lothar let the silence grow and deepen. Then: 'Do you still think of the war in this way, Lieutenant?' he asked softly.

'No!' cried Cowper with sudden emphasis. 'It's very different now. All those 1914 ideas, they're long dead! They looked back to the past, you see, and now, now it's the future we're fighting about. For better or worse, this war's changing things. Nearly every other war in history was a hiccough compared with this one. I don't know what it's going to be like, that future. But I do know that what we're fighting for now isn't any of the things it all started over. They're insignificant, puny, and no one should ever have been allowed to shed even a single drop of blood for them. But it's gone past that. It's the war itself that's important now. It's a great changing force, and it's been let loose, and there's no controlling it. So all we can do is sit on its back and make sure we're the ones in the saddle still when it finally dies away. It's too late to opt out.

You can't opt out of a hurricane, can you? You just can't!'

He broke off abruptly and sat with a faint look of surprise on his face as if taken aback himself at what he had just heard.

Lothar turned his back on Viney and stood facing the men.

'Lastly,' he said, 'I would like to read you something. It is a letter Mr Cowper gave me; asking me to post it somehow, if things went badly for him.'

'You bastard!' shouted the officer on his feet and flinging himself forward, 'You Hun swine!'

Lothar evaded the blind rush with a simple side step and a couple of men seized Cowper and dragged him, cursing and threatening, back to his chair. He waited till the man's protests faded into the expectant silence and then he started reading.

'My dearest one, If you read this, the worst has happened but I believe you will know that long before this letter can possibly reach you. Yet I do not believe the worst will happen, at least not here, not now. I'm with some chaps who seem to threaten all kinds of unpleasantness to me, but to be honest, the most of them are so like those decent, good-hearted, long-suffering soldiers I've been working alongside for three years now that I cannot believe they'll let the one or two madmen here harm me. In fact, this is probably one of the safest places I've been for a long time, and I myself must be a madman to want to get out of it and back to the fighting! So, no, I'm sure you'll never have to read this but instead will be able to hear rather than merely see these words of love and longing that I now send to you.'

Lothar paused, then slowly refolded the letter and replaced it in his tunic.

'There is a little more,' he said. 'But such things should only be heard by a wife from her husband's lips. I hope that is how they may reach Mrs Cowper.'

With a slight nod at Viney, he sat down.

The Australian's face wore a pensive frown. No one broke the utter silence of the chamber.

Suddenly the Luger butt crashed down on the table.

'Trial adjourned,' growled Viney, rising. 'We'll take an hour off for grub.'

It was in its way an admission of defeat.

But not, Lothar guessed as he made his way to the sleeping area to check on Josh, necessarily a final defeat. The 'hour for grub' would probably extend as long as Viney could drag it out. The Australian would know full well that time was the best dilutant for the sympathy which had been won for Cowper. Lothar resolved to press for the reconvening of the so-called court as soon as possible.

Instead, ten minutes later he found himself offering Viney an excuse to delay as long as he wanted.

'Disappeared? What the hell do you mean, Josh has disappeared?'

'He's not in the Warren, Viney,' said Lothar wretchedly. 'I think he must have gone out.'

His eyes met Taff Evans's and the Welshman looked away.

'Out?' said Viney, his eyes hard as slate. 'Who's on sentry?'

Coleport left the chamber and returned a moment later with Eddie Nelson, the bald-headed Londoner, who looked terrified. Before he could say anything Viney hit him high on the temple, sending him crashing to the ground.

'I've warned you what'd happen to anyone sleeping on sentry!' he cried.

His threat had a dreadful irony as Nelson's close associates knew that it was being caught asleep on duty which had precipitated his desertion.

'I wasn't sleeping,' protested the man, cowering away from the angry Australian.

'How the hell did he get out then?'

'I say let the silly little cunt go,' said Strother. 'He's doolally tap anyway.'

'Mebbe,' said Viney. 'But he's not so daft he couldn't lead the redcaps back here, Strother!'

'I don't think the lad'd betray us,' protested Hepworth.

'Don't talk wet, Heppy,' said Viney wearily. 'Half an hour with them bastards and Jesus Christ'd betray us! So let's get organized. We'd better find him quick!'

8

For the first ten minutes after he left the Warren, Josh had run as fast as the broken ground would allow him. He headed westward, towards the declining sun, towards the leaf-heavy trees he had distantly glimpsed from the escape hole of the ventilation shaft.

But as he approached the edge of the Desolation he began to slow down. It was over a year since the battle had moved on from this land. A harsh winter had balmed its wounds in snow, sudden spring had washed them in floods, and the heat of summer had tried to coax the dead earth back to life. The untouched countryside to the west, now beginning to ripen into rich autumn, had sent raiding parties of couch-grass and bindweed, of dock and vetch, into the desolation. Here and there a flash of blue or a smear of red showed where cornflower or poppy-seeds had been washed deep into the still living subsoil. But for this season they were merely a gesture, wreaths in a charnel-house, with the bare gaunt bones of the wrecked trees protruding from their own dry dust.

Somehow Josh felt that this was where he still belonged. To venture out of this dead forest was to risk once more the hot, violent pains of life.

He came to a halt a yard or two from the uneven boundary and sat down, his back against a tree, his eyes full of the slanting sun.

He felt no surprise as he saw the girl. She was moving parallel to the dead forest, picking up pieces of dry wood which she dropped into a pannier she carried over her left shoulder. She wore a simple blue dress which hung loosely over her slender figure and her long blonde hair was bound back with a green ribbon.

She moved among the stumps like the Spirit of Spring herself, come to promise that next year would be better. Such a fancy notion was far from being articulated in Josh's mind, but he felt her presence as a blessing, not a threat, and he sat quietly drinking in the total scene.

Nor did the girl have any such conceit of herself. She was enjoying the evening, enjoying the air, enjoying this excuse to absent herself a while from the daily round of hard toil.

But when she saw Josh, she felt no doubt that he was more likely to be a threat than a blessing.

With a cry of alarm she started back, the pannier slipping from her shoulder and falling to the ground where half its contents spilled sideways. This was the only consideration which inhibited instant flight. Her mother had warned her a hundred times, and without mincing her words, of the perils of close contact with soldiers of any class, rank or nation in these times of war.

That this was a soldier of *some* kind she saw at a glance. The boots, the worn and tattered but still recognizable khaki uniform; *un Anglais*, she guessed. It was partly the excellent camouflage of the man's clothes against the brown background which had kept him concealed so well, but added to this was his unnervingly absolute stillness. He sat unmoving, his back against a tree-stump, staring unblinkingly westward towards the sunset.

Slowly she reached down for the pannier. Her simple instinct was to pick it up and run. But her eyes, fixed on the soldier's face at first in terror, gradually admitted some diluting curiosity.

He was *so* still; and so young too. His face had something of that expression she saw every day in her own brother, Auguste, who had come back from the war a ghost in the ruin of his own mind. It was touched with unhappiness and suffering, but not aged by it. In fact he didn't look much older than herself, and she was only seventeen.

She postponed flight and began repacking her pannier. By the time she had finished this, curiosity had almost completely pushed out fear and was in its turn being joined by something like annoyance at the young soldier's complete lack of reaction to her presence. Her pannier replenished, she moved a few paces away and here paused, curiosity and pique finally compounding into boldness.

'*Bonjour, monsieur,*' she said in a clear, high voice.

For a moment there was no reaction. Then slowly the wide child's eyes turned towards her and seemed to see her for the first time.

'*Bong jewer,*' he said slowly, and smiled.

The smile dispelled what little remained of her fear. It was innocent and charming and free of any hidden intent, and it lit up his face with a glow that reflected the light of the setting sun.

She said, '*Qu' est-ce que vous faîtes ici, monsieur?*'

The smile remained but he shook his head.

'*Silver plate. Mercy,*' he said. Then he scratched his head with a look of comic concentration and finally added triumphantly, '*Van blonk. Bon!*'

She suddenly realized he was displaying his total French vocabulary for her benefit and laughed aloud. Eager to reciprocate she said slowly, 'Tommy. English. No Eggs. Cheerioh!'

He looked at her with such eager expectation that she racked her brains for more, but nothing would come.

Shrugging her shoulders, she said, '*Je regrette, c'est tout.*'

The two young people faced each other, smiling. Her shadow ran long before her, touching his body.

She said, '*Pardon, m'sieur, il faut partir. Maman . . .*'

He interrupted, saying, 'Josh.'

She looked at him in puzzlement.

He pointed his forefinger at himself and said, 'Josh.'

'*Ah! Vous vous appellez Josh?*' she cried. '*Moi, Nicole.*'

She imitated his gesture with the finger and repeated, '*Nicole.*'

'Nicole,' he said.

'Josh,' she said.

They smiled at each other in shared triumph as though the language barrier had been permanently broken.

Josh racked his mind in search of some way of extending the moment. Suddenly he thrust his hand into his trouser pocket, pulled it out and reached towards her.

Alarmed, she took a step back.

Then he said, 'Chocolate,' and she saw in his palm two

squares wrapped in silver paper. They were a present from Viney, given with a conspiratorial wink which hinted at a private horde, rather than the gentler truth that they were the Australian's own ration.

Slowly she reached out and took them. Then, as if fearing a change of mind, she rapidly thrust one of the squares into her mouth and chewed it with such an expression of delight that Josh laughed aloud in vicarious glee.

He motioned to the tree-stump inviting her to sit, but she wasn't ready for that. On the other hand courtesy as well as curiosity demanded that she did not eat this strange young man's chocolate and immediately depart.

'*Non. Promenadons*,' she said emphatically.

This was a word Josh recognized. He knew it meant *walk* but it also had such sexual overtones from its unique use in the always erotic invitation '*Voulez-vous promenader avec moi?*' that his stomach contracted for a second. Then he grasped that the girl's intention was to lessen the chances of the physical contact and he blushed as if she could understand his thoughts more easily than his speech.

They strolled along slowly with a good yard between them. From time to time each glanced at the other and from time to time their glances met and they smiled. She nibbled the other piece of chocolate, savouring its flavour. Some crumbs floated on to the front of her dress and she picked at them greedily, her hand against the loose fabric giving a hint of the slight breasts underneath.

'*Tu es soldat?*' she asked suddenly.

Josh nodded, delighted to have understood.

'*Oui, moi soldat*,' he answered.

She regarded him curiously and, looking down at himself, he realized what a strange kind of soldier he must appear. What remained of his uniform was tattered and stained. He wore no hat or battledress tunic and carried no weapons. And though his enforced daily shaving had never had much more to remove than what Wilf had unkindly termed 'bum-fluff', the total neglect of the last months had allowed a generous haze of blond hair to emerge on his chin and cheeks.

But he knew no way to explain, and was not altogether sure

he wanted to, that he was an ex-soldier, a deserter, a criminal in the eyes of his own country and probably hers too.

'*Soldat*,' he repeated, and performed a mime of shouldering arms and marching along with exaggerated limb movements. This made her laugh aloud and for a couple of strides she imitated him.

Josh racked his brains for something to say which would build upon this lightness of mood, but nothing came. So inexperienced was he in the making of small-talk with pretty young women that he would have had difficulty enough in English. In French it was utterly impossible. Fortunately the girl did not suffer the same inhibitions. She started to chatter away rapidly in French, seeming to find his looks of incomprehension amusing rather than frustrating. It crossed his mind that perhaps she was laughing at the nature of the things she was saying to him. He had heard British soldiers make the most outrageously obscene remarks to uncomprehending Frenchwomen, often accompanying them with serious and courteous expressions and gestures. It had usually seemed amusing at the time. Well, he couldn't believe that Nicole was saying anything obscene, but her delighted giggles made him guess that she was being a little cheeky.

But this cut two ways, he suddenly realized. There was no need to be tongue-tied as he would certainly have been with an English girl. Here he could say what he wanted, what he felt!

He began tentatively.

'I think you're smashing,' he said.

She paused in her chatter and looked at him in surprise.

'I think you're the prettiest lass I've ever seen,' he said with increasing boldness. 'I think your hair's lovely, and your eyes, and your face, and . . . and . . . all of you.'

They were walking parallel to the edge of the ruined wood. Nicole paused, regarding the fell of charred stumps distrustingly.

Then with a dismissive movement which set her skirt twirling to reveal brown legs, she whirled round to present her back to the ruined woodland and sat down on a gnarled root which had broken the surface like a bent knee. Large blue eyes were raised to him expectantly and she gave a little encouraging nod.

'I think you're smashing,' repeated Josh, suddenly as awkward again as if this had been Brack Wood behind St John's church and Nicole were an Outerdale lass he'd taken walking there after the service on a Sunday evening. The church, the wood, the whole valley came into his mind as clearly as though he were looking down on them from Outer Pike on a sunny day and with a poignancy that pierced him like a bayonet-thrust.

Slowly he squatted down on his haunches about a yard in front of the girl.

'Back home,' he said his eyes fixed on the ground between them, 'in Outerdale, this time of year, I'd likely be coming down from the fells at this hour. I'd be able to see our house in the distance and there'd be smoke from the kitchen chimney where me mam was making our supper. Dad'd be there already, he gives you a sharp look if he finds you've got home before he does, even if it's only by a minute. And likely as I got down to where Charter Beck starts to level out, I'd meet up with our Wilf . . .'

His voice trailed into silence. *Wilf.* He'd not mentioned his name since he and Lothar had walked off that mistily remembered battlefield. And though at first the horror of his dying had haunted his cratered mind like the cries of a wounded man alone and unreachable out in no-man's land, in the end the sound had faded or at least blended in with the background noise to become too familiar to be specially noticed. *Wilf.* He tested the name in his mind then spoke it out loud again. '*Wilf.*'

He could speak the name quite freely. Was it a betrayal that now so soon he could speak the name and feel nothing? The girl was looking at him strangely, at first he thought simply because of his sudden silence. But as he raised his head and tried to smile reassuringly at her, he tasted salt on his lips, and felt the evening breeze chill against his damp face, and realized that his eyes were running with tears.

Nicole made a small movement towards him. It would

probably have been easy then to precipitate what he so longed for – the close body to body embrace with his face buried deep in that golden hair. But he held back, made no confirmatory gesture; this grief and this girl had no acquaintance and to make one a bridge to the other would, he suspected, feel later like a betrayal of both.

'It'll be our Bert who works up the fells now,' he resumed in a stronger voice. 'He'd just left school when me and Wilf joined up. Next there's Annie, she's eleven; no, it'll be twelve now; and Ruth, she's seven; they're both still at school though soon it'll be the holidays and they'll be getting under Mam's feet. Then there's little Agnes, she's just a baby, always hungry. Three girls and three boys. Everyone said our dad was dead lucky when his first three were all lads. Then everyone had a good laugh when his next three were all lasses and said that God always evened things up. What they'll say now I don't know, now there's only Bert. They told them about Wilf, told them he was dead. Now I expect they think I'm dead too, I expect that's what they think.'

And now the tears which had been flowing so effortlessly were starting to force themselves out in huge body-shaking sobs. Through the swimming eyes he could see the girl's face twisted in bewildered concern. Once more she leaned towards him and this time he knew no strength remained in him to refuse her sympathy and whatever came behind it.

And then, like a waking dream, she retreated from him, her face contorted now in pain and terror. A rough hand was in her hair, twisting and pulling. Behind her, out of the coarse grass which had begun to make inroads on the still ashy ground of the Desolation rose a familiar figure with a triumphant leer on his face. It was Strother, and he was not alone.

9

Josh hurled himself forward but Strother merely used his free hand to strike him full on the nose, adding tears of pain to those of sorrow which still dampened his cheeks.

Nicole opened her mouth to shriek, but only one short, almost soundlessly high note emerged before Strother flung her into the arms of the approaching men, one of whom clapped his hand over her mouth as they bore her, kicking and struggling, back among the tree-stumps. There were about half a dozen of them, including the scab-faced Foxy, Nelson, the sentry who'd had to bear Viney's wrath, and the black Moroccan.

Strother himself followed up his blow to Josh's face by seizing the boy's right arm and forcing it so high behind his back that the pain of his nose was relegated by this new agony while Nelson stepped forward and hit him in the stomach.

'That's what I owe you, sonny,' he said viciously. 'No one drops me in the shit and doesn't get 'is card marked, I tell you!'

'Don't take on so, Nelson,' said Strother, forcing Josh to follow stumblingly after the others. 'Show a bit of gratitude. If the boy here hadn't of got out, think what we'd have missed! Josh, you're a revelation to me, old son. I reckoned you not to 'ave sense enough hardly to bash your bishop, and 'ere you was all the time, getting your leg over!'

The men ahead had stopped in a small clearing. Night was falling fast but there was still light enough to see their faces as they stood around the recumbent girl. Two were grasping her arms and legs and Fox was kneeling over her, one hand over her mouth and the other touching her body under the pretext of restraint.

'You fucking bastards! Let her go!' screamed Josh, trying to break loose despite the pain.

Fox looked up and the girl took advantage of the momentary relaxation to sink her teeth into the fingers squeezing her mouth. With a shriek, the scab-faced man released his grip and immediately Nicole began to cry out.

The Moroccan stepped forward.

'*Tais-toi*,' he commanded.

For a second, surprised by the French perhaps, she obeyed. Then a great torrent of words came gushing out. A knife gleamed in the Moroccan's hand. The point touched under her chin, forcing her head backwards.

'*Tais-toi*,' repeated the black man. And this time she obeyed.

'What's she going on about, Froggy?' asked Strother.

'Us,' said the Moroccan whose English was basic. 'Him,' nodding at Josh. 'She think he made trap.'

'Does she now?' laughed Strother. 'Well, in a manner of speaking, that's true, ain't it, Josh? You brought her to the woods 'cos you wanted us all to have a slice, isn't that right? Well, me for one, I'm not going to refuse such a kind and generous invitation. Here, Foxy, cop hold of our benefactor 'ere.'

Foxy, who was standing menacingly over the girl, nursing his hand, reluctantly stepped back and took over from Strother, forcing Josh's arm upwards till the pain was excruciating. The Cockney took Fox's place and, bending over Nicole, began to fondle her breast through her dress.

'Not much meat on this one,' he said. 'I've known young lads with as much to get an 'old of. I hope you've not been 'aving a bit of the Oscars, Josh. Viney wouldn't like that, not one little bit.'

Slowly he drew the girl's skirt up over her thighs. Underneath she wore nothing. An audible sigh went up from the men at the sight of the triangle of golden down on her pubic mound. Josh strained forward in anger and hatred, yet he was looking too. Apart from his pre-pubescent sisters, he had never seen a woman naked below the waist. To his other emotions were added shame and disgust as he recognized his own prurience and he screamed abuse at Strother till Foxy produced a bayonet and thrust its point so hard against his kidneys that it pierced both shirt and skin.

'Shut it, if you don't want gutted,' snarled Foxy in his ear.

'What did I tell you, boys?' said Strother, his hand caressing the girl's stomach. 'Nothing odd about young Josh, didn't I say that? Nice normal lad. Might be tempted by a sheep or two up in them mountains he lives in, but it'd always be the ewes he went after. Come on, girlie. Open up.'

He was trying to force his hand between her legs. The man holding her ankles drew them apart by force, the hand moved down, Nicole screamed, her eyes rolling round till they found Josh's and locked on them, agape with pleading or accusation.

'Dry,' said Strother. 'And tight. Very tight. Josh, you may prove a bit of a disappointment after all. Or 'ave you been saving it up for your old mates? Well, here's one that won't be disappointing the little lady.'

Slowly he began to undo his trousers, sliding them down over pale and flabby thighs to reveal his erect penis, short but enormously thick.

'Hold on, Strother,' said Nelson suddenly. He was kneeling beside the girl, holding one of her arms. 'I ain't sure about this. I owed that little shit one for dropping me in it with Viney but this ain't right . . .'

'No, you hold on, Nelson,' interrupted Strother powerfully. 'You'll get your turn soon enough, either with 'er or with me, take your pick. You, Froggy, tell the lady to stop thrashing about, else you'll slit her throat for her.'

The Moroccan looked down on the girl's madly convulsing body which the men holding her limbs were unable to control. Already she had moved her head with such violence that his knife had nicked the soft skin beneath her jaw and a trickle of blood ran down her throat.

Then the Moroccan's cold eyes turned to Josh who was struggling and screaming like a madman regardless of the pain in his arm and the prick of the bayonet in his back. After a moment he nodded as though at some conclusion reached, and stepped back from Nicole, returning the knife to its hiding place.

'No,' he said in his fragmented English. 'Not me . . . kill him also or he kills . . . not me.'

He nodded his head at Josh, but it was a gesture of neutrality not of support.

'Sod you, you fucking black Frog!' yelled Strother. 'I'll make the cow lie still.'

He knelt astride the twisting girl and raised his clenched fist.

'To hell with this,' said the man called Nelson and released the girl's right arm. Immediately she swung open-handed at Strother's face, dragging her nails across his cheek, leaving a quadruple track of blood in their wake. He cried out in rage and pain but was only momentarily diverted from his attack. His fist swung back once more.

Through the darkling stumps, unnoticed by the men in the clearing who were riveted by the drama going on before them, more men came running. First to arrive was Lothar with Viney close on his heels. The German caught Strother's upraised fist and held it effortlessly, pressing it backwards till the Cockney, his face bestial with rage rose to his feet. Dragging himself free, he plunged towards Lothar who swayed gently to one side and gave Strother an almost playful punch. Normally this would hardly have caused the man to stagger but his trousers were round his ankles and the one pace he took was enough to make him trip himself up. He fell on his face, his white buttocks in the air, and suddenly it was comedy. Someone laughed, others joined in. Lothar grinned broadly.

But Viney was not amused.

'What the hell's going on here?' he demanded. 'Are you jokers off your tiny heads, or what? Jesus wept! I bet they can hear the noise you're making at the sodding front! Strother, what the hell are you playing at?'

Strother got to his feet and began to pull his trousers up. He was no longer erect.

'It was him, that sheep-shagger,' he said pointing at Josh. 'We found 'im with this tart. We were just 'aving a bit of fun with her.'

'You were going to rape her, I think you mean,' said Lothar quietly.

'Shut up!' bellowed Viney at Lothar.

Encouraged, Strother went on, 'Christ, Viney, you know 'ow long it is since any of us saw a woman . . .'

'So you couldn't wait? You had to get stuck in right away, out here, only fifty yards from the edge of the Desolation?' said Viney. 'I'll tell you something, Strother, you want to come so bad you're going the right way about it. They say the last thing that an executed man feels is his cock spurting off, and that's the road you're taking all of us on!'

With sinking heart, Josh realized that Viney's anger was merely at the location of the proposed rape, not its morality. Nicole had been released, as the men holding her had decided that they didn't want leading roles in this business till they saw how Viney was reacting. She struggled to a sitting position and pushed her skirt down over her legs. Her eyes were flitting hither and thither around the surrounding men in desperate search of an escape route. Foxy still held Josh but in much less certain grip. It was easy for the youth to break loose and move towards the girl.

He dropped on his knees beside her and, not daring to touch, said, '*Pardon, pardon,*' trying to get her eyes to meet his.

Lothar came to join him.

'Tell her I'm sorry, Lott,' begged Josh. 'Tell her it wasn't my fault. Please!'

Lothar smiled reassuringly and spoke to the girl in rapid French. She regarded him distrustingly and he spoke again with greater emphasis.

'What the hell are you saying, Fritz?' demanded Viney.

'I was just assuring the young lady that no one would harm her,' said Lothar.

'Well, just you shut it, Fritz. I'll do all the assuring that's needed round here,' said Viney.

Encouraged by this reproof to the German, Strother said, 'Depends what he means by harm, don't it? All right, Viney, I was wrong to be so 'asty, but let's get her back to the bivvy and 'ave a bit of fun, what say you? Fair shares all round, I don't think we'll get many sitting this one out except mebbe the Kraut and his fancy boy.'

Lothar ignored the sneer which he had heard many times

before when his enemies wanted to try to set Viney against him.

He said, 'Let us try a vote. This is the English democratic way, is it not? Please, let everyone here who would like to rape this child raise a hand in the air.'

He turned slowly through a full circle. No one stirred.

'Mr Strother, please, you must raise *your* hand, I think,' said Lothar.

'This is bloody foolery!' yelled Strother. 'Viney, are you going to let this Kraut take control or what?'

But Viney was not paying attention to the Cockney's challenge.

'Belt up!' he said. 'Listen.'

They all listened. Straight away they heard it, a woman's voice, not too distant, calling, 'Nicole! Nicole!'

'Now see what you've done!' said Viney, rounding on Strother. 'Half the sodding countryside's probably looking for the kid now. Come on. Disappear!'

With a speed born of practice, the men faded into the undergrowth. Viney himself seized the girl, who had stiffened and turned her head at the sound of the woman's voice. Her mouth had opened as if to call out in answer but no sound had come. Her desperate dilemma showed in her eyes. Her terror urged her to call for help but her last reserves of control told her that all she would be doing was bringing someone else to share her fate.

Her silence was an act of courage which perhaps only Lothar of those present was calm enough to appreciate.

Josh jumped upright as Viney seized the girl, and looked ready to attack the Australian. But the big man only smiled at him half sympathetically and said, 'That's right, son. Give us a hand, will you?'

The voice called again, much nearer. Viney pulled the girl into the trees and Josh followed with Lothar close behind.

A few moments later a woman of about forty emerged from the dusk and stood at the edge of the clearing. She was dressed in peasant black, her thick brown hair was drawn back severely from her broad forehead and held by a twist of black cloth with ragged edges as though to proclaim its function was practical not decorative. She had a strong face, but hollow-cheeked and with a spottily sallow skin tanned to the semblance of health by sun and wind. Now her face bore an air of suspicious uncertainty, like that of a thirsty animal knowing there is danger near a waterhole but urged on by a need stronger than fear. As they watched, she stooped and picked up a heavy sharp-edged stone.

She knows we are here, thought Lothar.

Viney had reached the same conclusion. He released the girl and gave her a gentle push. For a second she hesitated, then with a cry of '*Maman!*' she rushed forward and flung herself into the woman's arms.

It was a clever move of Viney's. Flight was impossible for either woman in that moment of reunion, and by the time they gave it any thought, they were surrounded by an unbroken ring of men.

'Two of 'em now!' proclaimed Strother, smacking his lips appreciatively. 'We hang on much longer, boys, and we'll end up with one each!'

'Belt up, Strother,' said Viney. 'You've done enough damage. We could've made the girl look like an accident, but two of them's stretching things. You, Fritz, talk to them, find out what you can about them, where they're from, who'll miss them.'

Lothar approached the two women and began to talk in a gently persuasive voice. At first the older woman merely spat out what even the non-French speakers recognized as angry abuse, but after a while she seemed to calm down, either reassured by the German's manner, or because she decided this was the best survival tactic, and began to reply to his questions.

Viney typically became impatient after three or four minutes.

'Come on, Fritz!' he interrupted. 'What the hell's she saying?'

Lothar said, 'Please. A little time longer,' and returned to his questioning. Just when it seemed to everyone that Viney was about to explode, the German finished.

'*Merci, madame,*' he said to the French woman. To Viney

he said. 'Her name is Madeleine Gilbert. This is her daughter, Nicole. She says her husband was a farmer with a piece of land near here. More I think what you call a smallholder than a farmer. She says he had to go for a soldier and was killed at Verdun. Their farmhouse was not far behind the old front line when the French held it in 1915. The buildings were hit by our, that is, German shellfire. The stock was stolen by French troops. She does not care for soldiers, any soldiers. Nor any authority much. She has lived with her old mother since her husband's death and the farm destroyed. But such a life was hard for her and hearing that the war had moved away from her husband's land, she decided to come back to try to get things started again. The authorities said it was not yet permitted, but she came all the same. It is not easy for her. Two old people, Nicole here, and her brother, a young man made simple by the war but able to work, these are all the help she has. There are relatives of her husband's in Barnecourt, the village in the valley. The war has not touched them. They might help, but she is too proud to ask, I think. But she will be missed, she says. There will be searches for her.'

He paused. Viney said, 'Then mebbe we'd better make sure they're not found.'

Lothar said urgently. 'These people are no threat to us, Viney. They do not know who we are. I shall say we are a special unit of British soldiers in training . . .'

'Very special,' said Viney. 'With a black Frog. And a Kraut too. Your French may be good, Fritz, but I'll bet you've got a Kraut accent. In any case, we look just like what we are, no getting away from that.'

Lothar let his gaze run round the group of silent men. Viney was right. Even those he had come to know as susceptible to reason and reluctant to survive at the expense of looting and killing looked just as villainous as the rest, with their unshaven faces and filthy ragged clothes. Only Josh, because of his fairness and his youth, looked less than terrifying.

'All right,' said Lothar. 'Let us not deceive them. Let us rather trust them.'

'Trust them?' echoed Viney, screwing up his mouth like a nun forced to speak a blasphemy.

'Why not? What is the worst they can do? Tell the authorities we are here? They know that already, is it not so? How many of us do they see? Only a dozen. Where is our hiding-place? They have no idea. These women are no threat to us, Viney! I believe they hate the authorities almost as much as we do!'

Viney nodded abruptly, not in agreement but as a command for silence which Lothar felt it politic to obey. The big Australian stood deep in thought. Strother was obviously bursting to say something and Lothar willed him to speak before Viney was ready.

Finally the Cockney exploded.

'For fuck's sake, what are we waiting for, mates? Ere's a gift from 'eaven. We can take 'em back with us and shag them rotten, and who's going to miss 'em but a couple of antiques and a loonie, who shouldn't be where they are anyway from what the Kraut says? It don't need no thought that I can see!'

Thank you, Strother, said Lothar mentally.

Viney swung on the Cockney, his eyes blazing.

'You're the bossman now, are you, Strother? First the Hun here starts taking votes and then suddenly you've elected yourself king. Well, let me tell you something, the only way to get on top of this heap is by pushing me off and any time you fancy your chances at that, just let me know.'

He turned to Lothar.

'You're a clever cunt, Fritz,' he said. 'But in my experience, there's nothing more cunning than a pair of sheilas. What guarantees are there they won't talk?'

Lothar said, 'Their need to survive. If they complain, they will probably be removed. Also they are in great need. If we give them food, they will be grateful or at least dependent. Eventually they may have for us fresh milk, fresh butter. Sooner perhaps if you permit someone go from time to time to help with heavy work on the farm. But not Strother, I think.'

There was a small ripple of laughter at this.

Encouraged, Lothar added, 'They have contact with the

world outside. They can see and hear things we cannot. They could be our forward observation post.'

'All right, all right,' said Viney. 'Don't overdo it, Fritz. Say all this to them, see what the older one says.'

Lothar spoke to Madeleine Gilbert, who looked at Viney appraisingly before replying.

'What's she say?'

'She's guessed what we are. She says men who have left the war are in her opinion the only sane men in France. She will not do anything to harm them.'

'You believe her, Fritz?'

'Yes,' said Lothar. 'I do. Viney, they don't know anything that can harm us, not how many we are, nor where we are hiding. Let them go!'

Viney stood with his head bowed in deep thought for a whole minute. Could he really be contemplating the rape and murder of these two women? thought Lothar in horror. Why not? he answered himself. After the huge descent into the pit of war most other moves towards violence were a matter of degree.

At last the Australian spoke. 'It's your idea, Fritz,' he said. 'Tell 'em they can go, but we'll be in touch. I want 'em too scared to try anything funny.'

Lothar turned to the two women and told them they were free to go. The atmosphere relaxed generally. But Madeleine, who looked far from scared, now addressed her daughter who after a while shook her head, then pointed to Strother.

'She wants to know who the man was who attacked her daughter,' Lothar said to Viney.

The Frenchwoman strode across the clearing to Strother. The Cockney watched her approach with an obscene leer. She came to a halt about two feet in front of him, took a deep breath and spat full in his face. With a cry of rage he leapt forward at her but her right hand which had been buried in the voluminous folds of her black dress swung upwards. She was still grasping the stone she had picked up as a weapon on her first appearance in the glade. Now she crashed it with all her strength into Strother's crutch.

The Cockney collapsed like a pole-axed bull, and rolled on

his back, his knees drawn up to his chest, screaming in agony. Beside Lothar, Josh took a step forward, but the German caught his arm.

'Someone shut him up,' ordered Viney. 'And let's get out of here before they start selling tickets to come and watch us. Come on, move it!'

The men began to fade away into the night.

Josh went up to Nicole and anxiously, uncertainly he held out his hand.

She regarded it suspiciously, then slowly stretched hers out in return and touched his finger ends.

'*Au 'voir*,' said Josh, but she would not answer.

'Come on, Josh,' said Lothar. 'I think you have angered our leader enough for one day.'

He hurried the young man along after the others, talking gently into his ear in a manner both reassuring and distracting. He had noticed Patsy Delaney detaching himself from the group at a nod from Viney and guessed that he had been instructed to follow the women. Whatever his purpose, it was not in Josh's interests to be aware of it. But the boy was too elated with relief to pay much attention to the others.

'I thought that Nicole's mother had really upset things when she hit Strother!' he confessed. 'I thought I was going to have to fight everyone!'

'On the contrary, Josh,' laughed Lothar. 'It was the best move she could make. No need to assault your daughter's attacker, is there, if you propose to go immediately to the authorities and arrange to have him shot! No. Beating in Strother's balls was better than swearing an oath on the Bible, and Viney knew it! He has a good understanding of people, our leader, but not so good of himself. That's what makes him so good a leader, I think. And that's also what makes him so dangerous.'

10

In the days that followed, life in the Warren settled back into its old routine, but things were not the same. The trial and the episode with the Gilbert women had brought about a re-evaluation of relationships which produced some curious results. Josh had always been popular except with those who resented Viney's evident fondness for the boy. Now he was accorded an almost talismanic status. He was 'the still centre of innocence in a guilt-racked community', a phrase of Lothar's which had Taff Evans closing his eyes in concentration before nodding his head in agreement.

Nelson and Fox in particular seemed keen to dissociate themselves from Strother's attempted rape of Nicole. Nelson, a butcher in civilian life, who still plied his trade when theft or traps produced any fresh meat, went out of his way to slip Josh tidbits. Fox's method of making his peace with the boy was more indirect. His approach was via Lothar, whom he asked abruptly one day if there was anything he could do for his facial sores. Lothar washed them with makeshift disinfectant, mainly spirit and vinegar, and instructed Fox that this must be done at least twice a day.

Only Strother and one or two of the darker spirits continued to view the young Cumbrian with open resentment.

Josh himself was concerned with one thing only, the development of the Volunteers' contact with the family Gilbert. Viney just laughed at him when his impatience became too strong to hold and, ruffling his blond curls with his huge hand, said, 'Hold your horses, son. We need to move easy. But if it can be managed, I'll see it's done. That's a promise.'

Josh believed him and so did Lothar. Viney was a man outside most systems of morality and legality, but he would not lie to Josh.

The question which bothered Lothar most was the postponed decision on Lieutenant Cowper's fate. He didn't know whether to press for a reconvening of the 'court' or to leave well alone. He tried to keep the young officer cheerful by frequent conversations and he returned his letter to him with the assurance that it served no function now as his release was almost certain. But as the days slipped into weeks, he could see the dull despair of the prison cell beginning to steal over his spirit. What Lothar feared most was that Cowper would be driven to try to escape, thus providing the perfect excuse for killing him.

Finally he went to see Viney and urged a quick decision.

'Shooting the bastard, that's a quick decision,' mocked Viney.

'Release him. That is quick too,' said Lothar.

'And wait for him to bring the redcaps back? That'd be quickest of all, Fritz.'

Lothar said, 'He doesn't know where he is, Viney. If we blindfold him and take him many miles before releasing him, he can hardly lead anyone back to the Warren. The redcaps know we are somewhere in the Desolation, I am sure. All these raids you do are a very strong hint!'

'Too right,' admitted Viney. 'All right, Fritz, I'll think about it and give a verdict tomorrow.'

That night Lothar deliberately steered the conversation in the common chamber round to Cowper in the hope of recreating some of the sympathy that the trial had brought his way. It was interesting. Because they were talking about another man's fate, many of the Volunteers were much less inhibited about their own background and experience than hitherto.

Hepworth, it emerged, despite his bucolic appearance was a townie, a mill-hand from Bradford who had joined up with a lot of his mates in a local 'pals' battalion. He was adamant that he had had nothing against his officers.

'There were some grand lads among 'em. Grand lads. Especially the young 'uns. Just like us, they were. Up in the front line, same risks, same attacks, same patrols. Nowt wrong with them, and there doesn't seem much wrong with this lad Cowper to me.'

'Same as you, was they?' mocked Strother. 'Same grub? Same pay? Same leave? Give over, Heppy, you'll have me dead laughing.'

'Give you the chance at better grub or pay or leave, wouldn't you have jumped at it?' demanded Hepworth.

'That's why I'm here!' said Strother. 'Which brings me to the point: if you liked it so much back there, why'd you bunk?'

Hepworth took a long time to reply.

'Something gets stretched,' he said in a low voice. 'Something gets a bit tighter every time you go over the plonk, every time you hear a shell scream towards you or see a Flying Pig spinning over your head. I saw a lot of my mates get killed, but it weren't that. It were one chap who started talking of giving himself a Blighty through the foot before the next attack. Next attack came and bang! I heard his bondook going off down the trench a way. I went to see, but when I reached him, I saw he'd changed his mind. He'd blown his head off instead. I thought: That can happen to me. Hang on till you stretch too tight, and that's what you do. So I said, Sod this, and next chance I got, I was off out of it. But I'm not blaming any other bugger for it, leastways not anyone like Mr Cowper. Brass hats and base wallahs, yes, but in the front line, we're all in the same shit, that's how I see it!'

'No, no, it's not true. It's different for them, at least for most of them,' interrupted Fox vehemently. 'They think they're little tin gods. They've been to these fancy fucking schools and have got fancy fucking manners and they think they're better than the rest of you. They're just too thick to be afraid, that's the top and bottom of it. Too fucking thick!'

This rather odd intervention almost halted the debate for a moment, but Strother picked up the main line again, saying triumphantly, 'Anyway, he wasn't in the front line, this bastard, was he? He was riding around up there, wasn't he, this fucker, hunting down poor sods like us, just like we was fucking foxes or something, begging your pardon, Foxy. I say we should top him!'

Fox nodded his agreement.

'Why do you two hate officers so much?' asked Hepworth curiously.

Fox turned away, but the Cockney's face darkened and he began to talk. His story, carefully tailored towards self-justification, was that, while a cook with one of the London regiments, he had protested when he discovered an officer removing all the tins of strawberry jam from a supply truck, leaving nothing but the unpopular apple and plum for the men. All that he got for his altruistic moral stand was an immediate transfer to fighting duties and selection for all the most dangerous jobs thereafter.

'Victimization, weren't it?' concluded Strother. 'They were all in it together, the bastards. And they'd have seen me dead before they was through. So I said fuck it and slung me hook. That's why I hate fucking officers!'

Taff Evans caught Lothar's eye and grinned his recognition that the more likely version of the story was that Strother, having wangled himself a cushy job as a cook, was caught working a fiddle and returned to active duties.

You didn't contradict another man's story, however, and now the flood-gates were open and man after man gave his story as if in the telling they could purge themselves of fear and shame and guilt. Fox didn't offer to speak, Lothar noted, though even Nell, the younger of the two Indians, broke his customary reticence.

'They told to us that the King needed our help,' he said in a low, soft voice. 'They told to us that there would be battles to be won, an enemy to ride at with our lances, green fields to pitch our tents in after the battle. Instead there were holes filled with water to hide in, and no enemy to be seen but only noise and bullets and huge shells exploding, and soon our horses were dead or taken away from us to pull wagons. But it was not our officers' fault. They had listened to the lies too, they had been deceived like all of us.'

This simple statement produced a sympathetic silence broken by the irrepressible Strother.

'Don't take no notice of this black bugger!' he said scornfully. 'They're trained to think white officers are little fucking gods; they know no better. Anyway, he's bound to be all for that cunt back there, isn't he? They're both fucking cowboys, moaning 'cos they've lost their horses, aren't they?'

Nell looked bewildered at this but the West Countryman Quayle, who had appointed himself a sort of unofficial guardian to the Indians, flung himself forward and seized the Cockney by the shirt.

'Here, what's this?' yelled Strother as he was flung backwards against the wall. 'You want to watch it, Quayle. I'll have you, my son. I'll have you!'

'And you want to watch what you're saying,' said Quayle. 'You want to watch your tongue.'

He was a big man with a round, placid face, now contorted into an unprecedented anger. Usually a man of very few slow, drawn-out words, which had won him the ironic nickname of Quacker, he now spoke with a frightening vehemence.

'Leave him be, Quacker,' said Hepworth. 'You'll just catch summat if you get too close to him.'

Slowly the anger faded from Quayle's face to be replaced by an expression of faint embarrassment.

'Aye,' he said slowly. 'You're right there, Heppy. Just you remember, Strother, this lad's got his right to speak just like anyone else.'

He returned to his seat by the Indian and put a reassuring hand on his shoulder. The young man smiled at him in gratitude and relief.

'What about you, Quacker?' asked someone. 'You're on the trial board with Heppy. What do you think?'

'Me?' said Quayle, as if surprised that anyone should be interested in his opinion. 'I ran because I was frightened. If they catch me, maybe they'll shoot me for running. That seems stupid and wrong to me. But shooting someone for not running, that seems wronger and stupider.'

This felt like the final word. Many heads nodded in agreement. Lothar saw Strother shoot a glance of pure hatred towards Quayle but the Cockney was not prepared to risk another confrontation. He also saw Pat Delaney who had been sitting quietly in a corner rise and leave, doubtless on his way to report the general feeling of the Volunteers about Cowper's fate.

How Viney would react was hard to assess. Lothar felt he was reaching towards some kind of understanding of most of the men here, though he frequently warned himself against the arrogance of such an assumption in a man who could not begin to understand himself. But, arrogant or not, he had to confess himself beaten by Viney. Sometimes the man appeared a simple thug, sometimes a subtle and sinister manipulator, and sometimes (in very brief flashes, these) as a confused young man, bewildered and unhappy. *Young* man? How old was Viney? he wondered. Like everything else about the man, it was difficult to say. There was a barrier there against deep penetration, a darkness at the core which might conceal monsters, or nothing at all.

He rose and stretched.

'Sleep, I think,' he said.

As he went past Quayle, he said in a low voice, 'Be careful of Strother. He is a bad enemy, I think.'

Quayle smiled and said, 'No, it's mostly show, I shouldn't be surprised. Where I come from we take people as we find them and there's a lot of good even in the worst.'

'Yes,' said Lothar. 'All the same, be careful. We all must take great care.'

He went to bed.

11

Josh woke out of a deep sleep to the tail-end of a fearful scream.

All around was confusion and alarm as men started up from their pallets and grabbed for clothes or struck lights for lanterns. A hand grasped Josh's arm and, turning, he saw Lothar peering closely at him.

'Are you all right, Josh?' he asked.

'Yes, I'm fine. What is it, Lott? What's happening?'

Lothar gave him a reassuring smile. Inwardly, he was delighted to observe the normality of the boy's reaction to this alarum. The contact with Nicole and the near-tragedy which had followed seemed to have shocked him back to a much fuller contact with reality, if this strange troglodyte existence could be so termed.

'I don't know,' said Lothar. 'Let's go to find out.'

Together they set out to follow the others. There was not far to go and a corpse awaited them at the end of the trail.

It was Nell, the younger of the two Indians. His smooth brown body, completely naked, lay in a corner of one of the smaller dug-outs. His face was upwards and was composed into such an expression of peace that he might have been simply sleeping had it not been for the strange angle of his head to his shoulders.

In the other corner, also naked and with blood streaming from his mouth between broken teeth and crushed lips, was Quacker Quayle. His big brown eyes peered upwards with the incomprehension of a scolded animal at the figure of Viney whom rage seemed to have swollen to the stature of a Colossus. His right fist was clenched solid, with fresh blood staining the white knuckles, and his eyes were ablaze with mad fury. Blackie Coleport had placed a restraining hand on his shoulder and it was he who offered an almost apologetic explanation.

'He found them together. They were at it,' he said.

'I won't stand for it! I won't stand for it!' cried Viney in a terrible voice.

There was a disturbance of the press of onlookers round the entrance and Dell, the grey-haired Indian emerged. He took the scene in with an impassive glance, and, ignoring the Australians, knelt alongside his dead countryman and began to arrange his limbs.

'Oh Lott!' whispered Josh, his voice catching. 'So far from home. They're so far from home.'

Abruptly Viney shouldered his way through the crowd and disappeared. For a moment the rest of them just stood and looked at the dreadful scene, then abruptly Taff Evans stepped forward and bent to assist Quayle to his feet.

'Come on, Quacker, me boy,' he said. 'Let's be getting you out of here, shall we? Here, one of you lot, give us a hand.'

For a second there was no movement except for an involuntary back-stepping by a couple of men at the front. Then Lothar seized Quayle's other arm, slipped it over his shoulder and together they helped the half-stunned man along the passage to his pallet. Here Lothar bathed and cleaned his mouth as best he could.

'He needs medicines to make him rest,' he said. 'He is under shock, I believe.'

'Poor sod,' said Evans, then added with sudden ferocity. 'That Aussie bastard! He needs pulled down, that one. I don't approve nothing of these nancy boys, but he's not God Almighty to be dishing out such punishments.'

He raised his voice, but no one accepted the invitation to support – or oppose – him. When discussion did take place, it was carried on in twos or threes with the men whispering and casting fearful glances around as though anticipating Viney's sudden and violent interruption.

Josh was both horrified and fascinated by what had taken place. A virgin still, his inexperience of matters heterosexual was to some extent compensated for by natural inclination, barrack-room discussion and a country upbringing which made no secret of the basic facts of nature. But contact beyond the male–female baffled him completely.

Lothar tried to explain, but after a while Josh burst out in bewilderment, 'I still can't see why any man should want to put his cock up another man's bum!'

He was hurt and slightly indignant when Lothar's response was to smile and shake his head.

'Well, I reckon Viney's got the right of it,' he said sulkily.

'To kill a man for sleeping with another?' enquired Lothar mildly.

'No. I don't mean that, I'm really sorry about Nell. He was all right, was Nell. But I can see why Viney was, well, *outraged*, that certainly,' said Josh with conviction.

'Outraged?' murmured Lothar. 'Yes, it is true. Outrage can certainly rouse anger. But do you not think, perhaps, that it is fear that breeds fury?'

Josh couldn't answer this because, as with so much of what

Lothar said, he didn't really understand it. Strother, who was at the centre of another small group close by, suddenly raised his voice with the clear intention of being overheard and said, 'It's those Frog women I blame. A man's gotta 'ave 'is outlets. If we'd done like I said and all dipped our wicks, well, it'd have kept us going for a bit, like. Mebbe old Nell would still be alive and poor old Quacker'd 'ave all his teeth.'

Josh went pale and he would have risen and made for Strother if Lothar hadn't gripped his arm.

'Of course,' continued Strother, 'it'll be all right for *some*. I can guess what little creep and what fucking foreigner's going to be let out to sniff around that farmyard. They'll be *safe*, you see, dead safe. Only one knacker between the pair of 'em, and that's up the Kraut's arse!'

Again Josh moved, but Lothar's grip was like an MP's manacle.

'Sit still, Josh,' he said in a quiet but carrying tone. 'At home we have a saying: No need to fight a man with the smell of death upon him.'

This resort to soldiers' superstition rather than physical violence was nicely judged. Strother paled and his companions fell silent. The Cockney then began to assail the German with obscene abuse, demanding to know what the fuck he meant. Lothar sat meekly under this scurrilous onslaught, knowing that every soft reply and sympathetic smile rubbed salt into the other's raw fear.

'Can you really smell death on him?' whispered Josh in naïve awe as they finally settled down for the night.

'Only the fear of death,' murmured Lothar. 'He is a simple peasant, that one, and this is how the masters have always controlled the peasants – through their fears. I am ashamed to be doing the same.'

'Ashamed? After all them things he said about you? And me too!'

'Let us hope that he was in part telling the truth, Josh. Let us hope that perhaps we *will* go out to the farm.'

It was a hope Josh hardly dared utter. Everything depended on Viney, he saw that. His defence of Viney's action that night had been at least partially motivated by his awareness of how much his own happiness depended on the big Australian.

He fell asleep and tossed uneasily in an erotic nightmare in which Nell's naked black body turned into Nicole's, capable of arousing him even in death.

Close by him, Lothar did not sleep. He was certain in his own mind that Viney's discovery of the affair between Quayle and the Indian had been no accident. Strother had promised revenge and the outcome must have exceeded his wildest expectation. But his appetite was more likely to have been whetted than sated. Lothar had a sense of destructive forces being let loose which could turn the Warren into a microcosm of that greater destruction they had all fled.

The next morning Viney sent for him. The Indian's body had been removed in the night and now no trace of the violent incident remained either in the Warren or on Viney's face.

Before the Australian could speak, Lothar said, 'Last night, was it Strother who told you about Quayle?'

The cold green eyes regarded him. It was like being watched through ice by some alien creature of the deep.

'You'd better watch what you're saying, Fritz.'

'Why? Will I end like poor Nell?' mocked Lothar. 'It *was* Strother, wasn't it? He told you so that you would carry out his revenge on Quayle. You obeyed. Think about it, Viney. Monster you may be, but at least be your own monster!'

For a moment he thought he'd gone too far, but the Australian after a visible effort of control said, 'Lecture over, Fritz? Then let's get down to why I've sent for you, shall we? There's two things. First is, I've decided to let that Pommie officer go.'

The green eyes were focussed tight on Lothar's face as if challenging a reaction. It is a life for a life he's offering, thought the German. Cowper's for the Indian boy's.

But he merely nodded an acknowledgement.

'Second thing is, I want you to go out with Patsy here to take a look at that Frenchwoman's farm, have a talk with her, see how the land lies.'

Lothar was taken completely by surprise.

'You mean you wish us to check if she is to be trusted?' he said.

'Naw. I reckon she's all right. I've had Patsy and Blackie keeping an eye on the place for the past week or so, and no one's strayed even as far as the village, so she ain't in no hurry to talk with the authorities, is she? Time has come to remind 'em they're not forgotten. But it's no use him making contact seeing as he don't speak the lingo. He'll show you the way, though.'

When Lothar told Josh where he was going, the boy rushed straight to Viney and begged. 'Please, Viney, can I go as well?'

Viney laughed and went towards him and ruffled his hair.

'Not this trip, Josh. No room for billing and cooing this trip. We'll see what we can do later, mebbe.'

Coming from Viney, this amounted almost to a promise, and Josh waited in an agony of impatience for the recce party to return. The news was both good and bad. Both men affirmed that, in their judgement, Madeleine Gilbert had been as good as her word and had made no attempt to pass on news about the Volunteers to anyone else. What was bad was that the state of the farm and its buildings was very bad indeed and any hope of using it as an immediate source of fresh produce was out of the question. Indeed, Lothar argued that without help he could see no way for the inmates to be in a fit state to survive beyond the following winter.

'You mean you'd like *us* to support *them*?' enquired Viney.

'Why not? We have manpower without useful labour. And we have a plentiful supply of basic provisions.'

'At the moment, sport,' interposed Blackie Coleport. 'And we risk our fucking lives every time we go out looking for more.'

Viney relapsed into thought.

Josh held his breath and tried to will the big man into agreement. Not for the first time he wished that the constant and apparently instinctive antagonism between Lothar and Viney did not exist. These men were the most important people in his life and he wanted them united. Lothar he loved and respected and trusted. Viney was very different, much less close to him, but in some ways much easier to understand. The man's directness and power and even his use of physical violence were not without their appeal. Viney's behaviour could be frightening, it was true, but Josh had never felt himself directly threatened. Indeed, he'd always felt that Viney was very much on his side in most things, and this reliability added a pleasing physical dimension to the strong spiritual trust that drew him so close to Lothar.

Once again, Viney didn't let him down.

'Here's what we'll do,' he said. 'We'll lend a hand like you say, till we see how things get on. Three should do to start with. So let's see; who'll it be? Fritz, I think you'd best be one of 'em. Heppy, you're a big strong lad. You'll do. And for the third, let's see.'

He made a pretence of studying the assembled men closely, then his broad face split in a huge grin and he said, 'Young Josh, I suppose it had better be you, else you'll just be running off again.'

Josh answered the grin with a radiant smile and had to hug himself to hold in a squeal of joy. But others felt far from joyful and made little effort to hold in their emotions.

'Only three? Why only three, Viney?' demanded someone.

'Yeah. Or at least, let's 'ave a proper roster, so we all gets a turn,' cried Strother.

'We know what kind of turn you want, Strother,' said Viney. 'I said three and three's enough. We don't want the Desolation out there looking like Melbourne at eight o'clock on a Monday morning! As for a roster, that means starting from scratch each day. You get nothing done that way. Continuity's what you need. So I've picked my three and there's an end. Heppy's got the muscle. And Fritz speaks the lingo, and as we all know he's a clever cunt, so he'll get things sorted. And Josh here, now Josh is a country lad, born and bred, and I'll lay odds there's nothing he can't do around a farm, so that's all right too!'

It seemed marvellously and perfectly all right to Josh and he was surprised and distressed to discover that Lothar did not seem to share his enthusiasm.

'Nothing Viney does seems to be right for you!' he burst out. 'Why can't you give him credit?'

'I give him plenty of credit,' said Lothar mildly. 'All he says is true, I give him credit for that. Also, I give him credit for picking us because he knows he can trust us in this matter. The rest he sees either like Strother who would attack the women, perhaps, and certainly cause trouble; or like Arnold Tomkins who would certainly run off if he got the chance. Oh yes, I give Viney credit.'

What he did not add, because he had no desire to overtax Josh's mind or his loyalties, was that Viney's action had set up another barrier between himself and the men, causing resentment, reinforcing their sense of his 'otherness', and even putting him in a different time-scale by reversing the deserters' inverted day.

But it was worth almost anything to be going out into the fresh air in daylight, to enjoy a living countryside – not to mention the company of attractive women.

So he surprised Josh by laughing out loud and hugging the boy warmly, saying, 'Yes, you are quite right, *Knab*'! It is religion which examines motives rather than works. Here we are not religious and I am most truly and deeply grateful to our good lord and master for his excellent kindness!'

12

The next day Viney announced his decision about Lieutenant Cowper. There were no objections. Strother's malice seemed to have been temporarily slaked by Nell's death, and even Fox whose hatred of officers seemed almost pathological remained quiet. Blackie Coleport, Taff Evans and a taciturn Canadian called Pritchard were given the job of taking the officer to the other side of the Desolation and releasing him.

There were no fond farewells. The young lieutenant was merely told, 'You're going', and next thing he was gagged, blindfolded, with his hands bound behind his back, and stumbling up the tunnel to the surface.

Even under these constraints, the delight of breathing through his nose the cold night air was beyond compare.

There was a breeze which was in his face as they set out, and he tried to keep a check of their progress relative to it so that if it proved to have been constant that night, he might have a rough idea of backtracking their route. After ten minutes he gave up. Either the wind was gusting from all directions or their route was too circuitous to be remembered.

One man followed him close, his hand on the rope which bound his wrists. He guessed that one of the others scouted ahead while the third guarded their flank. Occasionally monosyllabic instructions were whispered in his ear and he quickly learned to obey them instantly. From time to time he was commanded to lie down and he lay with his face pressed close to the earth, sometimes for minutes on end, until the unrevealed danger was past.

But at last they came to a stop which had no sense of danger in it. This was confirmed when after a few moments he heard Coleport whisper, 'All right ahead. You all right, Taff?'

'Yes, boy.'

'Good-oh. This'll do nicely then. Well, Lieutenant Cowper, old mate, this is where you resign from the Volunteers. Give our love to the GOC.'

'Shouldn't we undo his wrists?' enquired Evans.

'Nah! He's got his legs, ain't he. We don't want him wandering around too fast, raising the alarm before we're safely away, do we?'

'But he's got to be able to see at least, Blackie,' protested Evans. 'Else he could kill himself falling down a shell-hole, and where's the point of it all been then?'

'All right, you soft-hearted Celt,' said Coleport. 'You and Pritch take off now. I'll loosen his ropes so he can get himself free with a bit of effort. Then I'll spin him around a bit, just to make sure he don't know what direction we're taking.'

'OK,' said Evans. 'Goodbye, Mr Cowper. Take care now.'

Cowper heard the sound of the two men moving away. And then he heard Coleport's voice soft in his ear.

'Sorry, Cowp, old son. I don't care what anyone says, you're too much of a fucking risk to be set loose.'

He tried to cry out against the pain of the knife which drove up beneath his ribs but the gag was too tight. Blindly he lashed out with his feet in the direction from which the blow came, but he kicked nothing but air. Curiously it was the blindfold which troubled him most. The sky and the stars were there to be seen just beyond that thin fold of cloth. He shook his head violently to free himself from it, but it was no use. He was falling into a blackness beyond all hope of light.

Coleport caught up with the others. Evans said, 'I hope he's all right.'

'Oh, he will be,' said Coleport cheerfully. 'Officer material, sport. Them bastards are always all right.'

He wondered as they made their way back if he could have got away with stealing the bastard's pocket watch. It was a fine kind of timepiece, silver-plated and elegantly engraved. He'd helped himself to it when they first caught Cowper but Viney had insisted that all the officer's possessions, revolver apart, be returned to him.

Nah! he told himself. He'd been wise not to take it. Viney would've spotted it for sure, and then there'd have been hell to pay. He reckoned Viney had made a mistake in wanting to let Cowper go free, but you didn't let Viney know you were correcting his mistakes, not if you had any sense you didn't! Anyway, what the hell did you want with a watch when time meant fuck-all to you!

Captain Jack Denial rubbed at the faint layer of tarnish which had formed on the case of the watch. Opened, it shone as bright as ever.

To Augustus, with love from his father.

The body had been dead only a few days when they found it. So they had kept him alive for several weeks before taking him out, bound, gagged and blindfolded, and slipping a knife under his ribs. Why?

In the same pocket as the watch there was a badly creased sheet of paper, covered with pencilled writing on both sides. One side was easy to read – easy in the mechanical sense, that was. It contained Cowper's letter to his wife. Did she still nurse hopes that he was alive? Denial wondered. Or had she accepted his death and was now to have him resurrected and killed again all in a moment? Perhaps the kindest thing would be to destroy the letter.

But Jack Denial was not in the habit of destroying evidence. And there might come a time when the faint and much less legible scribblings on the other side of the letter could be very important evidence indeed.

With infinite patience he pieced together the letters and words that August Cowper had scribbled both as a pastime and an aide-memoire in the days between his 'trial' and his hoped-for release.

30–40 men – Viney's Volunteers! – says he didn't mean kill nurse.

Denial paused here and closed his eyes. So it was confirmed. And Cowper had recognized him and challenged him. Oh, the brave, foolhardy young man.

Meant to kill me, I think. Thank God for the Graf!

What the devil could that mean?

Warren in old Hun command HQ. Location?

Some excitement. Contact with French peasants? Not too far from edge of Desolation?

Names: Coleport (Aus.) Strother (Cockney) Josh? (young lad – North Country) Hepworth (York.) Quayle (Devon) Do I want these men killed?

Screams last night. Someone dead? Can I trust them? Or try to escape?

And then the last entry of all.

Tonight Von Seeberg told me. But he's not coming – says it will be all right – but I wish

Jack Denial carefully put the paper into a cardboard file which already contained many scraps of paper and threads of information and went to report to his superior at Divisional HQ. Advancing as the front retreated, with the small gains of the Somme and the larger of the German withdrawal to the misnamed Hindenburg Line, Headquarters was now situated in an industrialist's country house, grandiosely called the Château d' Amblay, a couple of miles north of Barnecourt. It was from this same HQ that an RFC supply truck had set out across the Desolation. Its driver had been found by a cavalry patrol some time later with a bullet through his head. But of the truck, its contents, or its passenger, Corporal Tomkins, an HQ clerk, there'd been no sign. This too was all meticulously recorded in Denial's file.

His superior was an understanding man with a growing sympathy with Denial's view that the problem of disaffection in the British Army was not one which could be cured by pretending it didn't exist. If you made lunatic demands on the mental, physical and spiritual resources of men, in the end something must break, argued Denial. Desertion was the tip of the iceberg. Once let the feeling that a majority of men felt the same way spread and what you got next was mutiny. Denial had been regarded as slightly less of a crank as reports of what had been happening in the French Army since the fiasco of Nivelle's Chemin des Dames offensive spread. Over fifty divisions had been involved in a refusal to go up to the front and at one point early in June only two reliable divisions stood between the Germans and Paris.

The General Staff response was predictable. It couldn't happen to us, our chaps are good stolid Anglo-Saxon stock, none of your volatile Latins! Anyway, most of the trouble was caused by Bolshevik agitators riding on the purely temporary crest of the soon-to-be-reversed Russian revolution, and their filthy dogma would find no breeding ground in the minds of men brought up in the traditions of the world's oldest democracy.

This last argument Denial forcefully turned around. It was men brought up in these traditions who were eventually going to want reasons for risking their lives rather than obviously stupid orders to make impossible attacks. There had already been many instances of disaffection, quickly suppressed and subsequently lightly glossed over; and a legend was growing up about a whole lost battalion of deserters who inhabited the old battleground, sallying forth at night to take what they would from supply dumps and defying all official efforts to suppress or destroy them.

'From what you're saying, Jack, it sounds as if you think we ought to join them,' said his superior ironically.

Denial shook his head.

'We can't change policy, sir,' he said seriously. 'More's the pity. All we can do is help hold things together till the idiots at the top see sense, and hope it's not too late.'

'And capturing this bunch of deserters would help? I thought you objected to the execution of deserters?'

'I don't want them executed, sir. Killing the odd individual who breaks does nothing. It feeds disaffection rather than suppressing it. But this lot and other groups like them can become – are becoming – symbols of successful revolt. That's why they're dangerous.'

'And your own personal feelings, Jack?'

The question was put gently. The response was fierce.

'I've never allowed personal feelings to get in the way of my duty, sir. Either in the Army or out of it.'

Denial's gaze was unblinking, challenging.

'Yes,' said the other softly. 'I believe you. All right, Jack. I'll give you what rope I can. But remember. There's a bloody war being fought out there. Our job is simply to help keep our side of things running as smoothly as possible. Never forget that.'

'I thought that was precisely what I'd been saying,' said Denial.

On his way back to his billet, he made Sergeant-Major Maggs divert the motorbike combination along a skein of narrow sunken lanes whose coils took them ever eastwards. Finally he ordered a halt, got out and clambered over a tall hedge into a rough pasture. His instinct had been right. From here he could look down towards Barnecourt, its church, the bridge across the river. And the hospital.

He turned from that to look up a long shallow valley which rose gradually to where, a few miles away, a heat shimmer from naked earth marked the beginnings of the Desolation.

Up there; he was somewhere up there.

He felt nothing contradictory in the feeling which welled up through his entire being.
Sometimes hate was a kind of duty too.

PART TWO

THE FARM

In 1917, the Third Battle of Ypres (Passchendaele) began, like the Somme, in July, and lasted, like the Somme, till November. British losses were in the region of 300,000. Of the 346 men executed during the war, 266 were condemned for desertion, 37 for murder, 18 for cowardice, 7 for quitting post, 6 for violence against superior officers, 5 for disobedience, 3 for mutiny, 2 for sleeping on post, 2 for casting away arms. No Australian soldier was executed during the war.

1

Sunrise, and three figures crawl from the Warren and pause, as timid as any rabbits sniffing the morning air for danger.

The sentry does not share their caution. All he feels is envy.

'Lucky sods,' he says, adding as he makes an obscene gesture with his arm, 'give 'em one for me!'

The sun is more sudden in these rolling flatlands than it has scope to be among the foothills of the Harz or the high fells of Cumberland. As the three figures move westward it rolls swiftly up behind them, filling each dip and hollow in turn with an explosion of light. The thin mist has no defence and is blown away by the gusts of light rather than any movement of the air, as the three men move with swift care through the broken trees. They have gained their skill in different ways – game-stalking on a nobleman's estate, sheep-herding on precipitous hillsides, childish games of pursuit and flight through the ill-lit back streets of Bradford – but now their abilities meet and conjoin in a strange dance as each takes the point in turn, then waits and watches till the others have flitted by. They have been doing this twice, sometimes three times a week for over six weeks now and they are growing as used to each other's rhythms as old partners on a ballroom floor.

The huge sky is growing bright from horizon to horizon as the clouds which stem from the far western sea glow pearl-grey, then pink, then quilted white, and the three men increase their pace. They are out of the Desolation now and can take full advantage of hedgerows and trees as they hasten over the last half-mile to the Gilberts' farm.

The working day had already begun. Madeleine greets them with her customary brusqueness. Life has not treated her gently and it's going to take more than a bit of politeness, a few hours' work and a scattering of soldiers' rations to erode her high-piled suspicions of anyone and everyone outside her family.

Nicole welcomes them with bright-eyed delight, marred for Josh only by its apparently equal distribution among all three of them.

Madeleine's parents, Monsieur and Madame Alpert, come somewhere between the two greetings. Old Georges, half-crippled by arthritis in his knees after a lifetime as an estate gardener, acts as if he would not be surprised by the arrival of the Kaiser himself. His wife, a lively, rose-cheeked, bird-like seventy-year-old, smiles and nods and chatters happily to the newcomers as she bustles round the farm as though she had a large household to look after and keep tidy.

What she has in fact is the stone barn of her son-in-law's farmhouse, with a dilapidated but usable byre and stable attached. The farmhouse itself has received a direct hit from a German 'morser'. Part of one wall and a chimneybreast remain standing. The rest is rubble. Madame Alpert includes among her tasks the cleaning and polishing of the hearth which stands open to the elements.

Lothar was working among the rubble that morning when Madeleine came out with some coffee. Nodding at the old lady, Lothar said, 'She is very houseproud, your mother.'

He spoke with a smile, but there was no responsive humour from the woman.

She said, 'After he could no longer work, my father took my mother back to the little village where she was born. It was to the north, not far from Albert. It had gone. There was not one stone standing on another. It was impossible to tell even where the road had been. Here at least she can see something which might grow again.'

Lothar's mind was much exercised by the question of something growing again in every sense. Josh had taken control of the farming side of things, with willing but clumsy labour provided by Auguste, Nicole's shell-shocked brother. He was a man in his late twenties, appearing reasonably normal if rather shy and poorly co-ordinated, until startled by any sudden noise. Then he would tremble uncontrollably and seek out a dark corner to curl up in and Nicole would have to come and embrace him and talk soothingly in his ear for perhaps an hour before he would emerge.

He had a rapport with animals greater even than Josh's. Somehow the indomitable Madeleine had collected, as well as half a dozen scrawny hens, an old cow and a horse. Auguste was the only one who could coax more than a cupful of milk out of the cow. Josh had spent half a day digging a rusty old plough out of the rubble and the other half mending it. He and Auguste then coaxed the horse into a makeshift harness. Once harnessed to the plough, it took so readily to pulling it that Lothar guessed it was an artillery dray horse which had broken loose and strayed. On its rump there was a scar just about where a French army brand would have gone. He made some joke about this to Madeleine, but the only response he got was an uncomprehending stare.

With the horse and the plough, real work could begin in the fields. There had been the remnants of some grain seed-stock in the barn and to these Josh was adding what he could garner from patches of two-year-old vegetable crops which had survived and seeded themselves in the bomb-cratered fields.

But more important even than the actual farming, it seemed to Lothar, was the refurbishing of the farm buildings to provide a winterproof shelter. The barn was old and draughty, with sacking-patched holes in the roof and a floor of bare earth. In a dry, warm summer, it provided adequate enough accommodation, but once the rains, winds and frosts of the year's ending arrived, it would test the endurance of even battle-hardened soldiers.

He'd been labouring away in the ruined house for an hour, without feeling he was doing much more than redistributing the rubble, when he became aware of Josh standing behind him. The boy was grinning broadly.

'You didn't do much navvying in all them travels of yours, did you, Lott?' he said cheerfully. 'What's this? A wall?'

'How did you guess?' answered Lothar, straightening up.

Josh put his foot against the putative structure and pushed. The stones tumbled apart and gave up the pretence of being anything but a heap of rubble.

'A grand wall that was!' he laughed. 'Here; let the dog see the rabbit.'

As he spoke he began to select and place stones together and rapidly a structure recognizable as a section of wall began to take shape under his hands. He worked swiftly and efficiently. Stripped to the waist, he looked slight and bony, but Lothar could see there was a phenomenal chest expansion as he breathed slowly and evenly, and the sinews of his arms and shoulders slackened and tautened like the pulley-ropes in some cunningly designed lifting mechanism. Lothar regarded him with affection and admiration. The improvement in the boy, both physical and mental, was marvellous to behold. Fresh air, familiar work, relative normality, all these contributed, but the main cause was not far to seek.

She appeared now, blonde hair flying, her expressive face alight with joy, waving her left hand wildly over her head. In it she held a small brown egg.

'*Voilà!*' she cried. '*La p'tit', son premier œuf!*'

La p'tit' was the smallest and scraggiest of the hens whom Josh had opined was fit for nothing except soup.

As she rushed forward in triumph, Nicole stumbled on a mound of loose stones and almost fell forward. The egg flew out of her hand, and she screamed in dismay. Josh, who had looked up at her approach with a joyful grin which had nothing to do with successful hens, thrust out his right arm and plucked the egg from the air. He peered into his clenched hand and grimaced in revulsion as though the egg had broken as he caught it. Nicole made a sympathetic moue. Then suddenly Josh grinned and like a successful conjuror held up the egg between thumb and index-finger, turning it slowly round to show it was smooth and perfect.

Nicole did a little dance of relief, grabbed the egg from him and scampered off to show her mother. Josh stood looking after her, his face alight, and Lothar thought: He'll be pestering me for extra French lessons tonight!

It was unfortunate, but Josh seemed to have little talent for languages, and, to compensate, Lothar had set himself to instruct Nicole in English also and the girl was making much more rapid progress.

'Thank you, Josh, for the demonstration,' said Lothar.

'This? Oh, it's nowt. I help repair walls and sheepfolds up on the fells, so you pick it up. You should see my dad at it! He's a real artist. You'd need a direct hit with a "minnie" to knock down a wall he'd built!'

He referred to his father easily and naturally, but there was a shadowing of the eyes as his mind followed his words into that remote area of time and space which was home.

Lothar didn't give him time to brood.

'Josh, how high could you build something like that?' he demanded.

'A drystone wall, you mean? Oh, pretty high. I've seen 'em five or six feet. Why're you asking, Lott?'

'It's a matter of trying to get some kind of house back together for the family before winter,' he said. 'They can't stay in the barn.'

'Why not?' asked Josh.

'Because it is damp and draughty and dark and leaking,' said Lothar in his best magisterial tone.

To his surprise Josh's response was a loud guffaw.

'So you're going to build another house out here?' he said. 'It'll take ten years at least, the rate you're going! And it's a daft idea anyway! This isn't just a ruin, it's a wreck. And bloody dangerous too. One good gale and I reckon the rest'll come down. Look, there's four walls in the barn and a roof. It'd be much easier, and a lot safer, to work on that.'

'But the floor's just earth, it'll be cold and damp.'

'Spread a bit of rubble, lay a few planks,' said Josh promptly. 'You needn't do it all over, just in one corner. We could put up a couple of new walls. They'd be out of the weather so it wouldn't matter if they were a bit flimsy. Knock a hole in an outside wall for a window. Build a fireplace with a chimney. Patch the roof, or put another roof over the room we're building. You can do nowt out here, Lott, honest.'

Lothar opened his mouth to reply, but there was no reply possible. It was all so obvious, he couldn't believe that he hadn't seen it for himself! He had let his mind be seduced by that damned fireplace so lovingly cared for by old Madame Alpert.

'Josh,' he said, 'in future I will stick to repairing worlds and leave houses to you! Let's take a look inside and see what is to be done.'

But before they reached the barn, they heard a thin but carrying triple whistle drifting across the fields.

'Heppy,' said Lothar. 'Someone's coming.'

It had been Viney's express injunction that one of the trio should always be on sentry duty between the farm and the village, four miles away. In the event this was always Lothar or Hepworth as Josh was far too important to the farm work to be wasted in this way. The signal had been Josh's idea. He had a marvellous skill of mimicry of bird-calls and he had taught the others the redshank's song which would alert those on the farm without arousing the interest of non-ornithological visitors.

The two men made their way swiftly to the small orchard where a 'hide' scratched out against an earthen bank and camouflaged with straggly undergrowth had been their first construction on the farm.

Madeleine watched them go with her usual impassivity but Nicole's young face showed a lively alarm. Her mother did not give it time to grow but brusquely commanded her to check that no evidence of their helpers' presence had been left in sight. This was the first time that visitors had interrupted their work and Madeleine felt as anxious about their purpose and identity as she did about the possible discovery of the deserters' presence. This anxiety grew as the sound of a motor engine labouring up the long, winding and badly rutted farm track became audible. Only trouble came nowadays in motor vehicles.

The vehicle that finally bumped into view did not look substantial enough to hold much trouble. It was a motorcycle and sidecar combination, dust-covered as were the two riders who now got out and started beating themselves respectable.

The sidecar passenger was a small man in a dark civilian suit. The other was taller and dressed in British Army khaki. He removed his goggles and leather helmet and picked up a hat from the bottom of the sidecar. It was an officer's peaked cap and as he put it on, Josh peering through the concealing brushwood twitched nervously.

Madeleine had emerged from the barn and stood facing her visitors, not speaking, forcing them by her silence to explain themselves.

It was too far from the orchard for the hidden men to hear what was said and after a few moments the distant trio went into the barn so now there was nothing even to occupy their straining eyes.

Minutes ticked by. Josh shifted his cramped limbs and murmured, 'Lott?'

'Yes?'

The boy said no more, but his expression was agitated. Something strange was taking place in his mind. Perhaps it was caused by lying here, concealed and frightened, pressed close to the ground with the musty smell of excavated earth in his nostrils. All that was missing was the scream of approaching shells, the blast and heat of their detonation making the ground tremble in time with his fearful flesh and turning the skies red with blood and horror. It seemed to Josh – though he knew it only *seemed*, that it *was* a lie and a deceit – but it seemed to him that all he had to do was stand up and walk into the barn and explain himself to that British captain, tell him how everything had come to pass, tell him about

Wilf, and Lott, and how much he hated the killing and how much he loved Nicole . . . all he had to do was this and all would be well, all would be positively and finally and endlessly well. Not just well in the sense of preserving him safe from military discipline, but well as things had been well before . . . Before when? Before that moment, whenever it was, that the decision had been made by which he had given away all that was so perfectly and naturally well in his life that he did not know it. A second chance, there for the taking . . .

Then it came to him with the cold shock of experienced truth that all along that sinuous spoor of blood and slime which marked the war's trail across the face of Europe, millions of other men lay in earth and mud, beneath the blazing sun or the cold stars, and stretched back their minds to some recollected happiness, within reach, it seemed at times, of a simple act of will, but in reality unattainable now for ever.

He began silently to weep and Lothar's arm around his shoulders was a weight not a comfort. He had no idea how long it was before the sound of the motorbike startled the birds in the trees above his head, but their outcry and the rattle of the engine fading down the track and out of earshot was like a farewell to hope.

Lothar let him lie for several minutes. He sensed that this was no time for an intrusive questioning.

Then he said, 'Hey, Josh. Your face. You cannot risk to so worry Nicole, heh?'

The boy did not reply but spat on the shred of cloth which served him as a handkerchief and scrubbed at his cheeks. Lothar watched for a moment, then said, 'Perfect. Let us go now and see what news the great world has brought us.'

In the barn, Madeleine was already busying herself about her customary household tasks as though nothing had happened. She seemed inclined to answer Lothar's queries in monosyllables, but as Nicole was ready to remultiply these a hundredfold, there was no shortage of information.

The civilian had been an official for the local administrative department come to check on the Gilberts, and in particular on a rumour that there was a man of military age working on the farm. He had been satisfied with the explanation, supported by Auguste's medical discharge papers. But he had been very stern with Madeleine for failing to obtain official approval for re-entering what was still a militarily interdicted area. Using the stately language of French bureaucracy, he informed her that he could offer no guarantee that she would not be further contacted on this particular matter.

The British officer had also been unhappy, but merely because of a wasted journey. He said he was in charge of army billeting in the area and had accompanied the official in the hope of discovering an unused billet in the vicinity of Barnecourt.

There was news of the war too. Old Georges had been the one to ask how it was going, and the official, with the grave courtesy the rudest of Frenchmen reserve for the old, had explained in detail that the glorious French Army with Pétain as its new commander-in-chief had been holding firm against great Boche pressure along its section of the front. Russia, now in the hands of evil men who had overthrown their Tsar, was a spent force, leaving the fate of the whole of Europe in the hands of France whose main ally, Great Britain, instead of putting all its forces at Pétain's disposal here, where the real struggle lay, had been frittering away its energies in minor skirmishes across the border in Belgium around Ypres.

This last section of the analysis, Lothar assumed, had been given with the British officer out of earshot. Rearranging the pieces, he reckoned that Russian war-weariness was permitting more German units to be sent to the West. The French, regrouping under Pétain after their spring disasters, were in too poor a shape to do more than occupy their trenches, and the main effort of tying up the German forces by a large attack had fallen to the British.

His countrymen should have been attacking the French-held front with everything they'd got all through the summer, he found himself thinking. A great chance had been missed.

Then Josh said eagerly, 'What's it they're saying, Lott?' and Lothar was suddenly reminded of where and what he was, and stood amazed at himself and the ease with which that old nationalistic militarism could be rekindled.

'Nothing, Josh,' he said. 'Nothing important. There is no danger as long as we keep watch.'

'Why do I find it so hard to speak their lingo, Lott?' enquired Josh in angry frustration, looking longingly at Nicole. 'I must be really thick or something.'

'Do not worry,' said Lothar, smiling. 'It will grow easier, believe me. But meanwhile I shall continue to teach Nicole your language. She is a quick learner, I think.'

He spoke to the girl in rapid French and she laughed delightedly. Josh frowned sulkily and Lothar clapped his hand on his shoulder.

'Come, Josh,' he said. 'Show me where and how we should start our reconstruction in here. Let Nicole see that you are the master and I am the labourer, eh? Such language as that all women understand.'

He laughed as he spoke, but the young man's face remained serious.

'Lott,' he said, 'must we tell Viney that this officer came today?'

Lothar caught the drift of his thought instantly. Viney's prime concern was always for the security of the Warren and Lothar guessed that he had regretted almost at the moment of giving his permission for the three men to visit the farm. Josh had been the key. The youngster's delight had touched the big Australian in some obscure way and he was clearly reluctant to destroy it. But he had been adamant that they must abandon any visit if there was the least evidence of outside activity anywhere near the farm.

News of today's events would mean a week's embargo at the very least, perhaps longer.

'I think we *ought* to tell him,' said Lothar hesitantly. He watched Josh carefully as he spoke. His reaction took him by surprise, not because of its content but because of its manner. Resistance, argument, protest he had expected, but so used had he become to thinking of Josh in almost childlike terms that he had looked for these things wrapped up in emotional, even tearful terms. Instead the young man replied in a carefully measured tone, 'Lott, let's get one thing straight. If you feel you have to tell Viney about that officer being here, then I'm not going back with you. I'll stay on here by myself. You can tell Viney that even if I got caught I'd not tell them about the Warren or anything. I know he'd likely not believe you but that'd be up to him, and he'd have to come here himself to get me back, and I'd not come easily.'

The only sign of emotion was a slight breathlessness as he finished and a faint flush on his cheeks.

Lothar regarded him with a frown. The distressed boy he had been able to comfort, and to deal with, had disappeared. This was someone different.

He said, 'There is Heppy too. We must talk with him.'

The Yorkshireman saw no problem.

'What? Risk getting ourselves stuck back down yon hole permanent? Nay! I'll tell you what, Fritz, if you're going to say owt of this to Viney, I'm stopping here with the lad. In fact, it's not such a bad idea anyway. It's bloody daft us making the trip twice a day when we could doss down here and be on the spot to work in the morning. And a sight less dangerous too. Roaming around the Desolation, that's when we're likely to be spotted, not by some skiving, short-sighted billeting officer!'

Lothar sighed. Confrontation with Viney would mean more trouble than he personally felt able to deal with. Yet, in a way, the man had a right to know what had taken place.

He chose the lesser of two evils and said, 'All right. I will say nothing. But you must also say nothing about this business of spending the night here.'

'Why not?' grumbled Hepworth. 'Makes sense, doesn't it?'

'Of course it does. But it also makes sense not to risk provoking Viney in any way, doesn't it?' snapped Lothar. 'Now, leave it to me. Agreed?'

He waited till the other two nodded their heads, then turned and walked away, irritated at being once more forced into the position of leader and strategist. It was much easier simply to mock and make rude gestures at authority rather than get into harness with it. How this would amuse my father! he thought.

Josh overtook him in a few steps.

'Lott,' he said, 'Don't be angry. Like Heppy says, why spoil things for a skiving old billeting officer?'

It was hard to maintain anger long in the face of Josh's open young face.

'Yes, Josh. Probably you're right,' smiled Lothar. 'Why indeed?'

If he had known the identity of the 'skiving billeting officer', he would have felt very differently.

Jack Denial had learnt the hard way that information was the better part of detection. Information and a meticulous attention to detail. He had made it clear by repetition and by reprisal that he expected everything relevant to activity in or around the Desolation to be reported to him. By a mixture of flattery, threat and bribery, he had even won the cooperation of local officialdom and it was from the district *mairie* that word had come of a report that there were people working at this isolated farmhouse above Barnecourt, including at least one young man. He had asked permission to accompany the investigating official, and some innate caution had made him masquerade as a billeting officer.

Well, he thought as the motorcycle bumped back down the track towards Barnecourt, it had all been a waste of time. Yet there had been something, an atmosphere . . . no, not even that; a sense of constraint going beyond even the usual Gallic charmlessness in the face of officialdom. The girl . . . yes, it had been most evident in the girl. She was a pretty thing. His thought was quite without passion. He doubted if he would ever look at another woman with desire, not since Sally. Something of himself had perished with her; some long dormant spring which she had tapped had finally dried up.

But the girl had sent out a vibration which he had picked up. Probably simply fear that the official was going to insist that they move. She needn't worry. He had asked the man what he would do and received the answer, *nothing*. The winter would almost certainly do it for him.

Yet, though he was almost satisfied, Denial was not yet ready to relegate the Gilbert farm to that area of his thinking and of his filing marked 'closed'. He recollected Cowper's notes, that brief and puzzling reference to contact with the local peasantry.

He filed it instead under 'pending' and turned his mind back to the wider, day-to-day problems confronting an APM in this apparently unmoving and unending war.

2

'Well, if it ain't the kraut from Snowy River,' said Blackie Coleport, reclining on his pallet with a long thin wine bottle in his hand.

'Your pardon?' said Lothar at the dug-out entrance.

'Old Pardon, the son of Reprieve!' laughed Coleport. 'Have a snort, Fritz. You know what today is? Today is my birthday. Go on, have a snort!'

'My congratulations on your birthday, but no, thank you, for the drink. Viney, I would like to speak with you.'

'Go ahead, Fritz,' said Viney. 'The air's free.'

'The liquor's free too,' protested Coleport, 'but this fucking Hun's too stuck up to share it. I reckon he wants educated in civilized manners.'

'Shut it, Blackie,' ordered Viney.

Coleport blew a raspberry, echoed it with a fart, and took a long suck at the bottle.

'You did well last night?' enquired Lothar, looking significantly at the bottle.

'Pretty fair,' said Viney. 'The lads deserve a treat now and then. Man cannot live on words alone, ain't that the Good Book's teaching?'

It was early morning. Lothar had looked for Viney the previous night, but found him blacking up for a raiding party. They must have gone far afield to get within striking distance of wine. He had a nasty suspicion that Viney might have taken to raiding civilian targets, but this was not the moment to bring it up.

'Viney, we need more men to help with the work at the farm,' he said boldly. 'It is September already and there is much to be done to prepare for the winter ahead.'

'Is that so?' said Viney. 'How many did you have in mind?'

Lothar was taken aback by this apparently sympathetic reception. Several weeks had passed since the English officer's visit to the farm and there had been no more official contact. With everything quiet, Lothar had decided it was time to try out the idea that he and Hepworth and Josh should be permitted to stay overnight at the farm. His chosen approach was oblique. He fully expected his request for more men would be rejected out of hand, and then he would have put the other suggestion. But now Viney, relaxed perhaps by the drink, was apparently in the mood to give him what he really didn't want.

'Well, not too many. We must be careful always not to risk attracting attention either while we are there or by movement across the Desolation,' he replied, putting forward the arguments he'd expected to hear from Viney's mouth. 'Two or three good workers. Taff Evans perhaps. He is strong and hard-working. Quayle too.'

'Not Quayle,' said Viney, his mouth closing like a prison gate.

'But he is a countryman, big, strong . . .' protested Lothar.

'He's a poofter,' said Viney. 'A big soft bumboy. Quayle goes nowhere.'

Ever since Nell's death Quayle had retired into a shell of isolation, driven there perhaps by sorrow or guilt, but maintained there by the silent force of Viney's disapproval which made many of the others unwilling to be seen too close in his company. To work on the farm might be a therapy, but there was on Viney's face a look which did not invite argument. So, reminding himself that he didn't really want the extra men anyway, Lothar quelled his natural inclination to argue and said, 'One of the others then.'

'If you like,' said Viney as if suddenly tired of the topic. 'We'll talk about it later. How's young Josh?'

'He's fine,' said Lothar. 'Better all the time.'

'How's he making out with that young sheila?' demanded Coleport, grinning salaciously. 'Getting his share, is he?'

'Stow that, Blackie,' commanded Viney.

'Sorry, I'm sure,' said Coleport. 'What about you, Fritz? Getting your German sausage exercised all right? That Madeleine looked like she'd know a thing or two. What say you, Patsy?'

Delaney who'd apparently been asleep on the floor opened one eye.

'I heard there was a grandma,' he said.

'That's it then!' exclaimed Coleport in delight. 'My money's on Fritz fucking the grandma! So if Josh is fucking the girl, that leaves the mother for Heppy. Lucky bastard!'

Lothar did not speak and made to leave.

'Where're you off to?' demanded Viney.

'To collect Josh and Heppy. Soon it will be time for us to move.'

'Not today,' said Viney, looking at him broodingly from under the heavy shadows of his eyebrows. 'You'll stay here today.'

'Why?' demanded Lothar angrily. 'There is no reason that I know of. Why?'

'Because I say so, that's why,' grated Viney. 'Now sod off and tell them other two!'

'How strange,' said Lothar, fighting to control himself and knowing that his old duelling scar was probably standing out white and proud. 'I thought we had abandoned the Army and its stupidities!'

He left and went in search of Josh and Hepworth. He found them in the large chamber and quickly passed on the news. Neither concealed his disappointment.

Josh demanded, 'Why, Lott? He's got no right, no right! He acts like he's a bloody sergeant-major or something. It's time he was told we're out of the bloody army!'

Lothar was struck by the echo of his own response. He put

his arm round Josh's shoulders and urged, 'Leave it, Josh. Let us relax and join the fun, heh?'

Angrily the youth shook himself free but at least he remained where he was.

Lothar settled down beside him and looked round at the assembled men. There were several bottles of wine in evidence though not enough, Lothar assessed, to let anyone get more than gently merry.

'Here, Fritz,' called Evans. 'Come and have a wet.'

'That bugger probably sups a gallon of the fucking stuff every day,' growled the man with the nearest bottle, but he didn't resist when the miner plucked it from him.

'Thank you,' said Lothar. '*Prosit!*'

He drank and returned the bottle.

'I would like to speak to you, Taff,' he said in a low voice, thinking that if his diversionary proposal were in fact accepted by Viney, the Welshman ought to be forewarned. Not that he expected any objection. They were still the object of much envy, not all of it good-natured.

Before Evans could reply, however, someone shouted, 'Here, Taff, give us one of them songs of yours, get the party started.'

'Get it finished, more like,' grunted Strother. 'Fuckin' Welsh caterwauling.'

'It's a sight better than listening to them Aussie rhymes of Blackie's. Come on, Taff.'

'All right, boyo. If that's what you want.'

The Welshman took a deep breath and without rising from his place began to sing, not the rousing song with a catchy chorus his encouragers clearly wanted to hear, but a gently lilting, melancholy tune which he sang first in Welsh and then in English. It was a song about an ash-grove, an ironic choice, thought Lothar, for these men who inhabited a landscape of charred tree-stumps, but no one else seemed to feel the irony, only the deep silent pain which filled the chamber after the last words had died on the air.

  . . . the Ash Grove, the Ash Grove which sheltered my home.
Strother broke the silence.

'Well, that was real cheering, that was. Real fuckin' jolly.'

'Stow it, Strother,' said the ginger-haired Nelson, who since the incident with Nicole had been much less under his fellow Cockney's thumb. 'Come on, who's next? We ain't going to let a Welshman show us the way and not follow, are we? Heppy, what about *Ilkley Moor*, eh?'

Hepworth shook his head morosely.

'He's sulking 'cos his little trip out's been stopped today,' laughed Foxy, whose sharp ears picked up most things that were said within his vicinity.

Hepworth scowled and looked as if he might take this further, but Nelson went on hastily, 'What about you then, Fritz? One of them Jerry tunes. I 'eard 'em sing *Silent Night* in the trenches, Chistmas before last. It was Kraut words, but, fair do's, they made a bloody lovely job of it.'

'I am sorry,' said Lothar. 'I have no singing voice. I would be ashamed to follow Taff's beautiful song.'

'We don't want no Hun music anyway,' said Strother. 'Tommo, give us a tune!'

Arnold Tomkins, though obviously a recruit from terror of Viney rather than terror of the war, had proved an asset in at least one way. He turned out to have been a theatre musician in civilian life and in his hands an ancient mouth organ on which one of the others had wheezed out barely recognizable tunes turned into a musical instrument. The trouble was that Tomkins was sliding slowly into a deep depression and his choice of music tended to reflect his own melancholia. Tonight by request he had started with a foot-tapping rendition of *Hold Your Hand Out, Naughty Boy* but within seconds of finishing that, he was into *There's A Long, Long Trail A' Winding*. During the mood of introspection brought on by this mournful tune, Lothar studied the others. It was not that he was unaffected by the mood but simply that he feared its inevitable conclusion. For him all such melancholy trails invariably wound towards Sylvie and his rejection of her and the discovery of that dear, slight body drifting in the cold waters of the lake pool.

Anything was better than that image. He looked round the assembled men, saw their closed and inward-looking eyes, and guessed that for most of them their plight was worse than his. Their pain rose not from images of the unreturning dead but of the unreachable living. Wives, sweethearts, children, friends. At first, escape from the horrors of the front had seemed enough; then survival in the Desolation under the leadership of Viney had come as a blessing.

But now at moments like this they shifted their inner gaze from a joyous past to a hopeless future and tasted the bitterness of a new despair.

The tune ended and perhaps with some idea of cheering them up, Tomkins ran almost immediately into the lilting melody of *Après la Guerre Finie*. It was a bad choice.

'Give it a rest, Tommo, for fuck's sake!' growled someone and there were no protests as the music stuttered into a silence broken at last by Lothar speaking very softly.

'But what is it that we *shall* do, *après la guerre finie?*'

For a while the question seemed to have gone unheard, then a hesitant voice said, 'Mebbe there'll be an amnesty!'

'What the fuck's that when it's at home?'

'It means they'll say it's all right if you turn yourself in, no charge, like.'

'Oh yes? I can just see them doing that!'

'It wouldn't be for a long time anyway,' said Hepworth. 'Not for years, if ever.'

'At least once the fucking war's over, there'll not be all them fucking redcaps hanging around the ports. It'll be a bloody sight easier to find a boat to stow away in, and get back home!'

This was the first comment which produced any lightening of the atmosphere. It lasted only until Strother said, 'What makes you cunts think this war's ever going to be over?'

There was a mutter of unconvincing protest.

'It's got to finish sometime,' said Taylor. 'Here, Fritz, you started all this, what do you think'll happen to us after the war?'

There was an expectant silence.

Finally Lothar said, 'I do not know, my friend. I think that some of us may get home, but not all of us, and it will not be a happy homecoming for all who do.'

'Christ, you're the cheery one, ain't you!'

'I am sorry. I do not mean to be depressing, only realistic. It seems to me that the future is too dark to penetrate. But one thing is certain, what we are and what we do now is important. The future is built on the past, all of it.'

'What do you mean, Fritz?'

Lothar sighed and thought that probably it would be wise to stop here, but then he told himself angrily that if he couldn't practise what he was about to preach, he had no right even to think of preaching it.

'The way we live together, the way we regard each other and behave to each other, this must be better than in the life we have abandoned,' he said earnestly.

'You mean the Army? That won't be difficult!' said somebody.

'Yes, it is difficult,' insisted Lothar. 'In the Army, you have the rules to blame. Here, we cannot blame the rules for that is to blame ourselves. How shall we justify what we do here? Killing. Stealing. How shall we justify these?'

'You ain't including raiding for grub, are you, Fritz?' demanded Nelson. 'How the fuck are we expected to live?'

'Taking stuff off the Army ain't stealing anyway,' said Foxy. 'We all know that!'

There was a ripple of amusement.

Lothar held up a wine bottle.

'Did this come from the Army?' he asked quietly.

'Likely not, but I saw you drinking it happily enough!' cried someone.

'And these Frogs have been robbing the swaddies something rotten, so it's about time some buckshees came back our way. Any objection, Fritz?'

'It is not *my* objections, or *Viney's* objections, that are important,' frowned Lothar. 'What I have been trying to say is that each man must now make his own objections for there is no longer a force upon him to do things one way or another.'

'No?' said Evans. 'Have you seen Viney lately, boyo? Now there's a force for you!'

'Someone taking my name in vain, sport?'

How long had he been there, standing in the doorway, listening with frowning intensity?

'Sounds like someone's plotting a spot of revolution, Viney,' said Coleport at his shoulder.

'No need for a plot,' said Viney. 'I've always made it clear, anyone wants the top job, all they do is kick me off the heap. Right, Fritz?'

'That is force you talk of, Viney,' said Lothar. 'What we have been saying here is that what is possible for us is a society without force, so while the future may be black, we can at least live the present in dignity, in charge of ourselves.'

'That's what's worrying you, is it?' said Viney. 'The future! Well, I ain't promising and I don't persuade. It's up to you jokers to come along my way if you want. I don't give a toss!'

'Alternatively, you can follow Fritz here,' mocked Blackie Coleport. 'Give us your policies, Fritz. I like a good political meeting.'

'I have no policies,' said Lothar wearily. 'This is not a political meeting.'

'No? Well, I'd bet you'd like to make it one, mate. Listen, you jokers, ain't it ever struck you as odd that Fritz here didn't take off to his own country? I mean, the rest of us we deserted all right, but we didn't fucking well go over to the enemy, did we?'

There were some sounds of agreement.

Lothar said, 'I did not know that Viney's Volunteers were anyone's enemy.'

Hepworth intervened. 'Any road, we've been through this before. He did it to take care of young Josh. If he'd wanted, he could have been sitting the war out in a POW camp, right comfy. Well, at least as comfy as this.'

'You're right there, Heppy. Tell you what I think,' said Coleport. 'The reason he ain't is that he can do his little job better here. I reckon our German friend here is one of the Russki agitators, a fucking anarchist or some such thing. What he's after is recruits. Watch out, mates. He'll get you back home all right, long as you promise to blow up Parliament and hang the King!'

Strother and a couple of others made agreeing noises once more, but the majority were silent, except for Josh who said indignantly, 'That's daft! If Lott here could get the lads home to do that, don't you think every last one of them wouldn't promise to help him just to get home?'

It was a devastating piece of reasoning which took Coleport by surprise as much by its source as its logic.

'Oh, she's finding her tongue, is she?' he mocked. 'Mustn't let anyone say bad words about Daddie!'

Lothar saw Viney's face darken as it always did when anyone dared to speak ill of Josh, especially when sexual innuendo was involved. In another moment, Viney would put Coleport in his place. But suddenly Lothar was not prepared to let the big Australian always take the lead.

He scrambled to his feet, pushing down Josh who was also trying to rise, and said, 'Coleport, you talk too much, I think. In the East, their leaders, sultans they call them, also often have fat men to do their talking. Sometimes they are fat because they are pigs; sometimes because they are eunuchs, that means they have no balls. But in the end, a man's ears grow tired of the eunuch's squeaking and want to hear the sultan speaking for himself.'

The foot snaked out fast and vicious. Ready though he was for the blow, it had been the hands Lothar was watching and he was almost taken by surprise. But he twisted his body so that the heavy boot drove into his thigh rather than his crutch.

Pain shot up his leg, but he paid it no heed. His own attack was already prepared and he knew that speed was essential. Fat Coleport might be, but he was a vicious and experienced brawler. His very fatness made assaults to the body pointless and Lothar doubted if he could put him down with a single blow to the head, yet a single blow was all he might get in.

So the hand he darted forward was not clenched but had the first and middle fingers splayed rigid as though making the sign to ward off the evil eye. The evil eyes these fingers were aimed at belonged to Coleport.

The fat man screamed in agony as the finger tips drove against his eyeballs. Lothar skipped to one side in a scatter of men. Even half-blinded, this man could wreak terrible damage if he got you in his grasp. Clasping one hand to his eyes, Coleport rushed forward. Lothar came in fast from the side and, locking both his hands together, swung with all his strength at the base of Coleport's skull. The blow almost broke one of his own wrists. The fat Australian went down full length without even a groan and lay quite still.

Lothar spun round to face Viney. Patsy Delaney, his cold eyes suddenly ablaze with fury, was trying to push past to avenge his fallen friend, but Viney's huge arm was stretched out to restrain him.

The German and the Australian stood a couple of feet apart and locked gazes.

'For a man who don't approve of force, you've an odd way of working,' said Viney softly. 'Is this the way it's going on?'

It was a challenge; at least that was how it must sound to the expectant men; but Lothar thought it also sounded like a genuine query.

He stood lightly balanced on the balls of his feet and ransacked his mind for an answer. To back down was impossible, yet if physical battle were joined now between Viney and himself, Lothar could see no hope of victory.

Then the silence was broken, not from within the chamber but somewhere else in the Warren. A high keening note pierced the air, wailing up and down, up and down, an alien sound, almost inhuman, but full of human anguish.

'For Christ's sake, Patsy, go and see what that din is!' commanded Viney.

Delaney rushed off. He was back in only a few moments his face pale.

'Aw shit, Viney,' he said. 'Shit!'

'What the hell is it?'

'It's Quayle. Aw shit.'

There was a general move to see what had happened. Delaney did not follow but knelt beside Coleport who groaned and turned over. The others did not have far to go.

The older Indian, Dell, was squatting at the entrance to one of the smaller dug-outs, singing the high, keening song through his nose.

In the dug-out was Quacker Quayle, lying on his side, his hands still grasping the bayonet with which he had killed himself.

He had not made a very good job of it. The first thrust up through the belly under the ribcage must have started too low, but with a will as hard as the tempered steel inside him, he had wrenched the blade this way and that in search of the elusive vital thread.

Viney stood and stared down at the bloody mess as if testing himself against the wide, staring eyes of the dead man.

Finally he said in a terrible voice, 'Weak! These poofters are all the same. They break where a man should hold.'

And roughly shouldering his way through the press of men, he left.

'Poor sod,' said Taff Evans at Lothar's shoulder. 'Poor desperate bastard.'

'There was no need for this,' said Lothar. 'No need for this.'

'No need at all,' agreed Evans bitterly. 'He should just have told Viney he wanted to leave, and it would all have been done for him, nice and neat, like. Nice and neat!'

3

Outside the sun was already over the horizon, drawing up tendrils of mist from the dewy earth, but down in the Warren lights flickered low as men sought to find in darkness that solitariness which their close confinement made almost impossible in fact. A gloom of the spirit which matched this physical gloom was felt everywhere.

Lothar lay by Josh and brooded in silence like all those around. But when a hand gripped his shoulder, he twitched into life, ready to attack or defend as the case might demand.

The sight of Patsy Delaney's long features rendered even more cavernous by the dim light of a lantern did not reassure him.

'Viney wants a word,' the man said.

Lothar regarded him distrustingly. Suddenly a knife glinted in the man's hand.

'If I'd wanted to gut you, it'd have been done by now,' whispered Delaney.

The knife vanished.

Lothar rose and followed him from the sleeping chamber.

'Is Coleport all right?' he enquired.

'Right enough to tell me to leave you for him,' was the terse reply.

Viney and his two lieutenants shared the small dug-out which had once belonged to the officer commanding the HQ. It even had a curtain to pull across the entrance. Delaney motioned Lothar through but did not follow. Viney was sitting alone on a stool, leaning back against the boarded wall. On a makeshift table in front of him was a lantern turned up high as though darkness was no refuge to him. Alongside it stood a half-empty whisky bottle.

'Sit,' said Viney. 'Drink.'

Lothar obeyed. There are moments in life when even the most rebellious of spirits knows that unquestioning obedience is the only reaction. They drank in turns without speaking till the bottle was empty.

Viney knocked it over.

'Dead soldier,' he said. 'No more whisky. Brandy.'

He produced an unlabelled bottle which contained a raw spirit which was nearer brandy than wood alcohol, but only just.

'You got this on the raid?' asked Lothar.

'Yaw.' Viney's eyes challenged him to comment.

Lothar said, 'What am I doing here, Viney?'

'Getting drunk.'

They drank on. Lothar had acquired an excellent head for alcohol during his years of experimental debauchery, but the enforced abstention of recent months had left him vulnerable. Soon his mind was filled with shadows flickering like those of the lamp, but he drank on, certain that whatever lay at the end of this meeting could only be reached by following Viney's prescribed route.

The big Australian showed no outward sign that the drink was affecting him, but poured it down like a man trying to drown something trapped deep inside.

'You married, Fritz?' he asked suddenly.

'No.'

'Why not?'

Lothar, about to shrug and say he had never thought about it, was struck by the notion that any evasion would instantly be spotted by Viney and that would be an end of it.

An end of what? he asked himself. God alone knew. But Viney wanted to talk and his pride would only permit such talk to be straight trade; frankness for frankness, confession for confession.

He said, 'The only woman I would have wished to marry married my brother.'

'That's rough,' said Viney.

'Yes. Are you married?'

A long pause.

'Yaw.'

'Do you have children?'

A longer pause. Then an evasion.

'This brother of yours, he in the Army?'

'Was.'

'Dead?'

'Yes. Dead.'

'Were you sorry?'

'Yes. Very. I loved him dearly,' said Lothar, staring blindly into the flame.

'And his wife. Widow.'

'She is dead also,' said Lothar. He took a deep breath. 'I helped kill her.'

Viney expressed no surprise but nodded as though this was confirmation of what he knew. Or perhaps it was in confirmation of what Lothar now knew to be the unspoken and inviolable agreement between them, that nothing of this exchange would ever pass beyond the pair of them, no matter what state their rivalry should reach.

'She wanted, needed, my comfort. She came to me. I rejected her, out of guilt for my brother, out of shame at my own desire, out of pride and egotism and self-regard and all the other things that make men fight wars.'

He paused. Why had he said that? It was not a connection he had made before. But there was truth in it.

'So she sought comfort instead in death,' he concluded.

'Killed herself.'

'Yes.'

'Like Quayle.'

'Just like Quayle. All hope of comfort and love gone.'

For a second Lothar thought Viney was about to burst forth in another of his rantings against perversion, but the tremor of emotion which passed over the man's face vanished quickly and he took another draught of the raw spirit.

When he spoke, his voice was very low but controlled.

'My wife has a kid. Not mine.'

'You are sure?'

Viney considered this.

'Fair do's,' he said. 'Almost sure.'

Lothar did not speak but Viney seemed to hear a question.

'I said I'm almost sure,' he growled. 'It's like a firing squad, Fritz. One shot up the spout. You can't be sure your one shot hits the mark when there's a dozen others blazing away all around you, right?'

'One shot? You mean . . .'

'We were only married two days before I sailed,' said Viney. This was hardly an adequate explanation but instead of enlarging upon it he chose to go off at a tangent, or so it seemed.

'You had a lot of girls, Fritz?'

'Enough.'

'Me too. *Girls!* It was all right then. Young. They were young.'

'How old are you, Viney?'

'Me. Twenty-four, twenty-five. Fuck knows.'

Lothar was taken aback. He'd speculated on Viney's age but had never placed him as young as this. That brooding intensity, he must have been born with it! A man with his power over others imparted a sense of agelessness. Twenty-four! He was a boy!

And I am twenty-six and an ancient! thought Lothar with heavy irony.

'These girls, I've thought of them. Some were scared. It was the first time. They cried out. I told them to shut up. Even when it wasn't the first time, they'd often cry out and I'd shut them up. They'd lie there like I wanted them. They'd lie still till I finished. Sometimes on their backs. Sometimes on their bellies. But always still till I'd finished.'

Lothar watched him drink, took the proffered bottle and pretended to drink himself. A small deception, instantly spotted. He affected to sneeze to explain it, then drank deep. The bottle went down on the table with a crash which made the lantern jump and flicker. Viney did not react.

'She was different,' he said. 'She was the one they were all after, officers, swaddies, everyone. She was for me, it had to be, everyone thought it. I never looked for it, but she came right at me, and everyone got out of the way.'

Lothar could see it. That incredibly powerful physical presence, the men deferring – or rapidly dealt with if they didn't – the women fascinated. How could she not be? – used to being pursued and flattered and courted, suddenly coming face to face with Viney's massive indifference!

'I didn't want it, but it had to happen. I said *marriage or nothing*, but it made no difference. I didn't want it, but . . .'

And now there was uncertainty and hesitation which disturbed Lothar more than anything else he'd ever encountered in this man.

He demanded, 'But why, Viney? Why? If you didn't want to? If you *knew* you didn't want her?'

'But I didn't know! Don't you listen, Fritz? I knew *nothing*! I didn't know what was going to happen! I don't know what happened yet! Can't you tell me? You're so sodding clever, you're so sodding knowing, can't you tell me, you clever bloody Hun?'

Viney's eyes were blazing now and he thrust his face towards Lothar with a terrible mixture of threat and pleading written there.

'I do not know,' said Lothar helplessly. 'I cannot tell, Viney. You have not told me everything . . .'

'She wasn't like the others, she was quick, she was at me, she was saying things, she was hot and never still and pulling at my clothes. She wouldn't stay still, she wouldn't be silent. She got me in her mouth, she wanted me to do the same to her. She was soft and wet and warm and moving and moaning, I wanted her to be cool and still and firm and quiet, that's what I wanted, but she wouldn't be. I couldn't . . . she wanted too much . . . I couldn't . . .'

He looked at Lothar, all pleading now.

'All I want to do is know, Fritz. What happened inside me? A man has a right to know what's happening inside him, don't he?'

Lothar regarded the big man helplessly. Such confusion as this was beyond any remedy he could suggest. Such doubt and incomprehension at the centre of such massive, rock-like certainty was like molten lava at the core of some vast, flawed, granite mountain – a recipe for catastrophe. He sought for words to soothe, to help. All he really wanted to do was get out of this chamber, this situation!

He said, 'But you said that you did . . . once . . .'

'Fuck? Yaw, I said it. She didn't say much that first night. We slept apart in the end. Next day, I kept her in my sight. I wasn't going to have her running around, spreading the word, so I kept her on a short chain. I drank a lot. I wasn't drunk, I couldn't get drunk, but I drank a lot. She tried again that night, same approach, same result. This time she let fly. She said . . . well, I won't tell you what she said, don't need to, not with you being such a clever little Hun.'

He paused to drink again.

'A man can only take so much, Fritz,' he said. 'I hit her. I don't care for any bastard who hits a woman, but I hit her.'

He clenched his huge right fist and gazed at it as if he'd never seen it before. The butterfly on his biceps swelled monstrously.

'I just about knocked her cold. She went spinning away from me and fell across the bed. I went up to her. She was lying quite still. No noise except her breathing. I looked down at her. She seemed, I don't know, *smaller* somehow. That's when I realized I was getting bigger. That's when I fucked her, Fritz. She opened her eyes as I was doing it, but she didn't say anything. She didn't say anything afterwards either. Next day I sailed, and she never said another word to me, Fritz. Not another word. I heard nothing from her till I got this letter.'

He took an envelope from his tunic pocket and passed it over.

As Lothar unfolded the creased sheet of paper, he recalled that other letter he had read in the Warren, Lieutenant Cowper's, and felt the same frisson of revulsion at being forced to peer deep into open emotional wounds.

But this was no message of faith and love.

Dear Viney *(God! even his wife called him Viney!)* I've got one in the box – probably not yours – but you'll know that – anyroad it'll have your name – unless you say different – I don't want more trouble than I've got – I could do with some money though – thought I should let you know before some other joker does – hope this finds you well – Fran.

'Viney,' said Lothar helplessly. 'I'm sorry.'
He stood up, his head reeling from the drink.
'Nothing for you to be sorry for, Fritz,' said Viney, standing also. He seemed totally unaffected by the drink and as he rebuttoned the letter in his tunic pocket, it was as if he was buttoning away the last hour too. Lothar could only hope he didn't bring it out again later for resentful examination. But for the moment at least the old Viney was back as if their exchange of confidences had never taken place.

'You near on blinded Blackie, you know that?' Viney growled.

'I'm sorry,' said Lothar, relieved to be on other ground. 'But he would have near on killed me, I think.'

'You're not wrong there. But you won't pull that trick twice.'

Threat or friendly warning? It didn't really matter.

Lothar said, 'Viney. Why could we not go to the farm today?'

Viney yawned massively and said, 'You're always after

reasons, Fritz. Some day mebbe I'll come looking for *your* reasons. I hope you'll have 'em ready. Me, I've always got 'em ready. Last night we raided out of the Desolation, down to a Frog village. Barnecourt, it's called. There's soldiers down there. It's a rest area. And there's a hospital too.'

He paused as he said this and seemed about to relapse into introspective silence.

'You knew there were soldiers there, and still you went?' prompted Lothar.

'You can avoid what you know, Fritz,' said Viney. 'Why pick somewhere strange when there's a spot you've got all the drum on? We hit a farmhouse just outside. Christ, you should've seen the booze that old bastard had! We tied the old sod up, grabbed as much as we could carry and set off back. It was a long haul. We passed close to your precious farm, Fritz. Blackie wanted to lay up there for the night, but I said no.'

Lothar sighed with relief.

'That was kind of you, Viney,' he said sincerely.

'No, it wasn't,' said Viney. 'If they came on after us, I didn't want to be caught snoring in a fucking barn.'

'You think they might go to the farm?' asked Lothar in alarm.

'Wouldn't you, if they find tracks leading that way? And if they do, what'll stop them Frogs from talking?'

'They won't talk, Viney, believe me, they won't!' urged Lothar.

'No matter if they do,' said Viney indifferently. 'They don't know anything except that we're holed up in the Desolation, and the redcaps know that anyway. But you and the other two, you're a different matter, Fritz. I don't want the redcaps tickling you with their batons. So you'll keep away for a few days, you hear me? Then you'll go back nice and steady, take no chances. All right? You've got your reasons, so now fuck off. I need some sleep.'

But Lothar was not done. He scented an opportunity.

'Viney, when we go back, how if we stayed there overnight sometimes? The days are getting shorter. It will save much of our energy and we will get so much more done.'

'What's this? First you want more men, then you want to

stop there at night? What're you up to, Fritz? Setting up a rival organization?'

Lothar didn't reply, but waited.

'We'll see,' said the Australian finally. 'But you don't go near the place till I say, you've got that?'

Lothar nodded and turned to go, content for the moment with this small concession. But Viney's voice stopped him at the doorway.

'I've hurt a lot of men, Fritz,' he said softly. 'Killed a few too. But I was sorry about Quayle. Ain't that funny? I never touched the bastard and I'm sorry about him.'

Quietly Lothar slipped past the curtain and headed for a drunken, dream-filled sleep in which Viney and Sylvie and Josh and Quayle all rolled together beneath the water in a slow tangle of limbs, now sexual, now drowning.

4

When Jack Denial heard of the raid on the house on the outskirts of Barnecourt, he at first accepted the irate owner's version that he had been tied up and robbed by *des soldats anglais*. But a closer questioning of the man in an effort to get some hint of the thieves' unit produced disturbing information.

'Black faces, they had black stuff on their faces. And beards, they were not shaven like most of the soldiers,' said the old man. 'And their clothes were old and ragged. And they smelled. Phew!'

He pinched his nose for effect. Denial, who was not much impressed by French rustic hygiene, knew that this must indicate a very high degree of unwashedness indeed.

Physical description was vague, except in one particular. 'Their leader was a giant. Yes, I mean it. A giant!'

Denial talked it over with Sergeant-Major Maggs or rather he talked and Maggs listened, saying only after a long silence, 'So you think it might be our friend, sir?'

'It's possible.'

'What do you think then, sir? Mount a little expedition to go and find the bastard?' said Maggs eagerly.

Denial laughed without much humour.

'I don't think the General is going to be very keen on expeditions of any size going off in pursuit of an Australian deserter,' he said. 'There is a war on, as I dare say you've been told. We have more work already than we can decently manage.'

It was true and Maggs knew it.

'But in the circumstances, sir . . .' he protested.

'Circumstances?' said Denial mildly. 'The individual doesn't matter in this war, Sergeant-Major, can't be allowed to matter.'

'So there's nothing to do?'

'Not quite,' said Denial. 'There's one thing. Up there, four or five miles, close to the edge of the Desolation, there's a farm. Or the wreck of a farm. There's a French family there. I visited them, you recall. I felt . . . something. An atmosphere.'

'You think Viney's lot could be hiding out there?' said Maggs. 'Well, we can spare half a dozen lads for a quick run up there, can't we? It'll only take a couple of hours at most. Shall I set it up, sir?'

Denial shook his head.

'Not straightaway,' he said thoughtfully. 'If they are using the farm, I don't think they'll be there now. It's too obvious. They'll be safe in their hole. But give them a few days and they might start feeling safe again and drift back. Give them a few days . . .'

A few days passed. Viney had withdrawn into himself and hardly communicated even with his lieutenants. Coleport had recovered from Lothar's attack but his still inflamed eyes followed the German with a hatred which promised that the incident was not forgotten. An atmosphere of gloom and foreboding settled on the Warren, reaching its nadir when one night Nelson said to no one in particular, 'I reckon as how I'd as soon be back up the front!'

The remark seemed to act as a trigger to Josh whose despondency had seemed to be most intense of all. He jumped to his feet and rushed out of the chamber. Alarmed, Lothar followed.

He had no difficulty in finding his friend. His raised voice signalled where he was.

'It's no use, Viney,' he was crying. 'I know you're the boss, but this is daft. I'm not staying down this hole any longer, not for anyone! I've not run away from the fighting just so as I can live like a rat. I'm off back to the farm, no matter what you say, and I'm going to stay there, not come running back here every night. I don't care if they catch me in the end, I don't care. They can put me up against a wall if they like and shoot me. It'd be better than this!'

He was standing in front of the recumbent Australian who rose slowly onto one elbow as the youngster raged, then, when he had finished, pushed himself into a sitting position, in which he was almost level with the boy and put his huge hands on Josh's thin shoulders.

Josh did not flinch, nor did he show any fear as the big man drew him slowly down till their faces were barely six inches apart. There they stayed for a long moment, the young man's features stamped with determination, the other's inscrutable in their heavy blankness. Even when the fingers dug visibly deeper and deeper into the flesh beneath the shoulder-blades, Josh showed no fear or pain.

Then with an audible sigh, the Australian relaxed and pushed the youngster away.

'Off you go then, son,' he said.

Josh looked at him with uncomprehending doubt.

Viney said, 'Get your men together. You're the boss-man down there, so I've been told, so you give the orders. Fritz says you'd like Evans with you. Is that right?'

'Yes, Viney, he'd be a grand help . . .' stuttered Josh.

'Anyone else?'

Josh shot an agonized glance at Lothar who was standing in the doorway, then said, 'I'm not sure, Viney. Taff'll do for now, I think.'

'Right, sport. On your way. I'll be down to see how you're

shaping up soon, so keep them jokers hard at it.'

'Yes, Viney. Thanks, Viney!'

Excitement erasing all other emotion from his face, Josh pushed past Lothar without saying anything as though eager to be on his way before Viney could change his mind. The Australian's gaze touched the German's for a moment, then moved indifferently away.

Lothar went in pursuit of Josh and an hour later in company with Hepworth and Evans they were on their way to the farm.

On the evening of the same day, Jack Denial was holding a briefing session with Maggs and four of his men in Barnecourt. The summer was definitely over now and September was trailing its decay-scented mists over the rolling countryside long before the sun set and keeping them there for a good hour after it had risen. This suited Denial very well.

'Two things,' he said. 'One. We are acting only on suspicion.'

In fact in order to get authorization for this operation, he had had to go rather further than this in his justification to his superior. Well, that was part of the game. He'd done as much to get his own way at Scotland Yard. The thing was not to start believing your own exaggerations.

'Proceed therefore with the utmost care and consideration. These people are French nationals. Two of them are very old. Three of them are women. One is a shell-shocked man. Even if they are harbouring deserters, it may be under duress. In any case we have no direct authority over them.'

This was not quite true. In a war zone by agreement with their French allies, the British military had a quite substantial authority over French civilians, but Denial knew better than to make fine distinctions with the men in his command.

'Two,' he went on. 'These men, if they are there, are suspect criminals under the military code. They are not the enemy. Restrain them without undue force. Only use violent means if violent resistance is offered. But be alert. They may be desperate and ruthless.'

He paused and looked at the men. He had chosen carefully. On the whole the redcap soldier was no better or worse than his fellow in a fighting unit, but that was not enough. He ought to be better. He ought to be able to resist the constant temptation to abuse his authority. He ought to know the difference between brutality and strength. He ought to know when to push and when to bend. He ought to know when he was acting for the Law and when he was acting for himself.

Denial knew how hard *that* distinction was to make. Hence his scrupulosity in selecting the four most intelligent and mature of his men for this operation.

There was a knock at the door and an orderly stuck his head round.

'Sorry, sir,' he said. 'But you're wanted urgent.'

'All right,' said Denial. 'Mr Maggs, you carry on with details of timings and movement, will you?'

He left. Sergeant-Major Maggs waited till his footsteps faded out of earshot.

'Right, lads,' he said. 'You've heard what the officer said, and he's right. That's why he's an officer. What he can't say to you but I can is this. These laddos we're after, they're not just your ordinary deserters, oh no. Don't expect to find some scared kid, with tears in his eyes and shit in his britches, hiding in a haystack. These are hard men, real hard men. Like the officer says, they're not the enemy. Again he's right. They're worse than the fucking enemy! They've declared war on their own! They rob their own side, they kill anyone who gets in their way, they defile graves, I expect you've even heard it said that they'll eat their own old comrades' flesh if hunger really pinches! Well, I don't know if that's true. But I know this. Anyone who don't go in there tomorrow morning with his trigger-finger itching is a fool and likely to be a dead fool pretty quick. Hit hard. Hit first. There'll be lots of time for questions later. One more thing. If one of them's a big lad, six and a half foot minimum and broad to match, hit him harder and quicker than anyone else. Get that bastard laid flat and don't stop hitting him till he stops twitching. Understood? Right, let's get down to details.'

He had only been talking a couple of minutes when he too was summoned away, this time by Denial's batman. He found the captain in his room stuffing his belongings into a valise.

'Joplin, you carry on here,' he ordered the batman and took Maggs aside.

'I've been called away,' he said. 'I'm to go to Étaples. There's a little trouble there, it seems.'

'Trouble, sir?' said Maggs, instantly alert at the mention of Étaples. Here, only a short distance from the resort of Le Touquet, had been established a huge base for troops in transit from England to the front. It was in fact a vast complex of camps, the most famous of them being No. 2 Training Camp, set among the dunes within sight of the heartbreakingly normal roofs and chimneys of Le Touquet, and notorious throughout the British Army as the Bull Ring.

Denial had spent some time there on first returning to France as an APM. The conditions and atmosphere had alarmed him so much that he had written a report on them, stressing the virtual impossibility of effective policing in such circumstances. Like most of Denial's reports, it was put aside as exaggerated and alarmist, but the events of the past few days had caused someone to resurrect it and the result was this urgent summons.

Denial looked at Maggs assessingly, wondering how much to tell him. He'd already been assured that everything he'd been told on the telephone was top secret. This he took to be preliminary to the inevitable official cover-up. But he himself was resolved this time that he was not going to be part of any cover-up. And in any case, sooner or later every military policeman in France would have heard some version of what had happened.

'There's been a mutiny,' he said in a low voice. 'The camp's been taken over by the troops. They've been aided and encouraged by deserters living in the countryside round about. Officers have been locked up, even assaulted.'

'And our lads?' said Maggs.

'What do you think, Sergeant-Major?' asked Denial bitterly. 'An impossible job done badly. Yes, they've been abused, assaulted. And some of them have been killed.'

A light of fury blazed in Maggs's eyes. Denial saw it

unhappily. Those deaths at Étaples would be paid for a hundredfold by unfortunates all over the battlezone once the news spread.

He said, 'This means we'll have to cancel our visit to the Gilbert farm, of course.'

'No need for that, sir,' said Maggs. 'I'll take the lads up there and have a look round as planned.'

Denial shook his head wearily and sat down on his bed. He took a notebook from his pocket and began to scribble on a sheet which he then tore out.

'Here you are, Mr Maggs,' he said, handing it over. 'I've put it in writing so there can be no ambiguity. My orders are that the operation be cancelled and under no circumstances be renewed until my return. Do you understand that?'

'Sir!' bellowed Maggs, snapping to attention.

'We are policemen, Mr Maggs, policemen. Our task is to administer the law without fear or favour. Let's just do our duty, shall we?'

'Yes, sir,' said Maggs.

But as he moved away he thought savagely: Duty, you poncy bastard! Oh yes, duty! You want to have him down on the ground in front of you as much as I do, kicking his balls to a pulp and hearing him scream, that's what you really want, you bastard, and you know it. You try to hide it with all this talk of duty, but you know it, you bastard. *You know it!*

5

Josh was usually the third person up in the morning. There was no way to be first or second. Let what he would be his alarum – cockcrow, the first streak of light in the east, even the grey of the false dawn – yet Auguste would already have moved out like a silent ghost to milk his beloved cow and Madeleine Gilbert would be moving equally silently about her first tasks.

'*Bonjour, madame,*' said Josh as he passed her on his way

out of the barn to the pump where he performed his morning ablutions. There had been a heavy frost and it took all his wiry strength to persuade the frozen handle to move. The air was still and white with mist. He listened intently. No sound except the thin calls of a few birds and a shuffling of hooves as Auguste worked in the byre reached his ear. It was a blessed silence. For days now as the cold weather took its grip, the rumours of war which reached them from the east had grown less and less. Winter, the soldier's traditional enemy, in this most untraditional of conflicts had become his friend. It would nip him and bite him and starve him, but the tighter its embrace, the more it held him in his own trench line and kept him from the rapacious emptinesses of no-man's land.

Josh was no keen student of his own inner life, so he did not think it significant that he could now think about the front with something approaching objectivity. There was emotion there, of course, but no longer the emotion of horror, despair and panic. It was a more uncertain and manageable feeling, compounded of pain and pity and perhaps even the smatterings of guilt. But such detailed diagnosis was not Josh's way. He did not analyse. He *felt*. And the great difference was that those old feelings seemed burned into the living tissue of his being while these new ones, though strong, were easily sloughed away with the first shock of icy water on his naked torso, and even more by the sight of Nicole's face grinning at him through the doorway as he roughly towelled himself down.

She had disappeared when he re-entered the barn, which was now unlike any other barn he had ever seen. With Evans's and Hepworth's assistance, and especially since the growing coldness had begun to inhibit work outdoors, the rebuilding had advanced by leaps and bounds. A living-room had been built by the erection of two more walls against an angle of the barn. The new walls were solid for about five feet when they became wattle frames, but the effect was one of considerable cosiness, enhanced by the construction of a proper fireplace against one of the barn walls. This had been Taff Evans's device, the Welshman showing considerable expertise on the question of draughts and ventilation. And now a huge wood fire could burn there merrily with scarcely any driftback of smoke into the room. Madame Alpert was more delighted with this fireplace than she would have been with a million francs. She constantly drew attention to it, as if the others might have missed some of its excellences, and Evans paid for his expertise by becoming the object of her gratitude which was expressed in interminable and, to him, incomprehensible sagas of family history, from which he was usually rescued by Madeleine who was equally grateful to him for giving her mother this new lease of life.

At present, the living-room served as sleeping quarters for the women and old Alpert, while the men bedded down in the remaining section of the barn. Madeleine, who was the arbiter of all things to do with the farm, seemed content enough with this arrangement, but Evans felt differently.

'Got to have a bit of privacy, don't they? We live tight down in the Rhondda, got to, haven't we? But we like to be private and decent.'

And without more ado he'd set about constructing some small sleeping chambers alongside the living-room. Madeleine had shrugged, but offered no larger opposition, and now the work was well advanced.

Lothar had assisted to the best of his abilities but his lack of practical ability compared to the others had reduced him to the level of unskilled assistant, and since their return on a permanent basis to the farm, his old authority had suffered a similar reduction. When he tried to organize a sentry roster, this change became quickly evident. Hepworth, who seemed in some obscure way to blame Lothar for Quayle's death, voiced the new attitude.

'The weather will stand sentry,' he growled. 'If anyone wants to hang around out there in the cold, he's welcome. This isn't the fucking army, is it?'

'No,' said Lothar, frowning. 'It's not the fucking army.'

And Josh noticed thereafter that his friend from time to time drifted off to take a long look down the valley and soon his main job was that of almost permanent sentry.

This morning Hepworth was still lying wrapped in his blankets from which he rarely emerged till the smell of cooking blunted the sharp morning air. As Josh passed the entrance to the Gilberts' living-room, he glimpsed Nicole pulling a chemise over her head, slim brown legs forking provocatively beneath the turbulent garment. Involuntarily he paused, then gasped as something hard prodded him in the kidneys. Behind him stood Madeleine. Her weapon was the spout of an old tin kettle which she had just filled from the pump.

'*Vast'en,*' she ordered sternly. '*Vite, vite!*'

In his blankets, Hepworth sniggered and Madeleine went towards him and none too gently dug her foot into his ribs, rattling away at him far too fast for Josh's limited comprehension and for Hepworth's too, but it was clearly uncomplimentary. The Yorkshireman's hand snaked out from under the blankets and grasped the woman hard about the ankle so that she almost lost her balance. She re-doubled her abuse, and at the same time gently tipped the kettle so that a trickle of icy water fell on to Hepworth's face.

Swearing, he jumped upright. Whatever emotion their contact had roused in Madeleine, its effect upon Hepworth was quite evident in the stiff pyramid formed at the crutch of his long woollen underwear. For a moment the pair of them stood face to face like a pair of boxers preparing to fight. Then outside the stillness of the morning was disturbed by a distant rumbling.

'Listen!' ordered Lothar, turning at the doorway to strain his eyes into the mist.

'Thunder, is it, Lott?' whispered Josh in a futile effort to delay an obvious truth.

'No, Josh. It's a barrage,' Lothar confirmed quietly.

'Ours or theirs?' demanded Josh, forgetful who he was speaking to.

'Yours,' said Lothar certainly. 'There must be an attack.'

'Attack? It's nigh on the end of November!' exclaimed Josh. 'They'll not be any big push now until the spring, surely?'

His face was working as he demanded this reassurance. He had lowered his defences that morning with his thoughts of the quiet months ahead, for the men in the front line as well as himself, and this violation of *their* merited peace brought into doubt his own.

'Not a big push, no,' agreed Lothar. 'But someone has decided, perhaps wisely, that men should not be allowed too long a time to think. Come, everyone, soon it will be too cold to fight or work. Let us use the time as best we can.'

Evans had meanwhile headed out to the pump. Suddenly they heard him cry out in alarm. There was another voice and footsteps on the cobbled yard and the men in the barn moved rapidly towards their neglected weapons.

But they were too late. The new arrival was already at the door.

'Great lookout system you've got here,' said Viney, filling the doorway as effectively as a square of oak. 'What're you all looking so surprised for? I said I'd be along to see how you jokers were shaping up, and here I am.'

As he spoke his eyes were never still. Lothar guessed that he had taken in every detail of the barn and its inmates and stored them up for future reference. The Gilbert family were regarding him with grave suspicion. Lothar could not blame them.

'You heard the guns, Viney?' said Taff Evans. 'Some poor bastard's catching it. What do you think it is?'

'Fuck knows. Can't be anything big, not with winter coming on hard. It'll just be some brass-hat worried in case the poor bloody infantry get bored with having nothing to do.'

So the topic was dismissed. It was curious, thought Lothar. Viney spoke not with that dreadful pseudo-knowingness of the bar or barrack-room bore, usually found in a cushy billet miles behind the lines, or safely ensconced in a nice Home Front job, but with an irresistible, almost godlike certainty. He tried to imagine what it would be like if some freak of chance put Viney in total charge of the British and Allied armies. Instant peace or bloody devastation? Of one thing he was certain: the men would be out of their trenches and moving one way or another within a very short time.

Viney stayed all day, cheerfully giving a hand and taking

instructions without demur. One effect of his arrival was that all the men took their turn at standing sentry, even Hepworth, without being told. Josh, who viewed Viney's arrival with mixed feelings, tried to keep out of his way. The upset of that distant gun-fire had been bad enough, but the Australian's presence was even more disruptive. The guns only reminded him of the past, of what he had fled from. Viney reminded him of the present, of what he had fled to.

It would have been good to explain some of this to Nicole, but today she was more than usually uncertain in her moods, at one moment seeming to meet him as the closest, and dearest of friends, the next mocking and evasive. Lothar had suggested he should think of learning French as he did of learning to imitate bird-calls and this approach was showing results. But Nicole's shifts of mood would probably have left him tongue-tied even in his own language.

Late in the afternoon as darkness began to fall, Lothar said diffidently to Viney, 'Will you spend the night here?'

'I don't think so, Fritz. They'll be missing me back in the Warren, so I think I'll head on back. But I'd like a little confab with you before I go.'

He clapped his hand on Lothar's shoulder and led him out of the barn and away from the buildings. Sensing something important was going on, the others came and stood in the doorway and watched. At first it seemed friendly enough, though it was impossible to hear what was being said. Then suddenly Lothar broke away from the affectionate restraint of the Australian's heavy arm and it rapidly became clear that they were far from in accord.

For the onlookers, it was almost purely a visual impression, though the discussion increased in volume as it increased in vehemence, with the two men silhouetted against the western skyline like warriors in a shadow play, antagonism visible in every line and movement, and breath-clouds of anger lowering over them in the frosty air.

It ended with Viney turning away, taking a few paces towards the barn, then wheeling round and thrusting out an admonishing finger, stiff and threatening as a pistol barrel.

'You've got it cushy here, sport, and don't you forget it!

There's plenty back there that don't. So, either we play things my way, or we change the fucking game!'

He spun round again and strode towards the others who stood by the barn door, including Madeleine and Nicole, who grasped Josh's hand tight in defence against the Australian's anger. But his expression relaxed in a moment and he was smiling broadly as he reached them.

'*Au' voir, madame, m' moiselle*,' he said gallantly. 'Cheers, young Josh,' – ruffling the boy's hair and winking – 'don't you be overtaxing your strength. Heppy, Taff, behave yourselves, sports. See you soon!'

He made off at a rapid pace into the already bitter night. Evans looked enquiringly at Lothar who had walked slowly back to join them, but the German just shook his head and went into the barn. Madeleine, not to be put off by any strong silent man act, went after him, stabbing questions at his back, and Evans and Hepworth followed.

Josh and Nicole stood together and watched till the voracious darkness had completely swallowed Viney. Josh would have been content so to stand for hours, as long as the girl's hand was in his, but already he felt her fingers beginning to gently disengage. Impelled by an urge far removed from conscious decision, he turned suddenly to the girl and aimed a kiss at her face. He half missed her lips and found himself pecking almost fraternally at her cheek. But a bonus of his awkwardness was that his free hand, reaching up to steady the surprised girl, accidentally came to rest on her breast. Heat flowed through his body – sufficient, it seemed, to melt the frozen air for a mile around into balmy summer.

Then she twisted away and went after her mother.

Josh stood alone under the triumphant stars till the cold threatened to freeze his lips together, then went back into the barn.

Inside he found Madeleine and Lothar in a discussion as intense if not as antagonistic as that between Viney and the German earlier, while Evans and Hepworth stood uncomprehending but troubled on the sidelines. Josh by concentrating with all his power might have picked up enough words to get the gist, but most of his attention was focused on Nicole going demurely about her business, from time to time shooting him glances beneath lowered eyelashes, by turns accusatory and conspiratorial.

So it was only when the debating pair divided with not much evidence of accord that Josh was able to find out the cause of the agitation.

It was simple enough. With a long hard winter in prospect, Viney was concerned to build up a store of supplies sufficient to keep the Warren going to the spring. This was understandable. There'd already been several flurries of snow and when it came down thick, not only would it turn the trenched and cratered Desolation into a death-trap, it would also mean that any raiding party would leave tracks even a staff-officer could follow.

Viney was planning a series of raids to stock his larder and his strategy included, among the usual assaults on supply dumps in the support lines, another expedition into the living countryside behind the Desolation.

'Listen,' he'd said. 'What do we get from Army supply dumps? A lot of danger, that's what, and what for? Plum jam and hard-tack as often as not! There's real grub out this way. Fresh stuff, eggs, meat, fruit. Officers' grub. And civilians'!'

'You can't get away with it!'

'No?' Viney had smiled wickedly. 'We've already tried out our hand, you know that, sport. We got ourselves some nice vino, remember. But that was small-time stuff to what I've got in mind!'

Lothar resisted as best he could, protesting vigorously that innocent civilians were not legitimate targets even for a few desperate men.

That was when the discussion got lively.

'Listen, Fritz, you promised us fresh grub and Christ knows what else when you got us mixed up with this place. Well, it might be keeping *you* well fed, but it's doing nothing for the rest of us. So I don't see how you or these Frogs can object to helping your mates out in keeping body and soul together.'

'Helping?'

'Yeah. Oh, not with the dirty work. Your lilywhite hands'll stay nice and clean! But it's a long way from the Warren to the fat lands. What we need is a staging-post and storehouse for a couple of nights and this is it! Tell the Frog queen and tell her so she understands. There'll be no arguing! This is it!'

All this Lothar now told the three British soldiers. Hepworth and Evans received the news in silence, then turned away to digest it. But Josh was not silent, for once surpassing his friend on his outcry against Viney.

'They can't come here! They can't!' he protested. 'It's not right. Lott, you mustn't let them!'

It was curious, thought Lothar, how suddenly he was the leader again. From each according to his ability!

He said gently, 'What would you like me to do, Josh? Go after Viney and shoot him?'

For the first time in their acquaintance, Josh regarded him with anger in his eyes. Then, turning on his heel, he went to join the others, leaving Lothar standing by himself, apart, and suddenly once more foreign.

6

Ten days went by and nothing happened.

The intermittent rumbling of the guns told them that whatever 'show' had begun on the day of Viney's visit was still running. The weather remained hard and cold, and work within the barn proceeded apace. Sleeping quarters were completed, two small chambers, one for Madeleine and Nicole, and one for the old Alperts, but in the event Madame Alpert showed such reluctance to be removed from her new and dearly loved hearth, even for the hours of night, that she and her husband remained in situ and the Gilbert women had a chamber apiece.

Lothar tried to warn Josh that Viney's non-appearance did not necessarily mean he'd changed his mind, but at the end of a week, he could see the boy was thoroughly convinced the threat had been averted.

Then one morning as Josh was engaged in his customary wrestling bout with the frozen pump, figures began to loom through the sun-tinged frost mist, huge and menacing as the watery light diffused their shapes.

With a cry of alarm, Josh ran towards the barn, Madeleine appeared at the door before he reached it, with Lothar close behind. He joined them and gasped. 'There's men coming, Lott. Men!'

'I see them,' said Lothar, grasping his shoulder.

Josh turned.

Into the yard, shedding bulk but not menace as they emerged from the mist, came Viney and Delaney accompanied by about a dozen armed men.

'*Bonjour, madame*,' said the Australian, ''Morning, Josh. You want to get something warm on or you'll catch your death. 'Morning, Fritz. No sentries again? Could've been Haig himself, or even Hindenburg, with a couple of divisions at their heels. Still, no harm done, and I dare say you all need your beauty sleep here, working so hard all day. Patsy, post three of these jokers on lookout duty and work out a roster. Josh here will show you the best spots when he's made himself decent. Come on, Josh, let's be having you! Ladies present. I hope it's not like this all the time, Fritz. Very lax, very slipshod.'

There was something it was difficult not to admire in the smoothly efficient way in which Viney insinuated his men into the farm. Most of them Lothar was not surprised to see, but there were a few missing faces that he would have expected to find in the party.

As if catching his thought, Viney who had dropped his equipment off on to the barn floor and was warming his behind at the newly lit fire, said wearily, 'Now before you start, Fritz, you have a good listen to me. I'm here because I've had to come. The men would've come sooner but I paid heed to what you said. You're a clever cunt and you've made sense in the past, so I made them try the supply lines first. Result, two dead and three wounded and one of *them* like to snuff it any time. So, I've got jokers with me now that I'd rather stayed at home. Well, beggars can't be choosers, and we're beggars, don't mistake it. We're low on everything. It's a real bastard, but unless we start stocking up quick it's going to be a starving Christmas. Hello, lad. How're you doing?'

These last words were addressed to Auguste who had just come from the milking and was standing staring in some alarm at the sight of the newcomer. Nicole went to him and put her arm around his waist and led him aside, talking reassuringly.

Josh had not lingered long in directing Delaney to the best sentry-posts. Lothar guessed he was more concerned about mounting a personal guard on Nicole, and the German was aware of the youngster's mounting anger as he said to Viney, 'Exactly what is it you propose?'

'A couple of raids, that's all. Two nights, maybe three, four at the most. Then it's goodbye and see you next spring.'

'The whole countryside'll be up in arms!' protested Lothar.

'Mebbe. But that's the beauty of jumping off from here. We can go so much further afield. And in any case, it's the poor swaddies in the training camps and rest areas that'll get blamed! No one believes in us, Fritz, didn't you know that? Or at least, no one likes to admit we exist! Anyway, I'm afraid it's like it or lump it. We're here and here we stay till we're done. The more help we get, the sooner that'll be.'

'Help?'

'Sure. The Frogs can fill us in on the countryside round here. It'll be worth a dozen recces. And I'd like Heppy and Taff to help us out with the work. You too, if you like. I've got jokers here that I'd not trust to piss in a bucket.'

'I thought I saw Strother out in the yard,' said Lothar. 'Why have you brought him?'

'Needs must, sport,' said Viney. 'He's a hard man in a tough corner. I'll keep him out of the way of the ladies. I think Patsy'll have sent him off on sentry duty to start with. OK, Fritz. We're in your hands here. Show us where we can doss down. Any help we can give while we're here, you've only to ask. But I'd like everyone to get a few hours' kip before they go out tonight.'

There was nothing to be done. Viney now with great skill concentrated all his energies on first containing, then eliminating Josh's obvious anger. The boy let himself be soothed not by argument but by the Australian's frank and friendly manner. Watching Viney at work, Lothar sighed deeply. That there was genuine affection there he did not doubt. But that to Viney emotional manipulation was merely a sometimes preferable alternative to brute force was also clear. But it was not the power of strong amoral men like Viney that made Lothar sigh, it was the vulnerability of gentle and good men like Josh.

In the event, things went very smoothly to start with. Madeleine was a pragmatist and with the single exception of her assertion, too calmly stated not to be believed, that if Strother came anywhere near Nicole or herself, she would slit him with the pig-sticking knife, she made no protest over the raiders' presence. Auguste meanwhile had reacted to Nicole's reassurance by apparently censoring the newcomers completely from his mind. He went about his business ignoring their speech and their presence alike. It was eerily unnerving to some of the men but their sympathy for yet another victim of the war was too great for any adverse reaction.

Viney invited the men already at the farm to join the raids, though Lothar noticed he waited till Josh was out of earshot first.

Both Lothar and Evans refused out of hand, but Hepworth thought a long time, then said, 'I'll go.'

'Good man,' said Viney.

Lothar regarded the Yorkshireman in silent surprise and after Viney had left them Hepworth said, as if challenged, 'We've all got to earn our self-respect our own road, Fritz.'

The first night, only a small reconnaissance party went out, returning shortly before dawn with information which Viney did not share with Lothar but which clearly made him happy.

After a day spent in useful work under Lothar's direction, the raiders slept till ten o'clock, and then departed.

They returned shortly before dawn in a GS truck loaded down with all manner of goods, both military and civilian. The mood was jubilant as they offloaded the boxes into the byre. This work Lothar and the others willingly assisted, aware that the longer the truck stood in the farmyard, the greater the danger of discovery. The east, which had at last fallen relatively silent once more, indicating that after about a fortnight the recent battle had faded out, was lighting to grey as the truck was driven away. With the dawn came a flurry of snow, not deep, but settling enough to cover any tracks that might have been made by the laden vehicle.

The next night more than half of the raiding party were sent back under Fox's command to the Warren, bearing the loot.

'Haven't you enough?' asked Lothar, uneasy when he saw that it was the hardest of the men who stayed behind.

'Not by a long chalk, sport,' said Viney. 'But there's plenty more where that lot came from.'

'You will not raid the same place?'

'Why not? They won't be expecting us twice. Getting hold of that truck was a bonus. We'd be mad not to use it again.'

'I think you will be mad to do it.'

'Listen, Fritz,' said Viney, pointing upwards to the lowering cloud-cover. 'Any time now there'll be snow coming down from there like shit from a goose. Real snow. I want to be snugly tucked in for winter by then.'

His forecast proved too quickly correct. That night Viney and his raiding party set out to where they'd hidden the truck. There were six besides the leader – Coleport, Delaney, Strother, Taylor, the black Moroccan and Hepworth who was still working off his own strange sense of duty. Madeleine had glared angrily at him but said nothing, only shaking her head with great scorn. Four hours after they'd gone, the snow began to fall in big soft flakes; more, Lothar thought, like goose-feathers than goose-shit. Soon it began to dress the frozen ground in a soft white coat. There was a freshening wind blowing from the west which as it grew stronger would cause drifting. Lothar knew from childhood winters at Schloss Seeberg how quickly snow could change a landscape, turning the familiar into the foreign, and mocking even the most compass-like mind with lures of indirection.

Dawn approached, or what would have been dawn if light could have penetrated the drapery of snow, twisting sinuously now in the breeze which rose with the sun, and lying in folds and pleats like the train of some monstrous bridal gown.

In the barn, the women and old Georges slept, but the men kept intermittent watch and Lothar never closed his eyes. His wakefulness was initially due to a fear that the raiders might return with the authorities hot on their heels, but that fear was long past. Now he was beginning to wonder if they would return at all.

Darkness or light meant nothing to Madeleine, who rose at her usual time with the exactness of one summoned by a bell.

'They have not returned?' she asked Lothar. He shook his head and she turned away indifferently. He guessed her indifference was assumed as far as Hepworth went, but also guessed that she would not be displeased to learn that Viney and the others were frozen to death in the snow.

Lothar could not share her feelings, and not just out of concern for Hepworth. He suspected that if anything happened to Viney, the other Volunteers would rapidly deteriorate into disorganized scavengers, making inept attempts to reach the coast and smuggle themselves back to Britain. Capture for most would be inevitable, and inevitable too would be their rapid betrayal, deliberate or involuntary, of the Warren and the farm.

He went to the doorway. Josh was standing outside in the snow which was falling less thickly now, though the gusty wind was swirling it round in pulsating spirals that mocked the eye.

'Josh, you will freeze. Come inside,' he ordered.

Josh did not move but said, 'When I was a lad, I loved the snow. Our dad had made us a sledge and me and Wilf used to go up the fellside a way behind the house and come whizzing down. On the steep bits I was scared stiff, but Wilf would keep a hold of me and sing or laugh in my ear and down we'd come like lightning, and even if we fell off, he'd not let go, and we'd just roll over and over in the snow.'

Lothar put his hand on the boy's shoulder.

'Yes, Josh,' he said gently. 'I too used to go sledging with my brother. I too. Now come inside.'

Still Josh resisted, but this time for a different reason.

'Listen, Lott! I think I can hear something!'

Lothar stepped outside and cupped his ears, but heard nothing except the wind.

'There! Again!' said Josh.

Lothar shook his head.

'I can't hear anything,' he said. 'You lead and show me.'

His unquestioning trust of Josh's countryman's hearing brought a smile of pleasure to the lad's lips.

'Right,' he said, grabbing one of the long-handled shovels which stood just inside the door. 'Come on!'

The snow was calf-deep in the farmyard and up to their knees once they left the shelter of the buildings. It was obviously deeper still where the wind was drifting it, but Josh seemed to have an unerring instinct for finding the best way.

'There again!' yelled Josh after they'd gone a hundred yards.

This time Lothar too heard the cry and after a couple of minutes more they spotted the figures. There were four of them in two pairs about ten feet apart.

'It's Viney,' said Josh. 'Viney! Halloo!'

The big Australian was supporting his compatriot, Coleport, whose left arm dangled loose, streaked with blood through its bandage of snow. Both men were grey and exhausted.

'That you, Josh? Thank Christ!' exclaimed Viney. 'Here, give us a hand with Blackie, will you? I'm fair tuckered out.'

Lothar came up behind, glanced at Viney, then without speaking went on to the other couple which consisted of Hepworth, whose head was bleeding from a long gash along the temple, supporting Strother, who clutched his stomach and groaned and would have fallen to the ground without the Yorkshireman's arm to hold him.

This was no time for questions. Taking most of Strother's weight on himself, he set out after the other trio who were following the still visible track he and Josh had made from the barn.

Inside, the fire was roaring. Madame Alpert was still in bed, but old Georges was up demanding querulously to be told if the cow had been milked. Nicole was patiently assuring him that all would be taken care of in time and Madeleine was stirring a great cauldron of broth over the fire. But all conversation and activity stopped at the sight of the returned raiders. An indignant glance passed between mother and daughter when they saw one of the wounded was Strother, but any objections to his presence were quickly forgotten when Madeleine realized that Hepworth was hurt also. She scolded him anxiously till her careful examination revealed that his head wound looked worse than it was. Coleport's arm, however, was in a bad way. A bullet had passed clean through the upper arm, shredding flesh and muscle though miraculously it only seemed to have nicked rather than shattered the bone.

Worst of all was Strother. A bullet had entered his lower abdomen. There was no exit hole so it must be still inside. There had been a few packs of dressings among the last lot of loot, and Lothar who found himself cast once again as medical expert did what he could for Strother while Evans dressed Coleport's arm. As they worked, Viney told them what had happened.

'Bastards must've been waiting for us,' he said. 'MPs, I reckon. No warning, just started shooting, that's the kind of trick them cunts are good at. Two of our lot went down straightaway.'

'Dead?' asked Lothar.

'Oh yeah. They were dead all right. Heppy got creased and then we were off and running, right into another bunch of them. That's when Strother took one in the gut, and the other three got killed. Including Patsy.'

His voice broke slightly.

'Delaney's dead?' said Taff Evans.

'I said so, didn't I?' snapped Viney.

'And if you say so, it must be right,' said Hepworth in a strange voice.

237

Viney shot him a savage glance but did not reply. Coleport, however, roused himself from his wound-shock to say, 'What're you getting at, Heppy? Is poor Patsy dead or not?'

'Didn't you see him for yourself?' asked the Yorkshireman.

'What the hell could I see? I was lying on the ground fifty yards away, set to bleed to death myself from this fucking arm.'

'Count yourself lucky it wasn't your leg,' said Hepworth. 'I looked at Delaney before I went to Strother here. That's where he was shot, clean through the kneecap.'

'What're you saying?' demanded Coleport. 'You don't snuff it just because you've been wounded in the leg, not unless the fucking thing's been carried right away!'

'Well, this weren't carried right away, but nor was it going to carry Delaney right away either, that's the point!'

There was a silence as those present tasted the implications of this. Coleport broke it, his voice low and pleading.

'Viney, he's wrong, isn't he? Not Patsy! You'd not be worried about Patsy talking, would you, cobber? He'd not tell them bastards anything, not if they started hacking at his leg with a rusty saw!'

'I know that, Blackie,' said Viney. 'Don't you think I know that? He was wounded in the leg, sure, but he'd taken one through the chest too. He was dying when I got to him. I tried to lift him up, but he died in my arms.'

'Aye!' exploded Hepworth. 'What'd you try to lift him up with? Your jack-knife?'

Viney leapt to his feet, dragging his Luger out of his tunic. Hepworth, though unarmed, rose too and confronted him with no sign of fear.

'For the sake of God!' cried Lothar. 'Hasn't there been enough blood shed tonight.'

He stepped between the two men, the movement bringing Josh within his line of vision. The boy's face was pale, his eyes wide and unblinking with the look of one who sees more than what stands before him. Such an expression as this Lothar had not seen since the early days after their arrival in the Warren.

Viney too had noticed and he broke away from the threatened confrontation to go to the youngster.

'You all right, son?' he queried. 'Bit of a shock, all this. It'll be all right, you'll see.'

He reached out his arm but Josh stepped back, his eyes fixed with horror on the sleeve which was crusted with Coleport's blood.

Lothar picked up a rifle and thrust it into Josh's hands.

'There's no sentry outside,' he said in a peremptory voice. 'Josh, wrap a blanket round yourself and get out there and stand guard.'

The young man shook his head violently.

'Josh, we can't risk being taken by surprise!' insisted Lothar. 'They may have been followed. For Nicole's sake, go and stand watch!'

This seemed to get through. He turned towards the door. Lothar tossed a blanket over his shoulders as he went out, a whirlwind of snowflakes dancing in before the door shut behind him.

'What the hell's all that about, Fritz?' growled Viney. 'No one's going to come after us in this blizzard. They might even give the sodding war a rest till it's over.'

'Josh has had enough of this,' snapped Lothar. 'Of blood, of fighting, of violence. If you idiots are going to start killing each other, here, before his eyes, he will know for certain the world is mad beyond recovery, and he may run mad himself.'

'Yeah, well, all right,' said Viney. 'Let's forget what's happened, OK?'

He looked at Hepworth who shrugged and turned away. Coleport seemed to have relapsed into shock once more. Evans said, 'These poor devils ought to be in hospital, Viney.'

'Hospital? To be fatted up so they'll make a good target for the firing squad? Talk sense, Taff! They'll have to take their chances. We'll all have to take our chances. For the moment we're as safe here as anywhere in France tonight!'

And his confident assertion was still hanging on the air when they heard the shot outside.

Everyone froze except Viney who spun round to face the

door, his Luger in his hand, held steady as a rock at a man's chest height.

'No!' yelled Lothar, throwing himself between the Australian and the door. If they were being attacked, the slightest sign of resistance when their attackers came through the door could lead to a bloodbath.

No other shot followed and as the long seconds ticked by, some of those present began to doubt whether they'd really heard the first one. Some, but not all.

Nicole suddenly called out, 'Josh!' and headed for the door.

'*Attends!*' commanded Lothar, barring her way. Cautiously he lifted the latch.

Outside the snow was still falling, but gently now, a few big flakes feathering down to add the last touches to a scene which at first glance had something of the quality of a Christmas card illustration.

The farm buildings and the rubble-pile which marked the site of the old house were equally beautiful in their covering of white. In the middle of the yard crouched Josh. With his blanket draped round him like a Biblical robe, he might have been a shepherd kneeling in adoration outside the stable. Two things spoiled the image. The first was the Lee-Enfield .303 in his hands. The second was the half-open byre door. No shaft of divine light came pouring out. But on the whiteness of the snow which had drifted across the threshold lay a crumpled shadow.

Lothar went slowly forward and paused by Josh, his hand touching the young man's shoulder.

'I heard noises,' said Josh in a flat, dead voice. 'I thought they were in my head again. Then I saw the door opening in the corner of my eye. I thought . . . I thought . . .'

But what he thought was never expressed, for Nicole screamed a long wavering scream.

'Auguste!'

And ran across the yard and knelt in the snow cradling her brother's loose-lolling, open-eyed head in her lap, with her own hating, accusing gaze riveted to Josh's face which was now so white that it was scarcely possible to distinguish the pale skin from the snow which clung to it.

7

They laid the young Frenchman out in Madeleine's sleeping chamber, Madame Alpert and her daughter performing the task with the stoic expertise born of the peasant's long acquaintance with the realities of death.

Hepworth, his own hurt forgotten, went to Madeleine with clumsy condolence, but she only looked at him from the shadowy pain behind her blank eyes and spat out the single word '*Soldat!*' and went on with her work.

As for Josh, he too made an attempt to speak to Nicole. But when he stood in front of her no words would come. Slowly she raised her head and looked into his eyes. Then, with a visible gathering of strength, she drew in a deep breath and spat full in his face. Josh did not move, but the girl, as if this had exhausted all her reserves, now surrendered nearly all control of her limbs and would have fallen in a heap if Lothar had not caught her.

Madeleine came to his aid and together they laid her in the narrow chamber next to that in which her dead brother rested.

'It is shock,' explained Lothar. 'Like when a shell goes off near a man. She must rest and be warm.'

They piled blankets over her and left her in, though not at, peace.

Josh said, 'Lott, I heard a noise. I thought . . . I didn't think . . . I just turned and I thought it must be . . . the rifle went off . . .'

'Yes, Josh, yes,' said Lothar gently, putting his arm around the young man's shoulders and leading him to his bed-roll.

It was a tragic irony. The war had at last killed Auguste. In the shock of Viney's return with his wounded companions, it had not seemed possible for anyone's attention not to be focused on this drama. But Auguste, in whose war-crushed mind Viney and the others were not allowed to exist, had risen on this morning as every other and gone out unnoticed to milk the cow.

Viney said, 'Forget it, son. These things happen. Poor bastard's probably better off dead.'

It was kindly meant but Josh simply turned a comfortless face to the Australian and began crying. In helpless embarrassment Viney turned away.

Slowly the day dragged along. Strother and Coleport were made as comfortable as possible, but soon the Cockney's groans had become as much a part of the background noise as the wind which still rearranged the fallen snow outside the barn. Nicole did not emerge from her chamber. From time to time Madeleine who was seeking her own solace in constant activity went in to see her, taking bowls of soup and coffee, but always responding to Lothar's queries with an irritated shrug.

She did speak to him once, however, and Lothar approached Viney who was resting in monumental stillness in the furthermost corner of the barn, only stirring to dismantle and clean his weapons.

Lothar said, 'Madeleine says there will have to be a priest.'

'Je-sus,' exclaimed Viney. 'I thought those jokers were meant to be here before death, when they could still save your soul.'

'There is still work for him,' said Lothar. 'And there are the funeral arrangements to be made.'

'Funeral!' exploded Viney.

'Why yes, Viney. These people are not up to date with the modern custom of tossing the body into the nearest old trench or new shell-hole. They still expect consecrated ground, a coffin, and a ceremony.'

Viney thought about this.

'What you're saying, Fritz, is we'll have to be moving out of here pretty soon?'

'I fear so. In any case, once the snow goes, there may be men here looking for us.'

'Yaw. And Strother, he ain't going to be easy to move.'

'No. Perhaps we could leave him. He needs hospital treatment.'

Viney looked at him blackly.

'We leave no one, Fritz,' he said. 'How long do you think Strother'd keep his mouth shut? They'd be into the Warren in half a day.'

'The Warren is not the only place to hide in the Desolation,' said Lothar.

'You want to look for somewhere now? With winter all ahead of us? Don't make me laugh, Fritz! No, tell Madeleine she can fill the place with priests after we've gone, but that won't be till the weather improves!'

He reached into his pack and produced a bottle of spirit.

'Care for a snort?' he asked.

'Thank you, no.'

'Suit yourself,' he said indifferently. 'Suit yourself.'

It never really got light that day, but at last what passed for light faded, and gradually the activity in the barn died away and, except for Madeleine and Madame Alpert who were taking turns in keeping vigil by Auguste's body, everyone settled down for the night, though *settled* was not a word which suggested itself to Lothar. For the Volunteers, he thought, this was in many ways the low point of their existences, a nadir when all the horrors of the past and the uncertainties of their future ran into each other and showed they were the same. Strother, anaesthetized with brandy, tossed and groaned; Hepworth lay with his head between his hands and from time to time sighed deeply; Coleport stared unblinkingly into the darkness and moved his lips in soundless repetition of one of his favourite ballads; even Viney's indomitable spirit seemed subdued by the events and atmosphere of the day. Josh lay close by Lothar. He had spoken scarcely a word all day and now, when the German tried to put a comforting arm round him, he rolled further away.

Lothar lay in the darkness, more despairing now than at any time since Sylvie's death. Beside him he heard Josh's breathing at last lengthen into the steadiness of slumber, but for him no rest came.

How long he lay there he did not know, but when a noise came from one of the chambers, he was already fully awake. It sounded as if something had fallen over. Careful not to disturb any of the others, he rose and went to investigate.

The cause of the noise was soon found.

In the vigil chamber Madeleine, still at last as she sat by her dead child's bed, had finally succumbed to the fatigue of a day without rest or respite, and lay slumped forward in her chair with her head next to Auguste's. The movement had upset one of the pair of candlesticks at the bed head. Fortunately the flame had gone out as it fell.

Lothar picked it up, relit it from the other, and set it out of danger's way. So deeply fatigued was Madeleine that none of this roused her and he decided the kindest thing to do was let her sleep on.

He turned to the door and drew in a sharp breath. Wraith-like, a pale figure stood there. It was Nicole, clad in a linen shift, her white face set like a sleepwalker's.

Quickly Lothar left the vigil-chamber, drawing her with him.

She spoke in a voice so low he could not make out the words. He stooped so that his face was close to hers and she spoke again.

'Is he dead? Is he really dead?'

'Yes, my dear. He is. Come now.'

'No! I must be with him!' she protested.

'Your mother is with him. Come now. Come now.'

He led her back into her own narrow room. She was shivering with cold and with fear too, he guessed. As he tried to lay her down on the bed, she put her arms round his neck and begged, 'Do not leave me, do not leave me.'

'I must. I must,' he insisted.

'No. Do not leave me, please, please, I beg you, I beg you.'

In her voice was a fear and loneliness and despair which echoed his own. But worse than that was another echo, an echo from the lost, the dead time before his life had turned into this waste land. It was Sylvie's voice he heard, her blind pleading for warmth, for comfort, *Oh Lott, please, please.* He tried to break her grip again, but her strength was too strong now or his efforts too weak. She was a dead weight round his neck drawing him down to the narrow bed. He ceased to resist, but went with her and for a moment they lay there, perfectly still, body against body, separated only by the rough cloth of his threadbare shirt and the thin linen of her white shift which had ridden up round her waist. Then her mouth came in search of his.

Josh awoke from a dream of Outerdale with the snow crisp on the fields and the breath of carol-singers bright as their merry song in the morning air. Darkness and memory rushed back on him at once. Cold now, and in need of comfort, he rolled over and reached out his hand to touch the warm and comforting shape next to him. But his finger only encountered an empty roll of blankets.

Alarmed, he explored further, but there was no possibility of error. Lothar was no longer there. What was more, the nest of blankets was quite cold. This was no quick enforced trip to the farmyard for a cold-cursing piss. It might of course have been a more urgent call of nature which would have taken him across the yard to the privy behind the byre. Trying to accept this theory, Josh lay in the dark for some minutes. But his uneasiness was growing and as it was accompanied by his own desire to relieve himself, he finally rose, pulled on his boots and trousers, and began to move quietly through the darkened barn.

His conscious worries were all centred on Lothar's possible illness, or, in the icy conditions, even more possible accident. And his route towards the barn door was consciously dictated only by his concern not to wake any of the other sleepers. He reached and passed the entrance to the living-room. He reached and paused at the entrance to the first sleeping chamber in which Auguste was laid out. The door was slightly ajar letting out a glimmer of candlelight.

Slowly he pushed it open till there was a gap wide enough to permit vision.

The corpse lay on the narrow bed, lit by a pair of soft-flickering candles. Madeleine Gilbert's sleeping head rested on the pillow beside her son's. It was only her presence that gave him the strength, drawn by what impulse he knew not, to step into the room to look at the young man. Perhaps it was the need of one who had seen too much of limb-shattering, gut-ripping, flesh-searing death, to be reminded of the high valuation of life implicit in this washing and composing and dressing and decent laying out of the empty fleshly husk.

The face was calm and composed, but masklike – much deader in its way than the ruined head of some shrapnelled lad in no-man's land, one cheek ripped away to show the white curve of the jaw. No living flesh ever reflected the candle-glow like Auguste's waxen skin. It was touched with a light like that pale glimmer which steals across the high fells to mark what passes for day at an overcast November dawning.

Josh looked and shivered, seeing suddenly how death and its pains weathered a man's body like a landscape, taking soft richly rolling pastureland and reshaping it into a prospect of high bare mountains. With a strange flash of insight, it occurred to him that this in part was what brought all those odd tourists, whose motives had often puzzled him, to his own Cumberland. They were visiting and becoming familiar with the beautiful and fearful landscape of dying.

Then he heard voices and old Cumberland and new metaphysical speculation fled at once from his mind. It was just a low murmur, indistinguishable and almost instantly stilled. But his keen young hearing told him it came from the neighbouring chamber, where Nicole slept. And within the murmuring had been a man's voice. Yet in the very same instant as his ears registered this, his mind was telling himself he had been mistaken. It was a murmur of women's voices. Madame Alpert had gone to comfort her grieving granddaughter, that was all.

As reason strove with recollection, his body exerted its own will. He plucked the candle from its place at Auguste's head and left the chamber as quietly as he had entered. His mind called for him to pause and consider but his body knew that there was a cold waste of circumspection in which resolve would freeze and certainties be misted over with every future breath. Not hurrying, but not hesitating, he pushed open the door of the bedchamber and entered, holding the candle high so that its small light fell with all its force on the occupants of the narrow bed.

They pushed themselves half upright, alarmed, Nicole's face pale and small within the fanned-out tresses of her luxuriant hair which covered but did not conceal her naked shoulders. As his eyes were drawn resistlessly to her tiny but exquisitely rounded breasts, he recalled how even in that dreadful moment when Strother threatened what seemed inevitable rape, he had not been able to avoid the prurient thrill of his first glimpse of a woman's sex. And now again. Now he had seen all of her, but not as a man ought to take his view of the woman he loves; no, certainly not that.

Those should have been his arms round her, that should have been his chest pressed close against her bosom, his arms coiled protectively round her shoulders.

Now at last his mind confirmed what his eyes had been refusing to accept.

He threw back his head and screamed into the cold invisible air.

'*Lothar!*'

And flinging down the candle, he rushed out into the bitter anæsthetizing night.

8

The mutiny at Étaples had not lasted long. For eight days the camps had been in the control of the mutineers, prisoners had been released, deserters living in the woods and caves around the complex had mingled freely with their old comrades, and to wear a red cap within striking distance of the rioting men had been a provocation to assault if not to lynching.

But finally by a mixture of threat, promise, naked force and, above all, the deadening, self-defeating conservatism of the British working man, the situation had been brought under control. Now, except on a few bits of highly confidential paper in the most secret recesses of the War Office, Denial doubted if the mutiny was officially acknowledged to have taken place at all.

He had stayed on through October and November, helping to restore and maintain normality and also to track down those of the ringleaders who had gone into hiding. His work had been so impressive that it had been suggested that he ought to remain as permanent APM in the area. That this came as a suggestion rather than an order was a matter more of semantics than anything else, but Denial took it at its face value and produced a list of conditions for accepting the job.

'But these are absurd!' he was told by his baffled superior. 'These are to do with military policy, not with military policing! It's impossible – no, it would be grossly dishonest of me to even hint that they are acceptable.'

'Then let us hope the war ends within at most a year,' said Denial. 'I doubt if the idiots who are running it can trade on the tolerance of their soldiers much beyond that.'

'Jack, you almost sound like one of these new-fangled socialist communist fellows!'

It was meant as an atmosphere-lightening joke but Denial did not smile.

'I'm a policeman, sir,' he said. 'That means I believe in equality before the law. Tell me where I'll find *that* politically, and that's where you'll find me.'

He spoke vehemently and when he had finished, he sensed he might have gone too far. He had relied a great deal on his superior's willingness to let him have his head and this was ill payment. So now it was his turn to try a conciliatory tone.

'Oh, by the way, sir,' he said. 'That name von Seeberg in Cowper's notes. Did you ever find out anything?'

'Nothing very helpful. There was a Count von Seeberg, some big nob in Berlin, died recently. Had two sons, both killed in the war, poor devils. Win, lose, how do we replace the children, Jack? How do we ever do that?'

It was shortly after Denial's return to Barnecourt that the first of the Volunteers' raids had taken place.

'They got clean away!' Sergeant-Major Maggs reported. 'They stole a truck! What's it to be, sir? Go after them?'

Denial thought for a moment.

'No,' he said.

'No!' exclaimed Maggs. 'But what about that farm, sir? You really fancied that a couple of months back.'

'Yes, I did. But it's snowing, Mr Maggs. And it's going to get worse before it gets better. I could be wrong about that farmhouse and we could waste a lot of time and energy crawling around in the snow. So we'll just sit here and wait for them to come back, I think.'

'To come back, sir?' said Maggs incredulously.

'That's right. If I had got myself a truck, that's what I'd do. Wouldn't you?'

And so they sat and waited and if Sergeant-Major Maggs's scepticism had not conveyed itself to his men, they would have been a great deal more alert on the night of the second raid and might even have captured all the raiders instead of merely killing several.

Denial ground the point home, speaking out of a cold anger which took Maggs aback.

'In future you will follow my orders to the letter, Sergeant-Major,' he said. 'I have no desire that my men should be killed. Or lost to me in any way.'

Whether this last was an expression of concern or a threat was not clear, but Maggs was relieved when Denial went on to discuss their immediate tactics.

The blizzard now blowing fiercely made immediate pursuit impossible.

'It'll likely have frozen them buggers to death too,' opined Maggs gleefully.

'You think so?' said Denial. 'That's what you'd put your money on?'

He made it sound like more than money. Maggs thereafter kept quiet.

From time to time during the day there were lulls in the weather, but every time Denial began to think seriously about setting out, the snowstorm closed in again, and it was not till after nightfall that the winds finally died down and the sky became clear.

This brought its own problems. The carefully planned

flow of military traffic in the area had come to a complete halt during the blizzard. Now it would be starting up again at irregular intervals according to conditions and there was every prospect of tremendous confusion. Avoiding that confusion was the job of the Corps of Military Police – its main if not its only job in the eyes of many staff officers. Denial toyed with the idea of retaining a couple of his men to go in pursuit of Viney and his so-called Volunteers. He certainly had the authority to do it. But the provost department's hard-won autonomy was still an object of some staff disapproval and not lightly to be exerted on a doubtful priority.

He sighed deeply. For him to be debating with himself at all meant the priority was doubtful to the point of nonexistence. His first duty was always to the war effort. An efficient traffic flow was more important than hunting down a few deserters.

The snow had damaged communication lines too. In any case, first-hand reports from men he had trained himself were the only satisfactory basis of an efficient traffic plan. Before dawn he had dispatched all his men to check which roads were occupied by what traffic and to report the news back to Sergeant-Major Maggs as soon as possible.

Now, his sense of duty placated for the moment, he went out to the HQ stables. His small detachment of mounted policemen were all out on traffic reconnaissance, but there was a remount earmarked for his own use, though, being no horseman, he generally preferred the motorbike combination. This morning, however, legs were going to be better than wheels.

He told himself that all he wanted was a bit of exercise in the cold morning air to blow away the cobwebs of an almost sleepless night. But it was without debate that he set the animal's head eastward out of Barnecourt.

The wind had been set in the west during the storm, blowing up the shallow valley, with the result that at first the snow lay thin and the going was easy. But he could see how the drifts built up the further he went and after about a mile as it became increasingly difficult to steer a safe path through the high-piled snow, he let the horse come to a halt.

Up there another three or four miles was the farmhouse, and another mile beyond it, you were into the Desolation. From what he had seen of the farm buildings, it was going to be a desperate place for that family to winter in, but not as desperate as whatever hole those fierce and tattered denizens of the Desolation occupied. What had Cowper called it in his sad, last scribbles? The Warren. That was it. But rats, not rabbits.

Perhaps Maggs was right and the winter would do the job for him. He could almost feel sorry for these desperate, desolate fugitives. Then his mind went on further east to that other warren, that other complex of subterranean burrowings which was the front line, and he felt less sorry.

In any case, what had pity to do with a brooding, amoral, anarchic figure like Viney?

He was about to turn away when his eye was caught by a mark on the smooth blanket of the snow, a serpentine line as if some giant finger had idly started to trace a pattern and then quickly grown tired. It was about two hundred yards away, snaking out of a little wood whose own close-packed trunks created another blot of darkness against the white.

He studied the line with his field glasses. It was as if something, or someone, had come down through the wood and ploughed forward a little way further through the deep snow before . . . before what? Disappearing?

'Come on, boy. Giddup!' he said, urging his horse forward.

It took him ten minutes to force his way through to the end of the line. Dismounting a yard or two short with the snow now waist high, he forced his way through the remaining barrier.

He had been right. There on the ground lay a man, face down. Denial knelt beside him and turned him over. He was still alive, but with his ragged clothing sodden by exertion from within and melted snow from without, even a few hours' more exposure must surely have killed him.

'Come on,' said Denial, trying to raise the man into a sitting position. *Man!* He was more like a boy, with fine pale features and a shock of blond hair. Suddenly the eyelids moved, revealing light blue eyes which regarded him without comprehension.

'Let's get you out of here,' said Denial. 'We'll soon have you safe.'

Safe! It was a silly thing to say. If this boy was what he thought he must be, there was no safety where Denial was taking him.

He managed with difficulty to pull the youth to his feet, but he was a dead weight, supported almost entirely by Denial's arms round his chest. Even the few steps to the horse were beyond him. He'd have to bring the patiently waiting animal to the helpless man and try to throw him over the saddle.

Gently he started to lay the slack body down once more. As he did so he heard a noise. Startled, he glanced over his shoulder, looking along the track the boy had ploughed from the woods. He had a couple of seconds to glimpse a figure he had seen only once before but would never forget, huge and menacing and moving at surprising speed. He tried to straighten up and reach for his revolver but with his hands trapped beneath the boy's armpits, there was never any chance of success.

A fist like a club crashed against his temple, splitting the skin. The line of darkness through the snow suddenly ran wild and wriggled worm-like into his mind, and he saw no more.

9

Viney looked down at the two bodies. Pushing Denial aside, he knelt beside Josh and raised the boy's head. The eyes flickered open again, still blank, but the Australian sighed with relief. Reaching into the pack he was carrying, he produced a bottle of spirit and forced some between the blue-tinged lips.

'Come on, son,' he said. 'Let's get you out of here.'

His first thought was to use the horse, but the beast was unhappy at being urged to proceed further down this narrow corridor in the snow. It bucked as he lifted Josh on to its back and though the addition of his own weight stopped the bucking, it also made forward motion impossible. He drummed his heels savagely against its sides and it made an effort but he realized that within a couple of hundred yards he would probably have to abandon it exhausted anyway.

'Shanks's pony for you and me, son,' he said. 'For me, anyway.'

He stooped over Denial to check the man's condition. He was breathing shallowly. It would make sense to slit his throat here and now, and Viney got as far as gripping the shaft of his knife. But he didn't draw it. He recognized the face. This was surely the red-cap captain he'd left tied up in the hospital. That mountie lieutenant Fritz had persuaded them to let go, hadn't he said something about the nurse dying? And the nurse had been this joker's girl.

To another man these might have been simply extra reasons for killing him while he had the chance. But Viney's mind arranged the strangely shaped pieces of the moral jigsaw in a different way. A man whose relentless pursuit was based on desire for personal vengeance he could understand. But one who caught and brought to justice, meaning death, his fellow countrymen out of a sense of duty deserved to have his gut slit open by any right-thinking man.

He gave Denial the benefit of the doubt. The cold would probably kill him anyway. The clouds were building up in the sky once more as the wind began to reawaken. There would be more snow soon, cover for his tracks, and a pall for Denial.

He lifted Josh across his shoulders and set off back along the trail he and the fleeing boy had already beaten.

It was hard going, especially with the extra weight. He marvelled that Josh could have got so far. There was great strength in that slim body and of course he'd had the fuel of a burning emotion to drive him on. Fritz fucking his girl! Christ, there was a turn-up! Many another man'd have stuck a bayonet in his ribs there and then. He imagined himself in the situation and felt a surge of killing rage. Yes, that was right, that was what a man should do. But not him . . . he could never . . . There had been another man; there was the child – did it carry his name? He had had a dream some nights ago in which he found the man and rushed at him with his knife. But when he plunged the shining steel at his belly it turned to rubber and folded and bent and he could hear her laughing, derisive, dismissive.

He suddenly envied Denial back in the snow. Envied him his likely death, envied him his burning, driving need for vengeance. There was something not dissimilar in himself, he dimly comprehended, but it lacked that clarity, that simplicity. It was deeper and darker, and the only true object of its drive to destruction was himself.

He shook the darkness back into its place. It was time to think. To attempt to get straight back to the farm was stupid. It would try his own great strength to breaking-point and in any case he doubted if Josh could survive much more exposure in his present condition.

But there was an alternative. His mind had kept together when everyone else's was falling apart in the panic and excitement of that sudden awaking. Even Fritz, that clever, educated Hun, had been acting stupid, dragging on his pants and wanting to rush out after Josh in his bare feet!

Viney had calmed them down, not with reason, but with a single command which had cut through the confusion like a rifle bullet through cheesecloth. Once he had ascertained what was going on, he had made careful preparation for pursuit. Lothar had urged greater haste.

Viney continued methodically packing his haversack and said, 'I'll probably catch him quick, but there's always a chance not. Then, what I'm taking here could save his life. What're you doing, sport?'

Lothar was busy getting fully dressed.

'I'm coming with you, Viney,' he said.

'No, you ain't.'

'*No?* You aren't going to stop me!'

Lothar spoke challengingly, his fist clenched, ready to take his defiance to the utmost on this occasion.

Viney didn't even look at him.

'Two reasons,' he said. 'One is, I can maybe look after one idiot floundering around in a blizzard, but not two. The other is, it's you he's running away from, Fritz. What makes you think he'll be keen to let you bring him back?'

He stood up and put his pack on and looked round the lantern-lurid barn. Madeleine, roused from her slumber over her dead son by Josh's outraged cry, was in the sleeping chamber with a hysterically sobbing Nicole. Hepworth and Evans were standing around with the helpless look of the British male in emotionally fraught situations. Strother had fallen into sleep or unconsciousness, while Coleport sat bolt upright, his face flushed with fever, his eyes wild. But his understanding of the situation was not entirely deficient, for suddenly he cried out in a voice which shocked them almost as much as Josh's cry in the night,

'One is away on the roving quest,
Seeking his share of the golden spoil!'

When he fell quiet Viney had looked at him and said gently, almost affectionately, 'Yeah, that's the way of it, Blackie, old son. That's the way of it.' And left.

He had had little trouble in tracking Josh, but he had been surprised by the speed and distance the fleeing man had achieved. Now he was paying the penalty. Exhausted and in despair, his young body no longer had the energy to create enough heat to keep him warm beneath his sodden clothes. Shelter and warmth were necessary. Viney knew there was no building between Barnecourt below and the farm above, but he still strode forward with a purpose which gave strength to his own cold and aching limbs.

He had not too far to go and his keen instinct for direction had not failed him. Ahead was a huge snowdrift with an angularity of edge which a casual glance would have put down to a freak of nature. Laying Josh gently in the snow, Viney took out the trenching tool he had brought with him and began to dig. In a little while, he struck metal. He had remembered right even though the last time he had been here it had been night with the blizzard blowing and the groans of wounded men in his ears.

This was where the stolen truck in which the survivors had made their escape had finally ground to a halt in the impossible conditions, leaving Viney to lead them back to the farm by brute force and instinct.

Some snow had drifted in through the canvas curtain above the tailboard, but this was soon cleared. Big flakes were already beginning a new wind-tossed dance as Viney lifted Josh into the shelter of the truck.

Now he undid his pack and silently complimented himself on the care of his preparations. First he pumped, then lit a small primus stove and placed on it a tin of soup with holes punched in the lid. While this was warming, he forced another mouthful of spirit down Josh's throat and then began removing the youngster's soaking clothes till he was nude, his body slight and defenceless as a girl's. There were two blankets in the pack. Using one of them as a towel, he began to rub the boy's skin, at first to get it dry, and then for warmth. He paid particular attention to the fingers and toes which were a dead white. At first he was fearful that frostbite might have set in, but after a while Josh moved and cried out in pain as the circulation began to be restored. Encouraged, Viney poured some of the spirit on to his hands and continued his massage, going over and over the slim, sinewy body till it began to glow.

Josh now spoke for the first time.

He simply said, 'Viney.'

'Who else, sport? Let's get some of this soup into you.'

He poured the hot soup into a tin mug and when Josh's fingers proved still unable to grip the handle properly, he held it to the once more pink lips and encouraged him to drink.

Josh held the blanket close around his body and began to shake.

'That's good, son,' said Viney. 'When you stop shivering, that's when the cold's finally got you, did you know that?'

This was true. But there was no denying that it was still bitterly cold even in the shelter of the truck. The little primus stove gave out some heat, but not enough and in any case, they would need to conserve its flame for cooking. It would be easy to start a fire. There was a spare jerrycan of petrol in the truck. But that was a last resort as the danger of blowing the whole thing up would be immense. But something had to be done. Now that he was no longer in motion, Viney realized just how wet his own clothes were and how deep the chill had struck into his own bones.

Quickly he stripped off and gave himself a shorter, more violent version of the treatment he had just given Josh. Then he wrung out the damp clothing as best he could, but to put it back on would have been counter-productive. Wrapping the towel-blanket round his shoulders, he peered out of the back of the truck. The snow was falling fast now, not such a blizzard as had blown on the night of the raid, but bad enough to make any thought of an early escape impossible.

He turned back to Josh.

'More soup, son?' he said.

Josh shook his head.

'Right. Better get some rest then. We're going to have to keep each other warm.'

He took the blanket from the boy's shoulders and laid it and the other one on the floor. Then he put Josh on the middle and lay down beside him and drew the blankets around them in a cocoon.

So they lay intermittently in each other's arms for the next day, separating only for Viney to heat more of their small supply of food or to apply his violent massage to the boy's feet and hands. Gradually their breathing and their body heat and the occasional lighting of the stove raised the temperature of their shelter several degrees above that outside, but Viney was careful to preserve a good degree of ventilation by opening the flap from time to time and poking a hole through the fresh-piled snow.

This brought in fresh air but immediately lowered the temperature. Josh protested as Viney did this for the third time and the big Australian looked over his shoulder with a grin, delighted that the boy was recovered sufficiently to register the change.

'We don't want to suffocate, son,' he said. 'Looks like it's clearing up out there. We'll be on our way soon.'

'Where to, Viney?'

'Back to the farm,' said Viney.

Josh lay back in his blanket and closed his eyes and shivered in a way that was not altogether to do with the cold. Viney dropped down beside him and pulled him close.

'It'll be right, you'll see,' he promised. 'We'll just stay long enough to collect the others, then it's back to the Warren. OK?'

'You promise, Viney? I don't want to stay at the farm any more.'

'Promise,' said Viney. 'I reckon they'll be coming up after us once this lot passes, so it'd be stupid to hang around.'

'Viney,' the boy said, pressing himself close with his voice muffled by the proximity of his companion's chest. 'How could he do it? How?'

'Things happen,' said Viney. 'There's all kinds of needs, sometimes you can't help yourself. You'll learn.'

It was strange. Up to this point there had been nothing either in the sight or the touch of Josh's body which had touched him sexually. At least not consciously. But as he asserted the irresistible power of physical need, it was as if he punctured some thin membrane between the conscious decision-making part of his mind and the darkness which lay beneath. Desire swept over him in a long tidal wave. His penis, shrunken by the cold, arose, blood-gorged, against Josh's flat belly. The boy felt it and went tense. Horrified at himself, Viney tried to draw back and might have succeeded had not Josh, in a movement ambiguous of rejection or submission, rolled over so that his back was to Viney. Even now, had he spoken or resisted, the Australian's self-revulsion might still have been triggered to the point of breaking off from the contact. But as Viney's hated lust swivelled his hips forward bringing his steel-stiff member against the boy's skinny buttocks, Josh did not speak or flinch, but lay utterly still and silent except for one small cry of pain as Viney thrust deep in the final searing explosion.

Afterwards, without any spoken decision or agreement, they pulled on their still damp clothing. They had things to say to each other, but neither spoke for they could scarcely find the words to express these things to themselves. For Viney it had been the final open recognition that he was the very thing his whole idea of life cried out against, that he, of all men the strongest, was of all creatures the weakest. It was a devastating admission, yet one not devoid of comfort in that now, for himself at least, this one deceit was over.

For Josh it had been two things – recompense for what he had unconsciously acknowledged from their first encounter, Viney's love; and revenge on Nicole. Her body had been entered and sullied; so his too. Now they were even but not with an evenness which could ever match them together. It was a vengeance entirely devoid of comfort.

They set off later back to the farm, Viney breaking the path ahead, flailing his arms like a swimmer, or sometimes, where the snow was deepest, like a drowning man.

They arrived back at the farm in darkness. It was not a riotous homecoming. There was relief but no surprise. When Viney set out to do something, it was difficult to admit the possibility of failure. There was very little drama either. Lothar tried to speak to Josh but when the boy ignored him, the German simply shrugged helplessly and fell quiet. As for Nicole, she lay in bed within her sleeping chamber and whether her ears strained in the darkness or her eyes opened as she lay there, no one could know.

'We'll move out at first light,' said Viney after he had dried off and fed himself.

'Out?'

'To the Warren.'

Hepworth and Evans exchanged glances.

'But Strother can't move,' protested the Welshman. 'He'll have to stay.'

'He'll stay only if he can't move,' said Viney.

It was not a double-meaning. It was spoken in such a way as to be quite unambiguous.

Coleport laughed wildly but it was Hepworth who provided the real drama. As preparations for the move got under way, he said, 'I'm not going.'

Viney said, 'What's that, Heppy?'

'I'm staying. They'll not get through the winter 'less they have a man about.'

'How long do you reckon you'll be about, sport? They'll be coming up here, you know that? They'll likely clear the whole lot of them off, shove 'em in jail for harbouring deserters. Well, that's their bad luck. But *you* in jail, Heppy, that could be *our* bad luck, couldn't it?'

Hepworth regarded him calmly.

'I'll not promise not to talk, Viney,' he said. 'I know you don't rate that kind of promise very high. But I will promise not to get caught.'

'You mean you'll just keep out of sight when they start pushing Madeleine there into a truck?' said Viney disbelievingly.

'They'll not do that,' said Hepworth. 'I've spoken with her. As they come up, they'll meet her on her way down to Barnecourt to report there's been a bunch of wild men here, holding them hostage till they upped and left the previous day.'

'And you think the redcaps'll swallow that?'

'Aye, they'll swallow it. When she shows them her dead son, shot by one of these wild men, they'll swallow it.'

His eyes flickered to Lothar and he added softly, 'She can even tell 'em her daughter got raped by one of the bastards.'

Lothar went pale but did not move.

Viney said thoughtfully, 'You've got it all worked out, my son, haven't you?'

'I think so, Viney.'

Everyone waited.

'All right,' said Viney. 'It probably makes sense to have someone around keeping an eye on these people. We've invested too much in 'em not to take care of them now. See you in the spring, Heppy. Just don't get caught.'

'I'll not do that, Viney,' said the Yorkshireman.

A little while later as the sun's pink stain started to seep across the white snow, they left. Strother was carried on a rough stretcher between Lothar and Taff Evans. The Welshman still protested it would kill the wounded man.

'Then you'll not have to carry him far,' said Viney.

'But we can't even see the way we go to the Warren,' cried Evans.

'I'll find it,' said Viney. 'And Josh here says it'll thaw today. He's got a nose for the weather, eh, Josh?'

He smiled at the boy. He sounded almost light-hearted. Josh returned his gaze solemnly and Lothar noticed as they started the long trek away from the farm that the young Cumbrian did not once look back.

Jack Denial lay in his hospital bed. They'd found him in mid-morning stumbling along beside his horse, too weak from the blow to his head and exposure to get into the saddle. There'd been a threat of pneumonia but after three days in bed he had beaten it off.

Maggs had been to see him. They'd finally got up to the farm. On the way there, they'd encountered the woman, Madame Gilbert, who confirmed what he had suspected. The Volunteers had been there. Any doubts as to her own loyalties in all this were dissipated by the sight of her son's corpse, shot clean through the heart with a .303.

And now they were back in their hole.

But the war went on. No one could see any end to it. For the first time ever, Denial did not find this a terrifying thought. The end of the war might take him and Viney out of each other's reach.

But the longer it continued, the more certain he was that they would meet once more.

He closed his eyes and fell asleep.

BOOK THE THIRD

—◆—

Resolution

PART ONE

——◆——

ONCE MORE, THE SOMME

In March 1918, the Germans launched their last great offensive, 'die Kaiserschlacht', on the Somme front.

The British Fifth Army broke, and within a week the Germans had advanced up to forty miles, passing across the Desolation and penetrating countryside hitherto untouched by the War.

For the first time in more than three years, no-man's land ceased to exist.

1

It was a long hard winter. It seemed long even to the men in the trenches who knew that for them spring's renewal was merely a renewal of death. And for the men in the Warren, it was like a death itself.

Yet there was in some of the Volunteers also a desperate clinging to life, a sense that this subterranean existence was perhaps merely a finite purgatory rather than eternal hell.

By one of those massive ironies with which these carrion years were filled, their symbol of hope was the unlikely person of Dennis Strother.

Incredibly, he had survived the rough journey back to the Warren and eight weeks later he was still breathing, and still cursing with every breath.

He cursed his life and his luck and the lives and lucks of all those around him. He cursed all armies and all generals; all kings and all kaisers; all priests, potentates and politicians. He cursed England, France, and the whole world from pole to pole; he cursed the sun, the stars, and the entire universe.

His cursings were the unceasing burden of life in the Warren. Above all, he cursed Viney. He had in his possession a bayonet which no persuasion or threat could remove from his hand. It tented the bloodstained blanket like some grotesque erection as he lay (still feverishly muttering) in the semi-conscious state which passed for sleep. The bayonet was meant for Viney, he screamed, or for anyone else who came near him – except for Lothar. Lothar was the only one he trusted to tend to his needs. The others were spies for Viney. And Viney himself, he called out amid his roll-call of maledictions, would only come near him to kill him.

He was not entirely right in this, but not entirely wrong either.

The power structure of the Warren had changed. Viney's absence on the raids from the farm had not been long, but it had permitted a relaxation of atmosphere and a freedom of speech which some of the Volunteers were reluctant to give up. And though Viney had never gained his strength simply by using the strength of others, there was no denying that the losses in the raids had deprived him of a great deal of powerful support. In particular, the old triumvirate was now down to one. Delaney lay dead in whatever grave his killers had seen fit to scratch for him. And Coleport was now a withdrawn, inward-looking figure, sitting nursing his healing but almost useless arm in a corner most of the time, except for sudden manic outbursts of energy when he would rush around the passages and chambers of the Warren, sometimes talking to the dead Delaney, sometimes declaiming the sad and lively ballads of A. B. Paterson. To Viney he never addressed a word, and this visible disaffection told more heavily against the Australian than anything else. The evidence that Viney had actually murdered his own friend and lieutenant was far from conclusive, but it was strong. More importantly, it was easy to believe.

The opposition's weakness was lack of a leader. Hepworth might have fitted the bill. There was in his straightforward, down-to-earth manner a strength which could attract followers. But he might have been a million miles away for all the use he was to the men in the Warren.

Lothar was a possibility. His German-ness was a barrier which familiarity had almost removed. He had shown himself skilful in practical matters as well as debate and, above all, he did not seem afraid of Viney. But if the Australian suffered by the disaffection of Coleport, Lothar suffered even more by the more mysterious disaffection of Josh who avoided all contact or conversation with his old mentor and stubbornly took Viney's side in any matter of dispute.

Josh sat by the Australian's side one night at the supper table when the question of Strother came up again. His body had grown steadily weaker to the point where he was little more than a skeleton, but his voice seemed to increase in strength and could be heard clearly now, groaning and cursing in his pain.

Nelson banged down his tin plate violently and said, 'For Christ's sake, do we have to put up with that? Viney, ain't it time to put a muffler on that moaning bastard?'

Viney said mildly, 'Keep your hair on, Eddie,' producing a general laugh. Nelson had already been tonsured like a monk when he arrived in the Warren, but during the past few months, worry or diet had caused all his remaining hair to fall out, leaving him bald as a cathedral dome.

A Scot called Balfour said, 'It's true, but. If he's still making that row when the spring comes and there's a chance of someone wandering by, he'll be a real danger. I was on sentry last night, and I could hear him clear.'

Fox now joined in. He clearly felt little loyalty to his former ally, Strother, unless his words could be taken at face value.

He said, 'There's no hope for the poor cunt, is there? We'd be doing him a favour putting him out of his misery.'

All eyes were on Viney who turned to Lothar and said, 'Fritz, you're the next best thing we've got to a medical man. What say you?'

Lothar said, 'I agree. There is no hope. To me it seems a miracle he has not died weeks ago. Perhaps we should not interfere with miracles.'

Viney was not going to let him get away with verbal evasions. He said, 'So you're against putting him out of his misery, Fritz?'

Lothar took a deep breath.

'If Strother wishes to die, I will not stand between him and his death,' he said. 'But I make no decision on the life or death of anyone else, not any more. That is why I am here, Viney, not out there directing the guns of my battery.'

Viney didn't argue, merely nodding his head and saying, 'All right, if you want Strother to go on suffering and us to go on putting up with his caterwauling, that's just what we'll do, Fritz.'

He rose from the table, accompanied by Josh.

There were a handful of newcomers who had joined the Volunteers during the winter, bringing the numbers back up to around forty after the heavy losses at the end of the previous year. These had not yet seen Viney's strength in other than its passive mood, and though this was inhibiting enough, it was not totally intimidating. One of the newcomers, a Northumbrian called Groom with a cheerful wrinkled monkey-face, emboldened by Viney's assumed retreat, said to his neighbours, 'I bet there'd be no talk of putting him out of his misery if it was young Josh there doing the caterwauling!'

The older Volunteers froze into silence. Viney halted and turned.

'What's that supposed to mean, Groom?' he asked quietly.

Not yet alert to his danger, or unwilling to back down in front of his fellows, the Northumbrian essayed a conciliatory grin and said, 'Nothing, Viney, except that Josh is far too bonnie a laddy to let die.'

He was stupid enough to round off his remark with an appreciative smacking of his lips.

Viney said not a word but moved forward with tremendous nimbleness for a man of his bulk, smashing aside a table in a scatter of tin mugs and cursing men, and caught the half-risen Groom a blow which completed the process, lifting him clear over his stool and hurling him against the wall with a force that cracked the board-lining to reveal the bare earth behind. Viney went in pursuit, swinging his booted foot at the slumped body.

'Viney!' called Lothar in a voice which suddenly reminded

the others that he was or had been a German sergeant.

Responsive to the tone, Viney restrained himself with obvious difficulty.

Lothar said softly, 'Must everyone who offends you be killed?'

'Killed? Who said anything about killing?' snarled the Australian. 'I'm just going to teach this joker a few manners.'

'You may believe so. But in our present state, it is too easy to kill. Anything we cannot properly treat, no matter how minor to start with, can prove fatal. Only by helping each other, which means, to start with, by not harming each other, can there remain any hope.'

'Hope? Hope for what?' cried Arnold Tomkins in one of his rare contributions to debate. 'It's all right for you lot! *You* ran off because that's what you wanted. I never ran off at all. I was kidnapped! But who's going to believe that, I ask you? Who?'

'Not to worry, Tommo. We'll get Viney to sign you a chitty,' said Fox.

There was a general laugh which Viney joined in.

Tomkins said piteously, almost talking to himself now, 'I shouldn't have been in the Army at all, you know that? All the way along, they kept saying that. I was thirty-nine. *You* won't get your papers, they said when they brought conscription in. They'll never take *you* with *your* eyes, they said when I went for my medical. *You'*ll never get further than Aldershot, they said when I went to do my training. And when I got posted out here to Div. HQ, they said, well, at least you'll be safe *there*. No fighting for *you*. But here I am, aren't I? Here I fucking am!'

His voice had risen to a near-hysterical scream.

'Give it a rest,' snapped Nelson. 'At least they didn't make you fight. You had it cushy, mate, sitting in your little office, doling out leave passes for real soldiers!'

'I didn't want to sit in any offices,' yelled Tomkins. 'I was a musician, not a fucking clerk. They'll put you in a military band, they said! The nearest you'll get to France is sitting on the quayside at Tilbury Dock playing *Rule, Britannia*, they said! And look at me now.'

'No, thanks,' cried a wag. 'I'd prefer something more cheerful. Like a newly-dug grave.'

'Oh, you'll see that, likely enough,' said Tomkins viciously. 'You've all got a better chance of seeing that than you'll ever have of seeing the white cliffs of fucking Dover.'

This brought silence, broken by Evans.

'There's a thing, Viney,' he said. 'What do you think our chances are, really now, of seeing the white cliffs?'

'Don't ask me, sport,' said Viney. 'White cliffs mean sod all to me. Aussie's my home and that's ten thousand miles away. Though come to think of it, Dover might as well be that far, for all it matters. You'd better ask Fritz here. He's the clever one, the one with all the answers. Me, I'm just an ignorant digger. Tommo, you got a mo? I'd like a word.'

He left with Josh and a frightened-looking Tomkins following. To some of those remaining it might have seemed almost like an abdication. But Lothar, looking round at their uncertain yet expectant faces, knew that it was no such thing but a serious and deadly challenge to a duel for the primacy.

It was Groom who actually voiced the question, rubbing his jaw which happily seemed not to be broken.

'What do you say to that then, Fritz?' he mumbled. 'What *is* going to happen to us?'

Again the expectant silence. But this time the nature of the silence saved Lothar from immediate answer. Something was missing from it.

'That's funny,' said Evans. 'Strother's shut up.'

Pushing his way through the men, Lothar hurried to the chamber where Strother lay. It seemed unlikely that Viney would have gone straight from his recent encounter to do the man some harm, but not even the Australian's stability could be relied on after a year in these conditions. Strother was still alive, however, and he was alone. There was an expression of faint surprise on his face.

His eyes, huge in his wasted face, turned to Lothar and he said in wonderment, 'It's stopped 'urting. Just like that. Ain't that fucking marvellous?'

'Yes,' said Lothar. 'Marvellous.'

Something in his tone may have communicated itself to the

dying man, or perhaps he had reached the same conclusion independently, for he smiled wanly now and said, 'This is it, then? The big push.'

Lothar nodded. 'Yes, I think so.'

He sat on a stool close to the bed and bathed the man's forehead with a cloth taken from the bowl of water by the rough pillow.

'You're OK, Fritz,' said Strother. 'No bullshit. You were right about me when you said I had the smell of death on me. That's a sad talent you've got there, Fritz. Don't know as how I'd like a talent like that.'

He pulled his hand weakly from under the blanket. He was still grasping the bayonet. He looked at it thoughtfully for a moment.

'Won't need that no more, will I?' he said. 'Here, you'd better have it, Fritz. You keep a weather eye open for that Viney now. He'll 'ave you if he can.'

He dropped the bayonet into Lothar's hand and closed his eyes.

'I've had a funny life,' he murmured. 'Don't know whether it's been worth it or not.'

He then fell so silent and still that Lothar leaned close to see if he was dead.

But the huge eyes suddenly opened wide. In each of them a tear swelled.

'Tell you what, Fritz,' whispered Strother. 'I don't 'alf wish I was at 'ome.'

And, turning his head away, he died.

2

There was an atmosphere now of approaching crisis in the Warren. It was not necessarily coming fast, but it was coming certainly, and it gave a new vitality to their existence, a sense almost of urgency.

News had reached them via the newcomers of the war's

progress, or lack of it. They heard of the muddy horrors of Passchendaele; of the strange land dreadnoughts called 'tanks' used at Cambrai; of the first appearance of twangy-voiced, fresh-faced Americans in the Allied lines. But none of these things meant a great deal beside the tensions and rumours and groupings of the Warren.

Josh felt himself for the second time coming back to life. He was continually in Viney's company, but since that time in the abandoned truck, there had been no sexual contact between them. Had Viney made the attempt, Josh would not have resisted. The big Australian roused no desire in him, but he did rouse gratitude, and admiration, and loyalty.

And he was also a certain protection against Lothar.

Yet the new mood of confrontation touched him too. One day Lothar encountered him in the company of Arnold Tomkins. Their conversation stopped as soon as they saw Lothar but when the little clerk scurried off, Josh did not follow but defiantly stood his ground.

Lothar, desperate to talk to his friend, took this for encouragement not challenge.

'Josh,' he said urgently, 'I must talk to you before it is too late. Please, listen. You must understand. I did not seek Nicole that night. She found me. She was, I think, looking for her mother. She was in need, Josh, in need of comfort, of reassurance, of warmth. I was to her a source of these things; not me, Lothar von Seeberg, as a person, you understand, but as the headman of the house, its strength, its authority . . . these are strange things to explain Josh. Sometimes there are such needs . . . think of yourself after your brother died . . . No! Stay and listen Josh! I was not, am not, never can be Wilf. But in your need I was some of him. To Nicole I was grandfather, father, brother . . . and lover too. Yes! It had to be. I was there. In any case, how could she turn to you at that moment, you who had killed her brother? I took her to her bed. She clung tight, so tight. To make her let go would have been to let her fall into darkness. To have refused myself would have been . . . You must listen!

'I had a brother too, Josh. A twin brother, dear to me as Wilf to you. He was called Willi – almost like Wilf, is it not?

He died. This war . . . And after his death, his wife, my sister-in-law, Sylvie . . . she turned to me – for strength, for comfort, for reassurance. And I turned her away. Out of loyalty. And out of shame that I had always wanted her. She died, Josh. She took her own life. Perhaps it was inevitable, perhaps she was on that course beyond diversion, perhaps . . . But I can never forget that she turned to me and I turned her away.

'I could not do it again, Josh. Not again. You must understand . . .'

Now Josh interrupted.

'I understand,' he said in a gratingly ugly tone. 'You took her to bed and you fucked her. Like a big ram and a yow. You fucked her. Like a dirty Hun. They told us what you Huns were like and it's true. You bastards! Fucking, raping, murdering . . . I hope you die, Lott! I'm glad we killed your brother. I hope our lads are blowing all you bastards to pieces up at the front, all of you, every last fucking one of you . . .'

The voice had risen to a near-hysterical pitch. Lothar's fist swung at Josh's head. He had just enough control to open it out at the last moment and turn it from an angry blow into an anti-hysterical slap, but his face was pale with a sudden rage which made his duelling scar stand out like a jag of lightning in a grey sky.

'Now you really look like a Hun, Fritz,' said Viney. Behind him hovered Arnold Tomkins. The Australian came forward and put his arm around Josh's shoulders and urged him gently out of the chamber.

Josh resisted for a second, looking at Lothar with a curious compound of hate and regret in his eyes, then left.

Viney said, 'Don't ever lay a hand on him again, Fritz,' and walked away with Tomkins scuttling along in his wake.

The change in the little corporal-clerk was amazing. His slow slide into depression and despair had seemed unstoppable and his outburst during the discussion a few weeks earlier had seemed likely to have been his last flicker of desperate energy. But since that time, exposed to the sun of Viney's interest, he had blossomed into life again. Attempts to pry into the nature of his new closeness with Viney were met with a mysterious shaking of the head, often accompanied by a knowing wink. And if, as often happened, the questioner hinted physical back-up for his verbal probing, Tomkins would say, 'I'll ask Viney if I can tell you what he said, shall I?'

Supplies ran short as the winter wore on and when the harsh weather began to improve dramatically towards the end of February, no one was very surprised when Viney announced that he was taking a small recce party out. But several ears pricked at his next remark.

'I thought we'd look in at the farm,' he said. 'See how Heppy's managing. It'll be a good chance to catch up on the war. Might be all over for all we know.'

No one's hopes were raised. Though the winter battle solstice was clearly continuing, there was still activity enough on the front for anyone who cared to listen or go up at night and watch the distant flares.

In any case to the men of the Volunteers, the end of the war had ceased to be the unimaginably joyous goal it was to the men in the trenches. Eddie Nelson expressed this viewpoint tersely, 'War or no war, it don't matter to us, do it?'

'Why not?' asked someone naïvely.

'You don't think they'll put us on no troopship and ferry us home, do you?' asked Nelson with scorn. 'We're here for the duration, sonny. The duration of our fucking lives, I mean!'

There was a despairing silence, broken by Lothar.

'No!' he said with emphasis. 'The end of the war is as important to us here as to the world out there, believe me.'

'Why'd you say that, Fritz?' someone asked.

'The first and best reason surely is that it is important to men everywhere when senseless slaughter stops. Remember why we are here! Because we were sickened by that slaughter! Who more than us should be rejoicing if it stops?'

Another silence, broken this time by Viney with three slow handclaps.

'Bravo, Fritz,' he mocked. 'Only, while we're remembering why we are here, alongside all these noble motives you should put a few more. Like being shit-scared, like being lonely and unhappy and worrying about what the old lady back home's getting up to, like running doolally because of the noise and the stench and the mud, like putting a bullet up some bastard officer's arsehole and not being clever enough to do it while no one's looking. And a hundred more reasons along the same line. Don't forget to put these alongside your noble bloody philosophy!'

Lothar nodded reflectively. The atmosphere was curiously charged as if those assembled were sensing in this verbal fencing the beginnings of the head-on confrontation they had all been waiting for. Was the time right? Lothar asked himself. Right for what, he was not sure. A decision, a change of direction. *Something!* He took a deep breath.

'So. We are all men,' he said to Viney. 'We have human weaknesses as well as strengths. If sometimes our weakness leads us to do the strong thing, we should not complain, my friend.'

'I sometimes think your English ain't what it's cracked up to be, Fritz,' mocked Viney. 'What the fuck does that mean?'

'The strong thing would have been to say at the very outset, *"We will not fight!"* ' cried Lothar with sudden passion. 'To say, *"This is a war for money and for power and it should be fought by those who have money and power. We will stay at peace and live in unity with our fellows of all nations and claim no more than is our right as human beings!"* That would have been the strong thing, but we did the weak thing and let ourselves be tricked and lured and forced into fighting, till a greater weakness of weariness and fear drove us out. That was an act of strength by accident! And now we have the chance of making it an act of strength by will!'

He was on his feet addressing his audience with a strength and passion they had never seen before. Even Blackie Coleport was roused from his customary brooding isolation to pay attention.

'Think, my friends, what are we to do when this war is over? How are we to live? Not like this, for sure. We have been able to live like wild things because, in a country at war, wildness can pass almost unnoticed. But when peace returns, what then?'

'Mebbe there'll be that amnesty?' came the desperate reply.

'Yes, perhaps there will be,' said Lothar gently. 'But not for many years, I think. Even if they are not so keen to shoot deserters after the war, they will certainly give long prison sentences.'

'Aye, and it won't be in no comfortable civvy lock-up either,' said Groom. 'It'll be the glasshouse. No talking, doubling everywhere, and a dose of Mr Dunlop if you flicker an eyelash. I had a week of those murdering bastards once. Another day and I'd have been goaded into trying to kill one of them with my bare hands. That's what they want, man. Make no mistake, most of us here would be inside for ever!'

He spoke with a bitter intensity that made its mark.

Lothar spoke again, quietly now, but with no diminution of urgency.

'It is here in France that most of us must look to our future for a few years at least. We must try to live in peace and friendship, to work, to rebuild. It will be difficult and dangerous, perhaps, but it is possible. Those who have worked at the farm will tell you it is possible. Whoever wins this war, there will be men all over Europe who feel as we do, that it must never happen again. We are few here, but when the peace comes we shall be many! Some may be taken, that is inevitable. But when we are taken, what an opportunity there will be then to stand up before the world and declare our faith in the brotherhood of mankind!'

As he spoke, Lothar felt a pang of shame at what he felt must sound like empty rhetoric. And it was a pang that did not entirely fade as he saw the effect of his speech. Many of the men were gazing at him raptly. In the eyes of many of these filthy bedraggled scarecrows, he saw a desperate hope begging to be aroused. Almost guiltily, he looked away from them to Viney and waited for the Australian's response.

He expected mockery and was ready for it. Indeed, it seemed likely that it would be counter-productive at the present moment. But Viney was sharp enough to have felt that also.

He said, 'That's fine, Fritz. This equality and brotherhood, I'm all for it. Back in Aussie we've been practising something like it for years and it'd be good to see you jokers benefiting from it too. But be realistic. What's the future really hold here? There might be a chance if you speak good Frog like you, Fritz. Or if you've got your legs under the table somewhere, like old Heppy down at the farm. But most of the lads here are ignorant swaddies like myself. First gendarme who gets an eyeful of us will be blowing his whistle like it was full-time. And what about this brotherhood of mankind? I'll tell you what I think. After this fucking war's over, most ordinary folk'll want to forget it. Oh, there'll be parades and celebrations and Christ knows what, but after that it's forget-it time. And if they are reminded of it from time to time, the last thing they'll want to be reminded by is a bunch of cowardly deserters.'

He paused. Lothar could almost feel the spirit of hope going out of those around him. Perhaps he had been wrong even to attempt to awake it.

'Yeah, that's how they'll see us,' mused Viney. 'I mean, look around, for Christ's sake! What the fuck is there for anyone to see but a bunch of clapped-out no-hopers!'

'Now, I know you better than that. I know what you really are. And I know what you deserve. It seems to me that at the very least a man ought to be able to take his chances in his own country where men speak his own language and where he's got mates he can trust and family who love him. That's how it seems to me anyway. I'm not an educated man, like Fritz here, I'm no clever bloody intellectual, but that's how I see it and if I'm wrong, then just tell me here and now, without hard feelings, and I'll shut up.'

Taff Evans spoke first.

'No, you're not wrong, Viney, not in what you say, boyo. But you *are* wrong to be saying it! Don't we all want that? It goes without saying! But we can't have it. What Lothar suggests is bloody difficult, but at least it's not bloody impossible!'

'I thought you'd know me better than that, Taff,' said

Viney mildly. 'When did I ever suggest anything impossible?'

'What are you saying then, boyo?' demanded Evans. 'That we should head for the coast, maybe, and swim across the Channel? Or steal a boat perhaps and row across? Or stow away on Jellicoe's flagship and get across that way?'

'If that had to be the way of it, that's what I'd do to get home,' said Viney calmly. 'And I've got a sight further to swim or row than you jokers! No, here's my idea, and you don't have to wait till the war's over to do it. You go home now. You go on leave.'

'Leave!' said Groom in amazement. 'On leave? Who's going to give us our passes, then, Viney? You! I can just see those redcaps' faces when I show them a bit of paper with a little scribble on it saying, *"This joker's going on leave, signed Viney"!*'

There was a general laugh. Viney didn't join it but smiled in a puzzled fashion till it faded away.

'Yes, I'll sign your papers if you like. It don't much matter who's signed the sodding things, does it? I mean, you can't read most of those bastard officers' signatures anyway. As long as the details are right and they all tally, who bothers?'

It began to dawn on them that Viney was serious.

'But you need *real* papers with *real* stamps and all that stuff, Viney,' said Nelson.

'Of course you do. And once you've got them the world's your oyster. There's thousands of lads milling around at Calais and Boulogne and Dieppe at any one time. You'd probably not get a first look, let alone a second, if you've got the right bit of paper to wave around.'

'Viney,' said Nelson with incredulous hope, 'you ain't got no leave papers, have you?'

'No, I haven't,' said Viney. 'But I think I know where I could lay my hands on some! All this time, lads, we've had a little treasure in our midst and not appreciated it. Arnie! Where are you? Arnie, stand up and take a bow.'

Corporal Arnold Tomkins, late Headquarters Administration clerk, stood up with a self-important smirk.

And Lothar von Seeberg, late of the Imperial German

Army, felt his heart sink as he realized what Viney was proposing and knew that not all the political idealism in the world could resist the appeal of going home.

3

On the night of Thursday, March 20th, 1918, Jack Denial was dining in the mess at the Château d'Amblay.

He had been in the HQ all that afternoon giving evidence at the court-martial of a Service Corps lieutenant accused of organizing a petrol selling racket on the French black market. The trial would continue tomorrow, which meant he had to pass the night here. If the lieutenant were found guilty, he would probably be sentenced to be dismissed from the service. It occurred to Denial that France was full of ordinary soldiers who would have paid all they had for such a sentence but who for crimes much less heinous could only expect the humiliation of Field Punishment out of the line and exposure to all extremes of danger in it.

But he kept the thought to himself as he dined that evening with staff officers who clearly believed that the disgrace of cashiering was far more painful to a gentleman than anything the blockish and brutish private soldier had to endure.

As the wine relaxed those present, a sagging major puffed, 'Got those Diggers yet, Denial?'

The APM's long feud with Viney and his men was common knowledge. Denial merely shook his head and hoped the subject would drop.

'These Aussies are the living end,' drawled a cavalry captain. 'That mutiny down at Étaples last September was almost entirely due to them, I gather.'

A neighbour kicked his elegantly booted leg under the table as the senior officers at the head of the table paused in their conversation and turned frostily disapproving stares in his direction. As Denial had foreseen, the riots at Étaples had been officially described as small outbursts of unruly behaviour, limited in their effects and trivial in their causes. To use the word mutiny in this connection was a gross indiscretion.

'There were Anzac troops involved, yes,' agreed Denial.

'But not your chaps?'

'No,' said Denial. 'Not my chaps.'

'No? Well, good hunting, old sport,' said his interlocutor. 'I say, this claret's not a patch on last night's. I reckon someone's been switching labels again!'

Custom made it impossible for Denial to leave the mess before the senior officer retired and that night old battles were being re-fought and new ones which would bring the war to an end in a month if only Haig would listen, were being fantasized. The junior officers were infected by the game and a heated discussion began at Denial's end of the table. He sat in silence, his thoughts drifting to what waited for him at the war's end. And what might have waited.

At last, long after midnight, it was possible to get to bed. He took a look out of the window of the attic room he had been allotted. A heavy mist was draping itself over trees and buildings and as he watched the last star went out. He climbed into his restricting cot and fell asleep.

Outside and less than a hundred yards away the masking mist swirled around three shivering men. Two of them, Viney and Hepworth, were merely reacting to the dark chill which struck through every layer of clothing to the marrow. But the third, Arnold Tomkins, returning to the scene of his clerical triumphs, was shaking with pure terror. He hadn't stopped shaking since he left the farm several hours earlier. It was only the fact that of all the world's terrors, that inspired by Viney's wrath was the greatest, which kept him on his feet.

Hepworth's motives for being present were more obscure. He had greeted the appearance of Viney, Tomkins and Lothar on Wednesday night with watchful indifference. Madeleine on the other hand had made no attempt to conceal her irritation.

Viney had brought the German along for two reasons. Firstly to interpret and secondly because he did not care to leave him unsupervised in the Warren. With Coleport no longer reliable, there was no one even among his closest supporters whom he cared to trust to apply the ultimate sanction on Lothar.

But the promise of a return home with proper papers to guarantee a fair measure of protection had at least turned most of the opposition neutral.

As for Madeleine Gilbert and her family, Viney had urged Lothar to underline that it meant the Lost Battalion would be out of their lives forever. All that he wanted was to use the farm as a jumping-off point for this last raid.

'Are we asking or telling them?' enquired Lothar, knowing the answer.

Viney frowned.

'Don't get clever, Fritz,' he warned. 'You know that Taff and the raiding party will be turning up here on Friday night. We'll do the raid on Saturday. And we'll be off back to the Warren the next night and after that, it's every man for himself.'

Lothar regarded the Australian in bewilderment. For once he could not understand the man's motives. He was determined to show who was the leader, certainly. But success this time would mean the dissolution of Viney's Volunteers. And even if his grandiose scheme of providing every man in the group with genuine papers worked, it would benefit himself least of all. England was not home to Viney. And if he really wanted to get there, Lothar was certain he could contrive it without this dangerous and foolhardy raid.

Now he asked the question direct.

'Why are you doing this, Viney? Is it to get Josh home?'

Viney shrugged and said, 'Mebbe. If he wants to go. Or you might say I'm doing it for Patsy Delaney. In memoriam, Fritz. In memoriam. Now tell the good lady here what's going to happen. *Tell* her, not *ask*.'

Lothar told. Madeleine exchanged a glance with Hepworth but otherwise her response was the emotionless indifference he had come to expect.

'Where is Nicole?' enquired Lothar. 'I have not seen her?'

Madeleine shrugged and said, 'Busy.'

It didn't take any great powers of observation to see that Hepworth and Madeleine were now sleeping together. Old Alpert, Madeleine's father, had died shortly after Christmas, and as the sole male remaining, Heppy had become very much the master of the house, as far as anyone could be said to be master with a will as strong as Madeleine's to contend with.

Lothar spoke with him alone on Friday afternoon while Viney was sleeping in preparation for that night's recce.

'How are things with you, Heppy?' he asked cautiously.

'All right.'

'You will stay here after the war?'

Hepworth shrugged, an untypically Gallic movement.

'Madeleine will not, I think, leave here easily,' prodded Lothar.

'Then mebbe I'll stay here,' growled the Yorkshireman.

'But you are not a farmer. It needs expertise to work on a farm.'

The watchful eyes turned on him.

'Listen, Fritz, whatever a man can do, I can turn my hands to.'

Then suddenly the tone became more conciliatory.

'Not that you're not right, Fritz. I reckon in the long run, I can do it. But it really needs a lot more knowledge than I've got. Aye, and mebbe a bit more energy too.'

Lothar thought about this.

'It's Josh you mean, isn't it?' he said.

Hepworth nodded.

'Viney has promised they may all go home to England,' said Lothar.

'Home! That's pie in the sky, Fritz, and you know it,' said Hepworth. 'Any road, Josh has never seemed interested to me in getting home, not like them other bastards. You know that better than me, Fritz. How are things between you two, anyway?'

'Not good,' said Lothar. 'He hates me still.'

'And Nicole, how's he feel about her?'

Now it was Lothar's turn to shrug.

'Does she talk of him, Heppy?' he asked.

'Times, she does.'

Yes, thought Lothar. He could understand her pouring out her heart to this man. He knew well how much Nicole needed masculine strength to lean on and with Hepworth's developing relationship with Madeleine, he would make the perfect surrogate father and perhaps even confessor.

'And she would like Josh to come back here? To stay?' he asked.

He had still not spoken to Nicole, though from time to time he had glimpsed her, looking pale and preoccupied. Each time she had become aware of him, she had turned and hurried away.

Hepworth said, 'She needs someone, Fritz. If not Josh, you.'

Lothar smiled in surprise. He had plans and a mission far beyond a French farm once this war was over. Win or lose, Germany would be in a turmoil with a huge groundswell of anti-militarism fed by bereavement and waste and hunger. Hence would come the energies for the new socialist future which was the only thing that could make any sense of these years of carnage.

'Me?' he said. 'Why should I stay? How can Nicole need me?'

Hepworth stood up and said harshly, 'She's pregnant, Fritz. That's why.'

When the time came for the recce to begin, Lothar reiterated the refusal to go on it that he had made from the start. Viney came as near to pleading with him as he had ever done. Arnold Tomkins had already caused him considerable alarm by disappearing an hour earlier. They had found him shivering in the byre, his terror of the expedition having overcome his terror of the big gentle cow against whom he pressed himself for warmth and concealment. Now he was kitted up and apparently resigned to going, but Viney reckoned it was going to take at least two men to keep him in order once they got on their way.

It was Hepworth who ended the argument.

'I'll go,' he said.

Viney looked surprised but didn't argue.

'Hurry up and get ready,' he growled. 'We're late already.'

Under the pretext of helping Hepworth black his face, Lothar was able to ask him, 'Why?'

'I've seen how Viney takes care of the wounded,' said the Yorkshireman. 'But if Viney gets wounded, who takes care of him? There's no one here he'd be much bothered about protecting, is there?'

Lothar was in no doubt that 'take care' was a euphemism.

He gave it its proper inflection when he said, 'You take care of yourself, Heppy.'

'I'll do that. But if I didn't get back, you'd take care of this, wouldn't you, Fritz? Promise?'

His gesture included the farm and all in it.

'You love her?' said Lothar; then, embarrassed at this intrusion into the man's private life, he said, 'I promise.'

Madeleine approached at this point, in her hand a fresh baked loaf.

'Here,' she said. 'You will be hungry.'

'Aye. All of us,' Hepworth said.

He tore the loaf into three and tossed a chunk each to Viney and Tomkins. Then with sudden violence he grasped Madeleine and drew her to him in a fierce embrace which she returned with passion.

Five minutes later, the ill-assorted trio had slipped away into the night if Tomkins's reluctant stumble could be called slipping.

Lothar closed the door and said to Madeleine in a quiet voice which brooked no denial, 'Tell Nicole I wish to speak to her.'

Terror or time had blunted Arnold Tomkins's memory. All that he seemed to be capable of identifying through the night-glasses which Viney had brought along was the stable block in which he used to sleep. And as the mist blossomed and the château became a vague grey hulk he grew uncertain even of this.

Viney grew more and more impatient. He had made

Tomkins draw a plan of the château and its environs in the Warren, but the clerk's uncertainties had convinced him that a recce was essential. And now with the actual building within a stone's throw, the little shit seemed even less certain of the layout.

'We'll go in closer,' he whispered.

'No, please, Viney,' begged Tomkins. 'There's guards!'

'Of course there's fucking guards!' exploded Viney. 'But where? If you could tell me that, we'd know something worth knowing!'

He pulled down his woollen mitt and glanced at the wristwatch with the luminous dial he had 'won' from an officer in the Gallipoli campaign. A novelty still at that time, the officer had boasted in Viney's presence its convenience and its security from theft. But he had lost first his watch and not long after his life.

It was four-thirty, much, much later than Viney had planned. They should have been half way back to the farm now, but Tomkins' slowness of movement in getting here and uncertainty of identification on arrival had cost them a good two hours. It would be long after dawn by the time they made it back, so they might have cause to bless the fog after all.

He made up his mind.

'We'll risk the guards,' he whispered. 'Just a quick look, Arnie, old chum, just so's you can get your bearings, then it's home to a hot breakfast and a twelve hours' kip. All right, mate?'

His friendly approach did the trick. Tomkins nodded and the three men began to edge cautiously forward. Soon they had moved out of the concealment of the shrubbery and were making their way across a formal garden. It was sadly neglected, but the long, lank grass gave little protection. Fortunately it was criss-crossed with a pattern of slightly sunken paths flanked by miniature ornamental hedges and when they hit one of these they were able to move at a crouch with some small degree of concealment. Eventually they arrived at a central junction of paths marked by a small marble fountain choked with weed and debris. To Viney's delight, this seemed to trigger off a positive response in Tomkins's memory.

'Here, listen, I know where I am now,' he said. 'Right ahead there, there's a bit of a terrace, see?'

Straining his eyes through the mist, Viney could just about make it out.

'Right in the middle there's the main door, now that's useless as there's always a sentry in the box outside . . .'

'Well, for Christ's sake keep your voice down!' ordered Viney.

'Sorry. There's three big windows either side, then if you go round the corner, another four big windows down the side of the house. The third of them down the left-hand side, that's the room you want. Leastways it was when I was here.'

This, if accurate, was just the kind of information Viney wanted.

'The window fastenings,' he said. 'What about them?'

'It's a simple catch, I recall. Easy to prise open with a knife. But at night they usually close the shutters and I doubt you'd get through them without using an axe. But if you go round the back of the house, there's no shutters there any more. Got used for firewood, I think. And you could easily . . .'

He choked as Hepworth's hand closed sharply on his mouth.

A moment later they heard footsteps on a path. The mist made distance and direction difficult till there was the soft scratching of a match and the glow of a cigarette being lit in cupped hands about twenty-five yards to their left.

Voices spoke low. It sounded like two men. Tomkins stirred in fear, but Hepworth's hand remained on his mouth and Viney had his left arm in a grip like a vice.

The men came strolling towards them. After a few paces their shapes became clearly visible through the mist. Helmets and slung rifles. They were patrolling sentries, but fortunately, as Viney had guessed, rendered complacent by too much time in a cushy billet.

It seemed that they were going to walk straight towards the crouching trio but within a dozen feet of the fountain they suddenly veered off right, towards the house, obviously following the line of an intersecting path. A moment later and their shapes were almost invisible.

Slowly Viney let his breath out and, releasing Tomkins, took another quick glance at his watch.

It was four-forty.

He tapped Hepworth and jerked his thumb towards the east to indicate it was time to move out. The result was devastating. As though his gesture had stabbed a detonating button wired to a line of charges fifty miles long, the whole horizon exploded with a brilliance that lit up the fog and a couple of seconds later a dreadful pulsating but unremittant thunder came rolling over the trembling land.

Tomkins's teeth closed on Hepworth's fingers and the Yorkshireman jerked his hand away with a cry of pain. But his cry was drowned in the little corporal's shriek of sheer terror. He leapt to his feet and stared in horror for a second at the livid east, then, still shrieking, he turned his back on it and ran for the familiar safety of his former workplace.

The two sentries had spun round to look in wild amazement at the tremendous firework display which was ripping the night sky to shreds. Out of the mist, rushing at them, screaming like a man, or indeed a platoon of men, on bayonet practice at Étaples, came a demented figure.

One of the sentries tried to stutter a challenge but the other had a greater sense of his own worth. The rifle came off his shoulder with a speed that would have delighted his old training sergeant and the first bullet hit Tomkins when he was still several yards away.

The next half-dozen bullets took him in the back. They all came from Viney's Luger. His spine shattered by one of them, he went down with that unmistakable loose-flapping slump which proclaims a dead man.

Viney and Hepworth paused no longer. Turning and still crouched low, they sprinted away in the mist towards the distant crescent of flame.

The corridors of the château were rapidly filled with men in various stages of undress.

'It's the big one,' chortled the cavalry captain. 'I almost

hope they manage a little breakthrough. Then we'll see what the gees can do!'

Elsewhere a white-faced senior officer was saying to his ADC, 'I warned them about this. Oh Christ, with the state of our reserves, if they break through they could keep going to the sea!'

But Jack Denial looking out of his attic window dropped his eyes from the flaming sky to the garden below, where dimly through the mist he saw two men with rifles kneeling by a figure on the ground.

Quickly he dressed and descended. By the time he got downstairs, the body had been carried inside. He sought out the guard commander.

'Funniest thing,' he said. 'He came charging out of the dark like a crazy man. He wasn't going to stop so they fired off one round at him. Down he went. But look, you can see, his back's riddled with bullets. Oddest thing of all is, some of the lads recognized him.'

'Recognized him, Sergeant?'

'Yessir. It's a man called Tomkins. Used to be admin corporal here till he went missing about a year ago.'

'Tomkins.'

The name rang a bell. Denial had a card-index memory. He shuffled through it.

'Tomkins. He was in the RFC supply truck with that Scottish driver, Shawcross, the one they found shot out in the Desolation,' he said.

'I don't know nothing about that, sir. If you'll excuse me . . .'

Denial walked out into the night and stood in deep reflection. His mind was toying with three names. Shawcross . . . Tomkins . . . Viney?

And so he stood, completely oblivious to the cold, the mist, and, along the eastern horizon, the most concentrated barrage of artillery the German army had ever launched, the prelude to the *Kaiserschlacht*, the last great German attack by which General Ludendorff hoped to win the war.

4

Lothar had not slept that night.

If Viney should return with a platoon of British soldiers hot on his heels, he wanted to be ready for a rapid departure. Hepworth he was certain would not bring any danger back to the farm, but in the event of a policy dispute between the two men, he was by no means sure who would come out on top.

Besides, he had other worries to keep him awake.

His talk with Nicole had left him full of perplexity. She had answered him monosyllabically at first, but once her initial reluctance was overcome, the difficulty had been to interpret the torrent of words and tears.

The facts were simple. Yes, she was pregnant. And yes, he was the only possible father. But how to deal with these simple facts, *that* was not so simple.

His one source of consolation was that though his role as father was beyond doubt, he suspected he would be Nicole's own second choice as protector. At the end of their conversation she had asked almost inaudibly, 'Will Josh be coming with Monsieur Viney's friends?'

Unfortunately the answer was no. The raiding party consisted of half a dozen men, the best fighters remaining in the Volunteers. Viney was not going to risk Josh in this number, and the boy himself had shown no desire whatsoever to return to the farm.

So Lothar sat and mused till four-forty a.m. when the thunder of the German artillery brought him rushing to the door.

The fog was thick here too, but the farm was several miles closer to the front than the château and the display of firepower was proportionately more impressive. Lothar's expert ear attempted to identify what was coming over, but the continuous belt of noise defeated him. All he could be certain of was that practically every piece of ordnance from the lightest to the very heaviest must be in use.

Soon Madeleine and Nicole came to join him, their faces strained with new-awaking and fear. Madame Alpert, sadly declined in strength, remained in bed apparently unmoved by the fearsome thunder.

'What is happening? It sounds so close! Are the Boche coming?' demanded Nicole, clinging to Lothar's arm, oblivious of the fact that here was one of those Boche.

'Don't worry,' said Lothar. 'It is a long way off still. I think they are probably shelling the area well behind the front line to cut communications and damage supply routes. Come, let us go inside. Out here it is too cold.'

He ushered the unresisting women back into the barn where Madeleine stoked up the living-room fire and started boiling water for coffee.

'Lothar,' murmured Nicole, her young face troubled, 'this place out there in the dead lands where you all live, the Warren you call it, will the shells reach to there?'

'No,' said Lothar, smiling. 'They are still well behind the lines, many miles. They will be quite safe.'

What he said was true, but he also realized the devastating effect this bombardment might be having on the Volunteers. The Warren must be trembling in the shock waves of these explosions, and among Viney's Volunteers were many whose apparent normality was quickly destroyed by any sudden violent noise, let alone such an eruption as this.

He looked at his watch. It was after five o'clock. Viney should have been back by now.

At five-thirty there was a slight check in the bombardment though it was quickly resumed with a subtly altered timbre. Lothar guessed that sights had been lowered from the rear zone to the British infantry positions. At roughly ten-minute intervals after that there were other small hiccoughs in the distant bombardment.

'What are you listening to, Lothar?' asked Nicole.

'Just to my old comrades,' smiled Lothar, trying to reduce the girl's anxiety by adopting a light tone. 'The guns pause now and then, do you hear it? What they are doing is making range checks, I think. You see, with the first bombardment it didn't matter too much because they weren't aiming at anything in particular. But now they must be precise.'

'So that they can kill the Tommies in the trenches, you mean?' she said seriously.

'Yes, my dear,' he said softly. 'That is what I mean.'

He could have said more. The military part of his mind told him that this need for range checks during the bombardment meant that the checks had not been made, as was usual when a battery took up a new position, by firing a few shells and seeing where they landed. This desire for secrecy plus the sheer intensity of the bombardment told him that this was no mere local show to gain half a mile of no-man's land but a major push, perhaps a last desperate throw by Ludendorff in an attempt to break the Allied line, turning the British flank north of the Somme and rolling them up towards the sea.

But such matters were not for Nicole's ears. Nor indeed were the possible local implications of the attack. If there was a breakthrough, the British could be pushed right across the Desolation and the untouched pastureland between the farm and Barnecourt could become a battlefield.

But the front was many, many miles away. On past performance, a couple of miles in a day was an incredibly rapid advance. The great Allied push on the Somme in 1916 had gained at most five miles in as many months.

No, he told himself. Forget the madness going on over there. The real worry was, what had happened to Viney.

It was after seven o'clock when he at last heard sounds in the swirling mist, bright now under the touch of the invisible sun. He picked up a rifle and worked a round into the chamber. Then angrily he ejected it, and rested the weapon against the wall once more. If trouble was coming, it was not going to be the kind that could be turned away by bullets.

He stepped clear of the barn door and strained his eyes into the mist.

The touch of a hand on his shoulder made his heart twist and gape like a decked fish, and he turned round so quickly that he cricked his neck.

It was Hepworth, scouting ahead to check that all was well.

Viney now joined him, his huge figure damp with mist. Both men looked weary and depressed.

Lothar said, 'Tomkins?'

'Dead,' said Hepworth.

'The raid, it will be cancelled then?' said Lothar to Viney.

The Australian said, 'Why? I got what I wanted.'

'But they will be alerted,' protested Lothar. 'A man is shot – a man they know, remember that. They will draw conclusions, take precautions . . .'

Viney yawned massively.

'They'll have other things to bother them, I guess,' he said. 'Or haven't you noticed? Jerry couldn't have picked a better moment – Christ, but I'm whacked.'

He went into the barn.

Lothar and Hepworth stood and watched the flame-lurid horizon. The mist was lifting a little and the curve of fire was clearly visible.

'Tomkins was definitely dead?' queried Lothar.

'Oh yes,' said Hepworth. 'Back and front dead.'

'Viney?'

The Yorkshireman nodded.

'But getting that bugger back or front isn't so easy,' he suddenly added.

Was he describing his own failure? Lothar wondered. Or inviting a conspiracy?

He considered the matter, then shook his head in self-irritation. He had not come so far along this road from senseless killing to signpost it with a murder.

Madeleine came out of the barn and stood looking at Hepworth without expression or comment. Lothar felt a sudden envy of the man. He had found something in this war which to nearly all the others had simply meant loss.

He went into the barn and left the man and the woman to their private greetings.

5

The bombardment had continued for another two hours, rising to a climax at nine-thirty. Even at this distance it pulped the nerves; and lying underneath it must have been like having your soul flailed out of your body, leaving behind a gross dull incomprehending animal hulk.

Then it stopped.

And slowly all along the front, most of those near-lifeless hulks would be arising, and reassembling their manhood, and straining fear-dilated eyes into the mist and the smoke to discern the advancing enemy.

Lothar sat before the fire and wept silent tears for what man is and what he could be; for what nobility he wasted in what vile ends.

Then he stretched out on his pallet and joined Viney in sleep.

Noise of the distant battle rumbled on for the rest of the day and rumbled on far down the valley till it reached Château d'Amblay, where amidst the endless comings and goings and the hurly-burly of tactical conference and strategic assessment, despatch of orders and receipt of intelligence, Jack Denial could find no one of senior rank interested in the death of Corporal Arnold Tomkins.

Even among the other ranks, he got only the mockery of attention due to his three pips. The exception was Sergeant-Major Maggs, whose eyes lit up when he heard the magic word, Viney. But even he was distracted as the day wore on and reports of massive enemy pressure at key points along a fifty-mile front began to trickle out of the staff conferences.

The court-martial had been prorogued, but Denial had not returned to his unit in Barnecourt. Instead he had ordered every item of Tomkins's equipment and clothing removed and brought to him. Unaffected by the bloodstained tunic, the back of which was almost entirely ripped away by Viney's bullets, he went over the assembled material painstakingly, until early in the afternoon the door opened and Maggs appeared, just having time to say, 'Colonel Forbes, sir,' when he was pushed aside.

The Colonel looked at Denial in amazement.

'What the devil do you think you're doing, man?' he cried.

'It's this fellow who was shot last night,' began Denial, eager to try out his conclusions on anyone ready to listen, but the colonel was not on this very short list.

'Shot last night? We're all liable to be shot tonight, or tomorrow night or the night after that! For God's sake, man, don't you know what's happening? We're going to need every bit of our reserves, I tell you, and even then they may not be enough. The roads are going to be chaotic. My God, it's like Piccadilly Circus outside the château already. That's where I want to see you and your chaps, that's your damn' job, isn't it? This is no time to be playing Sherlock Holmes, Denial. Get out there and start sorting out the traffic!'

He left. Maggs kept his eyes fixed on a diplomatic space a foot above Denial's head and waited for the captain's orders.

'I think that Viney is planning a raid,' said Denial thoughtfully.

'On HQ, sir?' said Maggs disbelievingly.

'Why not? He has delusions of grandeur, I believe. And there must be any amount of stuff here these deserters would find useful. But I don't think he could possibly do it out of the Desolation in a single night. So it must be like last time, when Delaney was killed. They'll be using somewhere as a jumping-off point. Look.'

He held out a boot. Maggs looked.

'Smell it.'

Maggs sniffed reluctantly.

'Cow shit, sir,' he opined.

'Right. And fresh too. Like this.'

He held out some crumbs of bread removed from Tomkins's tunic pocket.

'Try some,' he said. Maggs looked at the bloody tunic and shook his head. Denial nibbled a crumb.

'It's very fresh. Recently baked. So. Cow dung and fresh-baked bread. A farmhouse, wouldn't you say, Mr Maggs?'

'That one at the top of the valley again, you mean, sir? Could be. But look, sir, no one's going to make a raid, not with all this going on.'

Denial said in surprise, 'Why on earth not, Sergeant-Major. Viney and his gang have opted out of the war, remember? The bigger the confusion, the more they'll like it.'

'Sir!' said Maggs, unconvinced. 'You may be right sir . . . meanwhile . . . ?'

He held open the door.

'Very well, Sergeant-Major,' said Denial resignedly. 'Let's go and direct traffic.'

The raiding party was supposed to journey from the Warren the following night, but when dawn came it had not appeared. The fog was very thick again and Lothar suggested that if, as was likely, it was even worse out in the Desolation, probably Evans had decided it wasn't worth risking the trip in such conditions.

'Or perhaps the stupid bastards set off and got themselves lost,' snarled Viney. 'Leave them alone for ten minutes and they're useless!'

Hepworth raised his head from the gun he was cleaning and said, 'Sounds to me like that battle out there's getting closer.'

'Yes,' agreed Lothar. 'It does. Perhaps Taff felt it wiser to keep underground.'

It was strange to see the same look of alarm leap into Nicole's and Viney's faces. The girl's English was still not very good, but she could follow simple statements well enough.

'The Boche couldn't have got that far. Impossible!' protested Viney.

'I believe so too,' said Lothar. 'All I meant was, if the front has moved, then everything will move. Perhaps there is new activity with reserves and supply lines closer to the Warren. In this fog then, it would be wiser to stay still.'

His explanation repeated in French seemed to allay Nicole's worries, but Viney was preoccupied all the next day.

As evening approached he said, 'Perhaps I should go to meet them. Or go all the way to the Warren.'

There was an uncharacteristic uncertainty in his voice and Lothar's reply was cautious.

'Perhaps, but you might miss them.'

'The raiding party? It's not them that's worrying me, Fritz. They can just about look after themselves. It's the other jokers . . . there's no one left to take proper charge.'

'There's Coleport.'

'Blackie? I'd have said so once, but he's been crook in the head since Patsy died. No, I've slipped up for once, Fritz. I should've made sure there was someone back there I could rely on. Like you. Mebbe you should have stayed back there, Fritz.'

He said it almost accusingly. Lothar forbore to point out that he had been practically ordered to the farm, and said, 'There's not much can go wrong, Viney.'

'What if the Boche break through? What about that, Fritz?' said Viney angrily.

It was a question which had already begun to trouble Lothar. Judging the progress of a battle at a distance was almost impossible. All they could say for certain was that it was still raging fiercely and that, as Hepworth had pointed out, from time to time it sounded closer. But if there *was* a breakthrough, and Josh should suddenly find himself swept back into that maelstrom of horror from which his mind had so recently been rescued, then . . .

He caught Viney's eye and saw in the big Australian's expression that exactly the same thoughts were running through his mind.

The next twenty-four hours passed in much the same way as the last with the sounds of battle now perceptibly closer and Viney growing more and more anxious.

Lothar took Hepworth aside in the middle of the day and suggested to him that it might be wise for the womenfolk to pack up and make the trek down the valley to Barnecourt.

Hepworth replied, 'The old lady's too weak and Madeleine'll not leave her.'

'They can use the horse and cart,' Lothar insisted.

The cart was a ramshackle affair which was the best that even Josh's skills had been able to construct from the debris of old implements. One wheel was missing a small segment, and the best that could be done was to give it a straight edge at that point. The result was a curious limping motion which, added to the effect of the deeply rutted tracks, made riding on this conveyance a nauseous and dangerous business.

Hepworth put the suggestion to Madeleine who shook her head vigorously.

'It will kill her,' she said. 'She is old and very frail. Since Papa died, she has lost the will to live. Well, at least I can see she dies in her own bed and not in a muck-cart.'

Hepworth shook his head in a gesture which conveyed perfectly to Lothar his belief that women were undeniably crazy but also crazily undeniable.

But Madeleine was no fool, and the judgement of these two men both of whom had won her not easily given respect was not to be ignored. If they felt the Boche might come, there must be something in it.

She said, 'Nicole can go. She can take the horse and lead the cow. You, Heppy, can see that she gets there safely. It is settled.'

Lothar regarded this indomitable woman with unstinted admiration. At a blow she was consigning to safety all that was valuable to her both emotionally and materialistically. Her daughter, her lover – and her livestock!

Hepworth began to protest, but she brushed him aside and called to Nicole. Lothar stood back to enjoy the clash of wills between the determined Frenchwoman and the stubborn Yorkshireman. It would be a battle whose ferocity would fall little short of that great battle raging to the east. But there was a surprising turn of events. Nicole on hearing her mother's instructions burst into tears and announced with a near-hysterical intensity that she was definitely not leaving. Madeleine frowned menacingly at this unexpected opposition, then equally unexpectedly did not press the point but went back to her work.

She shared with Viney, Lothar thought, the great strength of knowing when to bow.

The night came and went in the same pattern of fog and anxiety as those before.

As the morning sun began to draw up the mists, Viney said, 'That's it. Something's happened. I'm going to find out what.'

'Wait till noon,' urged Lothar. 'They may have set out and got lost. It would be stupid to miss them by a couple of hours.'

Viney agreed with surprising readiness. There was a look about him which Lothar recognized but could not identify. Then it came to him, and with it the explanation of its difficulty of identification.

It was fear; as unlikely a feeling to shape Viney's features as lust on a Mother Superior's.

It took a little while longer for Lothar to work out that it was not fear of the journey but what he might find when he reached the Warren. The battle-line – if a line actually still existed – must be snaking even closer. There could be no doubt that the bruised and ravaged earth of the Desolation which had lain fallow for nearly two years was now experiencing yet again the ripping, gouging plough of war. Sound told of its approach, and sight too, as fists of black and yellow and grey smoke clenched themselves in the air, then opened hazy fingers and hurled hot seeds of metal between the rambling, ugly furrows below.

The sky seemed full of aeroplanes today also. Lothar did not believe he had ever seen so many, not only nimble little biplanes swooping and diving over distant targets, but wide-spanned bombers and slow observation planes lumbering to join them, unnaturally low. It could mean only one thing. The British were desperately throwing everything they had into their efforts to inhibit the German infantry advance. After more than three years, war had finally broken out of the trenches and suddenly the old snail's pace of movement to which they'd all become conditioned was a thing of the past.

Viney was reaching the same conclusion.

'Christ Almighty, Fritz,' he said, looking haggard. 'I think I've left it too late. Who'd have thought the bastards would get off their arses at last and start making a real fight of it!'

Nevertheless, he was still set on returning to the Warren and seeing for himself when Hepworth came into the barn and reported movement to the east.

Alarmed, the men grabbed weapons and rushed out to see.

It was Viney, using his binoculars, who made the first identification.

'It's them!' he said. 'Leastways, it's Evans. And there's Blackie. He shouldn't have come, he's not in the raiding party!'

But his indignation that his orders had been disregarded changed to relief, as he saw the extent of the disobedience.

'And there's Josh too,' he cried out. 'Thank Christ! I think they've all come!'

In fact as the men started trickling into the farm it became clear that while this was certainly not the raiding party, it was far from all of the Volunteers.

Taff Evans explained.

'First night, fog was so thick, it was getting *inside* the Warren. "Bugger this," I said, "we're likely to end up breaking our necks or getting bogged down in an old shell-hole." So we stayed put, see.

'Meanwhile all hell was breaking loose out there. Next day we realized it was getting closer. Second night was the same, and now we were getting really worried. Some were for going, some for staying. It wasn't just a question of the raiding party any longer, Viney, it was the lot of us. One or two of the lads were already in a bit of a state, what with the guns and all. Anyway, the vote was for hanging on a bit longer.

'By the time we got to last night, I knew we had to be off out of there. The bloody war was coming right back to us, see! So some of us set off out. Well, we hadn't got far before we realized the place was crawling with bloody troops! Lucky for us, there was such a state of bloody confusion that we managed to scramble back to the Warren without been taken notice of. We lay doggo for the rest of the night, but come this morning I went up top to take a look-see and what I saw made me realize it was time to be off.'

He took a deep breath both out of necessity and for dramatic effect.

'I saw a whole fucking company of riflemen *running*! I mean, running *that* way.'

He pointed down the valley.

'They had their officers with 'em and they still had their weapons, but that was enough for me. If the British Army was on the retreat at the double, then Taff Evans wasn't going to hang about!

'I went back down and told the rest. Them as wanted came with me. There were some who took off in other directions. And a few there were who reckoned they were as safe down the Warren as anywhere, and they've stayed. Well, they may be right! We've got through to here only because we were fucking lucky and there's chaos out there. But whether we've run far enough or fast enough, I've got my doubts.'

He came to a halt. Lothar looked around. There were between fifteen and twenty men who had come with Evans. They included Fox, Nelson and Groom. Blackie Coleport was crouched in a corner, a private little smile on his lips and his unblinking gaze never leaving Viney. Close by him sat Josh. Lothar had observed a look of sheer delight and relief bloom on Nicole's face when she realized Josh was in the group, then it had gone out like a star swallowed by drifting cloud and she had disappeared from view.

Lothar felt angry with her, but knew there was little he could do. A spontaneous show of affection as Josh arrived might have done something to cut through all the knotted emotions that separated the two young people, but the moment was past. At least it was past for Nicole, but there was still time for himself to practise what he preached.

He went to the corner and squatted down beside Josh. The boy had been looking fairly animated as Evans spoke, but at Lothar's approach his face had resumed its old closed-in look and he fixed his eyes on the floor.

'Josh,' said Lothar. 'My friend, I am more glad than I can say to see you.'

He put both his arms around the boy and drew him close. The slim frame tensed and stiffened within his grip, the head sank obstinately lower.

'Josh,' he whispered urgently. 'Listen! Now you are here, there is something you must do. Madeleine wants Nicole to go down to the village with the livestock. It will be safer for them there. But she will not go. I think she would go with you, perhaps. Will you take her? I beg you.'

There was no response. Lothar tightened his grip on the boy.

'Josh, you must!' he pleaded. 'Josh, she is pregnant. She is with child.'

The boy's body started in his grip like a shocked heart.

'It's your kid,' he snarled. 'You fucking well take care of her!'

Viney had risen and come to stand close by them, a menacing figure. But he made no effort yet to interfere.

'Yes,' Lothar said. 'It is my child. But it is not me she thinks of, Josh. It is not me she puts all her hope of happiness in. Josh, I am nothing to her. She would, I think, gladly kill me if it would gain you for her once more.'

Slowly the boy's head came up to look at him.

He met the questioning gaze unflinchingly, but he suddenly felt like a shabby betrayer before those desperately probing eyes. Was he not simply trying to slip out from under his own responsibilities by this show of noble altruism? What future could there possibly be for a young French girl and an English deserter?

But now seeing the hope beginning to swim up to the troubled surface of Josh's face, he felt no doubts. No matter how short a future might appear, it was never too short to be important.

'Remember how you felt when Wilf died,' he said. 'Remember that others feel too. And fail, also. Go and see her now, Josh. She is outside, I think. Go to her. Talk, at least.'

Slowly Josh rose, breaking Lothar's grip, but holding his hands for a moment as he stood over him.

'Thanks, Lott,' he said.
He moved swiftly away.
Lothar looked up at Viney.
'It is for the best,' he said.
'Best for who?' said Viney. 'Tell me that, Fritz. Best for who?' And turned and walked slowly away.

Josh found Nicole in the byre. He had somehow known she would be there, standing close to the very spot where Auguste had fallen with a bullet in his chest. It came to him with a shock like an explosion that what he felt she had done to him was nothing compared with what he had done to her. The greatest, the most unbearable pains can be caused by accident, without malice, without design. He had scarcely thought about her dead brother during the months in the Warren. All his attention had been taken up by his own pain, his own despair. Lothar had reminded him of his own unreal, uncontrollable existence after Wilf's death, yet he had felt himself unique. In this world of death and horror, he had had the arrogance to feel unique!

She showed no surprise when he cautiously pushed open the door and stepped inside.

'*Bonjour*, Josh,' she said in a small, low voice.

They stood and looked at each other in the dim light which stole in behind him through the half open door. Through the aperture also came the crash of bursting shells and the rattle of automatic fire, distant as a land-rail's chatter in a neighbouring stubble field, but always coming nearer. This they did not hear but strained tongues to shape and ears to catch the words which could heal or destroy.

There was no chance of Josh rifling his small store of French and finding the right words and even his English vocabulary seemed empty of the subtle phrases needed to express the turmoil of feelings inside him. As for Nicole, fear of saying a wrong thing, of taking a step in the direction of recrimination and accusation, kept her paralysed.

Then the cow moved, swinging her heavy hindquarters in response to some minor irritation of the skin, but the result was for all the world like some kindly old aunt, impatient at the indecision of these silly youngsters, nudging her niece into action.

The broad hock touched Nicole and she stumbled forward. If she stumbled rather further than either the blow or her own powers of recovery made really necessary, then this was the prerogative of a pregnant woman in her own country who has made up her mind what she wants. As for Josh, he stepped forward to steady her, but found instead he had caught her in his arms, and he did not even have time to be surprised at how easily the right words came and how simple and few they turned out to be.

6

The parting was too swift to be sorrowful. That would come later. The afternoon light was fading fast and the distant glooms were continually ripped by jags of fire.

Josh, dressed in an ill-fitting assortment of Auguste Gilbert's clothes, mounted the cart with the old artillery 'hairy' before and the bewildered cow behind. Nicole crouched at the bottom of the cart, between a small bundle of her belongings and a roughly assembled crate containing the few hens they had been able to round up.

A few last words, shaking of hands, close embraces. Viney thrust a wash-leather purse into Nicole's hands. It was bulging with coins and so heavy that some at least must be gold.

'It's a war loan,' he said. 'Hide it. Take care of him.'

'*Merci, monsieur, merci!*' she replied, thrusting the purse down her blouse next to her skin.

Then Josh slapped the reins on to the rump of the horse, clicked his tongue, and the ramshackle equipage began its halting move down the valley.

They had an escort, both official and unofficial. Most of the men had decided too that their best bet was to head out before the war overtook them. Lothar, who did not trust them all equally, had viewed this with some concern till Taff Evans had announced he was going too.

Privately to Lothar he said, 'Don't worry, boyo. You're thinking some of these lads are capable of knocking those young kids over the head and stealing the horse, isn't it? Well, mebbe you're right. It's every man for himself now, that's the way of things. But I'll keep close till we get near the village. They'll come to no harm.'

Viney had made no effort to prevent this division of his force. Why should he? thought Lothar. This was no time to be thinking in terms of strength of numbers. Once the Warren was abandoned, numbers merely drew attention. It would be the wily – and lucky – individual who survived.

Lothar was concerned that, if stopped, there was very little chance of Josh passing himself off as a Frenchman. It was Nicole who supplied the answer. The girl was a transformed being, full of life and energy, and that this was a mental as well as a physical revitalization soon became apparent.

'He has Auguste's clothes,' she said. 'So, let him have Auguste's papers too!'

It was so simple it might work, thought Lothar. Of course, most of the villagers of Barnecourt would know that Auguste Gilbert was dead and in his grave. But they were not the ones who'd be asking questions. Indeed, with the German advance audibly getting nearer, they'd probably be too concerned with their own safety to worry about anyone else. As for the military, what would they see? A young man, so shell-shocked during his service that he could no longer speak, and with the discharge papers to prove it.

Still, as he watched the cart bump away down the potholed track, his heart was full of misgivings.

'Perhaps I should have gone too,' he said aloud.

'No,' said Viney, who had been absent during the farewells but was now standing beside him.

'No? Why not?' Lothar demanded angrily.

The Australian let out a single explosive bellow of laughter.

'Do you think if I didn't go, I'd have let you go off with

them, Fritz?' he asked. 'For a clever man, you're a real thick cunt sometimes.'

He went back into the barn. Lothar stood a long time, watching till the cart was out of sight and realizing for the first time just how much it had cost Viney to let Josh go.

It gave Jack Denial no satisfaction whatsoever to know that he had been right and the vast majority of his fellow officers had been wrong. They had relied too heavily on a forecast of the enemy's tactics which proved inaccurate; a scheme for the movement of reserves and supplies which proved inadequate; but above all, they had relied on the fortitude and altruism and strength of will of the British soldier who, badly treated, badly fed, badly clothed, and badly led, had held the line across Belgium and much of northern France with indomitable courage for three years.

And now the British Fifth Army was in full retreat. Not an organized withdrawal permitting the enemy to recapture a few hundred yards of no-man's land, but in many places a full-scale flight leaving holes in the line through which the German army was pouring like water through a dyke.

Since leaving the Château d'Amblay which was already being hastily evacuated, Denial had had no rest. The roads were chaotic and his men were almost at the point of despair in their efforts to establish some kind of control of the violent and conflicting flow of retreating soldiers, advancing reserves and fleeing civilians.

He moved from one key point to another, directing Maggs to take their dust-covered motorbike combination across fields and through woods to avoid the cluttered roads.

He had just used his APM status to bring peace and common sense to a head-on collision between a pair of majors who'd got to the point of arguing seniority, when Maggs who'd gone a little further down one of the roads returned.

'Hope you didn't leave anything valuable in your billet, sir,' he said.

'Why's that, Sergeant-Major?'

'I just saw one of the civvies from Barnecourt in that mob back there. He says Jerry's come down the river valley from the north, reckons they'll be in the village by now.'

'As far as that? My God!' said Denial.

He stood in thought for a while, then said, 'Sergeant-Major, round up all of our lads that you can. Start *them* rounding up any odd bods and stragglers they come across. Also all the guns and ammo they can find.'

Sergeant-Major Maggs caught his drift instantly, but said, 'Sir, what about this lot?'

He indicated the congested crossroads at which they stood.

'Leave one man here,' said Denial. 'But let's face it, Mr Maggs. It's all right directing traffic, but if we don't all do something about stopping the enemy, it'll be German traffic we're directing in a few more hours, won't it? So get to it, quick as you can.'

'Yes, *sir*,' said Maggs.

For Josh and Nicole, the journey down from the farm had been slow and perilous. The cart had almost lost a wheel at one point and it was here that the remnants of the Volunteers, far from being the menace that Lothar had feared, had proved invaluable.

They had shown a strange reluctance to split up, though what safety they could hope for certainly did not lie in numbers. Indeed they seemed almost pathetically keen to keep within view of the cart, as though its presence conferred some kind of civilian respectability upon them. Thus it was easy for Taff Evans to summon a gang of willing helpers to hold up the cart by main force while the repair was effected.

The delay meant it was thick twilight by the time Barnecourt came into sight. But the village itself was not in darkness. Shafts of light spilled out of doors and windows and there was a noise which sounded almost like drunken merriment audible in the interstices of the gusting wind which had blown up at the fugitives' backs.

'What's happening?' wondered Josh.

'Perhaps the people are packing up to leave, frightened that the Boche are close,' suggested Nicole.

'Mebbe,' agreed Josh and urged the horse forward.

The wind was blowing stronger and stronger, still carrying

the noise away from them as they approached, but there was certainly nothing in it of the sound of battle and they did not slow their pace till they were almost at the outskirts of the village.

The main part of Barnecourt lay on the other side of the tiny river and the leaders of the group of deserters were already crossing the narrow bridge.

Evans went running up the road behind them shouting to them to wait. It seemed to him that whatever was happening in the village, the sight of a gang of ragged, wild-looking men was not going to be greeted with universal delight. As he ran, the wind suddenly dropped for a moment, and the nature of the hubbub over the bridge became clear.

It was indeed drunken singing.

And, devastatingly, the words were German.

Even as the significance of this hit Evans a crackle of small-arms fire punctuated the serenade. Two of the men on the bridge screamed, twisted and fell, the others scattered, diving for cover down the river bank or among the nearest houses.

Taff turned and ran back towards the cart.

'Get back! Get back!' he screamed. 'They're here, the Huns are here!'

Josh did not wait to ask questions. The old horse had not been asked to make many violent manœuvres since he had last been harnessed to a gun-carriage, but all the old skills remained and in a few moments the cart was facing back up the valley.

By now the Germans who had fired on the deserters were up at the bridge. They loosed off a volley up the track which drink and darkness did not make very accurate. None of the bullets troubled Josh and Nicole, but behind them Taff Evans threw up his arms, stumbled and fell.

Nicole, seeing the Welshman fall, let out a cry of alarm but next minute he was on his feet, waving them on. It was a foolish thing to do, but it was not foolishness that gave him the strength, unless courage and self-sacrifice are the prerogatives of fools. Josh must not be allowed to stop for him, this was his only thought, instinctively according the terrified young man the same impulses of loyalty and bravery as he found in himself. The first bullet had ripped through the fleshy part of his thigh, but the next two caught him in the middle of his back and this time he went down into a darkness like that of the eighteen-inch seam he had once worked deep beneath the ravaged ground of the Rhondda.

Nicole did not see him fall again. A jolt of the cart had thrown her flat on her back. She had a sense rather than a sight of something falling out of the star-torn sky. Then, close by, there was a violent explosion and the cart halted with a suddenness which sent her sliding against the hencrate with a sickening crack of her head.

When she regained her senses after only a second or so though it seemed infinitely longer, she struggled upright and found herself looking by chance rather than design out of the back of the cart.

The sight she saw was horrifying. The cow had been hit by the full force of whatever shell or bomb had caused the explosion. Its belly was ripped open, spilling blood and entrails on to the ground where it lay, a dead weight which had acted like a sudden anchor to the moving cart. Only its head moved, rising to let out a cry of pain and terror and to fix huge brown eyes on Nicole which to her spinning mind seemed both pleading and accusing.

Turning to shout her news at Josh, to her horror she realized he had disappeared. Scrambling to the front of the cart, she looked down and saw him lying on the ground between the legs of the old horse, who fortunately seemed inclined to regard this unexpected halt as a rest stop and was peacefully grazing.

Imagining simply that the violence of the stop had thrown Josh off his perch, Nicole scrambled down to tend him. He was lying on his face but he half rolled over on to his side at her touch. Instantly she saw that things were far worse than she'd thought. A fragment of hot metal had caught him on the side of his neck, wounding and half-cauterizing in the same agonizing moment. How deep the wound went she could not say, but at least there was no great arterial pulsing of blood and his limbs seemed to have some co-ordination and strength as she dragged him sideways away from the horse's hooves. The worst thing was when he tried to speak and all that came out of his mouth was a strangled groan, fading into a harsh stutter like her grandfather's death rattle.

But this was no signal of death, she realized, just an effect of his neck wound.

'Josh, please, can you stand?' she said urgently, first in French, then in slow English. Another death-rattle. Unknowing whether this was *yes* or *no*, she put her arm over her shoulder and began to lever him off the ground.

There was little strength in his muscles but what there was came to her aid; his other arm flailed out and caught the side of the cart. Half carried, half dragging himself, he got round to the back and let the top half of his body fall across the tail of the cart. Sobbing with effort and fear, Nicole lifted his legs and pivoted so that they too were in the cart. Then with nervous fingers she unpicked the knot on the rope by which the cow was tethered to the axle. The animal had kept up its piteous mooing all this time. Nicole's upbringing on a French farm had left her little sentimentality as far as the farm animals were concerned, but now, even though every instinct in her body was urging her to the speediest of flights, she found it hard to leave this creature in its agony.

In the cart Josh rolled over. Whether by accident or design, Nicole could not say, but his scrabbling fingers plucked at the length of sacking which covered the .303 Lee-Enfield rifle Lothar had placed there. Nicole had never fired a gun, but she had seen them used and cleaned and checked. Now she took it and released the safety-catch and worked the bolt and put the muzzle to the animal's head.

At that moment, that the Boche should be making her do this seemed their gravest crime against humanity in the whole of this monstrous war.

To the sound of the shot ringing in her ears she added her cries of rage and hatred against the filthy Huns and all their works as she took the reins and sent the old artillery horse labouring back up the wind-steepened track down which he had so recently pulled his hopeful burden.

In fact, Nicole was doing the Germans an injustice. It was not a German shell but an English mortar bomb which had blasted the cow and wounded Josh. And the noise of gunfire she could still hear from the village was not the Boche firing after the cart, as her fevered mind was firmly convinced, but the beginnings of a fierce struggle as the British forces counter-attacked.

The Germans were in many places the victims of their own success. Conflicting orders were being received from different sources by the British front line units. Some, ordered to hold out, were offering bitter resistance to the last man; others, instructed to withdraw, had set off with a will, and at a gallop, in face of the huge enemy onslaught. Through these breaks in the line poured the Germans, Ludendorff's crack troops, tough men, battle-hardened men, but men who had been existing on the basic rations which were all that their war-starved fatherland could provide, and men who had long looked out from the fastnesses of the Hindenburg Line across the desert they themselves had created in their spring withdrawal the previous year. Across this desert they had fought till finally they entered that other great desolation, the ripped and broken landscape of the great Somme battles of 1916.

And finally, unexpectedly, just when it began to seem to them that the whole of France from Picardy to the Atlantic coast must be a desolate wasteland, they were out of it, into wooded and verdant highland, and rolling and cultivated pastureland, and villages and settlements untouched by shellfire, whose hastily vacated houses contained fresh food, comfortable beds and sometimes cellars full of wine.

To ignore such luxuries was beyond human fortitude. In Barnecourt, the officers of the regiment of Württembergers who had spilled down the little river valley through one of these sudden breaks in the retreating British line had not been aware of the dangers of success until too late. Within a couple of hours, from being the potent spearhead of a devastating assault, the troops had become a happy, undisciplined rabble, still ready and able to loose off a fusillade at any suspicious newcomers such as the fleeing Volunteers, but in very poor state to withstand a properly mounted counter-attack.

Not that the counter was very properly mounted in military terms. The British force which flung itself upon Barnecourt that night was an ill-assorted bunch of batmen, cooks, pioneers, and a handful of exhausted and demoralized fighting troops separated from their units. On having them paraded before him, Sergeant-Major Maggs had expressed grave doubts about accompanying them on *any* kind of operation let alone on a night attack, but when Denial offered him the serious choice (for the captain was little given to humour) between drunken Germans by night or sober Germans by day, he quickly chose the former.

He was quite right. By dawn the only Germans remaining in Barnecourt were neither drunk nor sober but dead. Whether the village itself had suffered more from its attackers or its rescuers was hard to say. After the initial stray shots with their mortars, one of which had wounded Josh and killed the cow, Denial's men had got range and considerable damage had been done. Also there was a slight tendency among some of the men to carry on where the Germans had left off, so far as the wine went anyway. Denial's orders soon put a stop to this and he also sent out parties of fire-fighters to control the blazes in a couple of burning houses. This was in his nature and even if he could have been aware that within forty-eight hours a German field-battery would reduce Barnecourt to a spoil-heap of smoking rubble, he would probably have done the same.

He stood now and watched the wounded and the dead being brought in. The corpses were being laid out in regimental lines alongside a wall which, appropriately enough, ran round the village cemetery. Denial's attention was caught by a trio of bodies whose scruffiness of clothing and unshavenness of feature was outstanding even in their present company.

'Who are these?' he asked the lance-corporal he'd put in charge of recording the dead.

'Don't know, sir,' said the man. 'No dog-tags, see. They was lying on the bridge back there. I think they're ours.

Leastways from what I can make out of their uniforms, they're ours. There's another one of the same with the wounded. They put him here first off, but I saw him twitch a bit and got him shifted. Mind you, I hardly think I need have bothered. He looks like he'll be back here before breakfast, whenever that is.'

Ignoring this hint, Denial made his way to the house where the wounded were being tended. The nearest thing they had found to a medic was a sergeant-cook who had volunteered gloomily, saying that after a long career in Army catering, at least there wasn't much that could turn his stomach.

He directed Denial to the resurrected corpse, confirming the diagnosis that the resuscuscitation was likely to be short.

As Denial bent over the pain-etched bearded face, the eyes opened and stared full into his.

'What's your name, soldier?' asked Denial.

The pale lips hardly moved but a wisp of sound trailed from them.

'Taff Evans,' he said.

'And what's your unit, Evans?'

There was no answer.

'Come on, man,' urged Denial. 'It'll help us inform your next-of-kin.'

Now the lips moved again, the words painfully emerged.

'Aren't you the cheerful one, then?'

'I'm being realistic,' said Denial, to whom this was the most powerful of arguments. 'Now, tell me. What's your unit?'

The eyes which had closed again now opened wide. What looked like a smile but what Denial felt certain could only be a rictus of pain twisted Evans's lips.

'Why, man,' he said in almost a normal voice. 'I'm one of Viney's Volunteers, aren't I?'

And he died.

'Sergeant,' called Denial. 'He's ready to go out with the others now.'

Behind his back, the sergeant-cook made a nauseous face. At last he'd found something to turn his stomach.

But Jack Denial did not notice and would not have understood if he had.

He walked through the village to the bridge and let his eyes run across it and along the road, between the few houses on this side and, using his binoculars now, up the track which wound through the fields, passing what seemed to be a dead cow on the ground, till his gaze reached the ridge which marked the near horizon.

Up there. He was somewhere up there.

7

That night those who remained at the farm had retired early. An impulse to huddle together round the fire in the livingroom was douched by the needs of old Madame Alpert. After a troubled and distressed day, she had finally fallen into a shallow sleep and Madeleine was determined to avoid any risk of disturbing her.

As well as Viney, Lothar and Hepworth, the remaining Volunteers were Fox, Nelson, Groom and Blackie Coleport. Coleport fortunately was not in one of his wild, excited, verse-reciting phases, but sat quietly as if in a trance. It was Viney who seemed the least disposed to sleep. Since the young couple's departure, he had been in a dark and dangerous mood and the others all moved cautiously around him, as though he were some huge, unexploded shell fallen in their midst, threatening to detonate at the slightest disturbance.

He had rifled the small store of liquor which Madeleine kept in her larder. No one had attempted to join him in his drinking, or to stop him either. Lothar hoped that the alcohol might anæsthetize his huge frame, but two bottles of wine and a good quantity of brandy had disappeared with no discernible effect. When the move to bed had come, he had arisen and made for the door, clutching the brandy bottle in one huge hand, but still steady-gaited.

'Should we post sentries?' suggested Lothar mildly.

'I'll watch,' said Viney. 'You sleep.'

There was no discussion invited – or offered.

Lothar lay awake a long time. Things were coming to an end. Curiously, it was a long time since he had envisaged a German victory. Now suddenly it was a real possibility. It gave him a strangely mixed feeling.

He fell asleep at last, but hardly seemed to have closed his eyes when he was awoken by a tremendous crash as the barn door was thrown open.

'For Christ's sake, help me!' roared Viney's voice.

It took some time to adjust his eyes to the dim light. Then he saw standing on the threshold the big Australian, his face twisted in grief.

And in his arms like a child's doll lay Josh, his limbs slack, and a scarf of scarlet blood around his throat.

It took several minutes to persuade Viney to put the boy down. Lothar then quickly established that he was alive and breathing regularly. His eyes opened and there was recognition in them. He tried to say something but no words came. Lothar guessed at his meaning and said, 'Nicole, come where he can see you.'

The girl who was sobbing in her mother's arms came forward and knelt beside Josh and took his hand, and he visibly relaxed.

Lothar set about cleaning the wound.

Viney demanded impatiently almost as soon as he had started, 'How is it, Fritz?'

'It could be much worse,' said Lothar. 'I don't think that any metal slivers have stayed in there and it missed the big vein, thank heaven. But it needs proper medical attention. Quiet, Josh. Don't try to speak. Just rest.'

The boy was opening his mouth once more but nothing was coming out but a rasping croak. Lothar pressed his hand to his shoulder till he subsided into stillness. Then the German turned his attention to the girl whose distress was now under control.

'He'll be all right, Nicole. Believe me. Now tell me what happened,' he commanded.

They all listened in silence.

Hepworth said, 'Bloody hell. Does this mean we've lost t'war?'

'I doubt it,' said Lothar. 'It's too early to talk of losing and winning. If they've reached Barnecourt without coming past here, that must mean a breakthrough to the north. They probably followed the river down its valley. It may be just a few to start with, but once they start coming in any number, the British will have to fall back on either side or else risk having their flanks completely turned.'

'What's that mean to us?' demanded Groom.

'It means that in not many hours we could see British troops retreating through here with German troops not far behind,' said Lothar grimly.

'What's best to do then?' asked Hepworth. 'Viney, what do you think?'

It was an instinctive turning to the man of action in time of need, but the big Australian who had slumped against the wall just shook his head.

'Best listen to Fritz,' he said indifferently. 'This is his show.'

The heavy irony which would normally have underlined such a remark was completely absent. For once, Lothar wished it hadn't been. To have an alternative, or at the least to have a source of mocking objection, was suddenly very desirable. But this was no time for indecision. He got control of his thoughts by pushing them to an extreme of cold logic in the light of that political idealism which the last couple of years had internalized from mere gesture to motive. As a group, they were lost. As individuals, of the men only he and Viney had any real chance of survival. And in the real European conflict which would be raging long after this present useless carnage was at an end, he alone might have a usefully active part to play.

Ergo he should leave now and devote all his energies to saving himself.

Sic probo!

And he smiled at the impossibility of it all, smiled at his own ineffectuality as a revolutionary idealist.

He spoke carefully, first in English, then in French, to make sure that everyone understood.

'The most, the *only* important thing is the protection of the women. Ideally they ought to move from here, but Madame Alpert is too ill to move. Madame Gilbert will not leave without her. And Nicole here will not leave without Josh who is also at present not well enough to move.

'So they must stay.'

'And what happens when the troops arrive?' asked Groom.

'If the British simply withdraw past here, that is well. We hoist a white flag so that the Germans may at least not shell the farm.'

'And if the British want to use the buildings as a stronghold in a rearguard action?' asked Groom.

'I will try to dissuade them,' said Lothar. 'At the least, they may have medical help and some mechanical transport that will be able to take the women to safety.'

It was a forlorn enough hope for the women and no hope at all for the rest. Lothar concluded, 'There is, of course, no need for others to stay. I speak French, English and German, so whatever negotiations are needed, I can help with. Josh must stay and would I think wish to, even if he were not wounded. But the rest of you . . . there will be much confusion everywhere, I think. It may be a good opportunity to make for the coast. Or Paris. A man can vanish in Paris, even an Englishman, perhaps. Heppy, will you perhaps help me get Josh on to his bed while the others decide?'

The Yorkshireman nodded and smiled his acknowledgement that Lothar had not deemed it necessary to state that he would also be staying.

Fox, Nelson and Groom and three other Volunteers who had made their way back after the massacre at Barnecourt went into a muttered conference at the far end of the barn. Viney did not join them. Instead, he watched the two men carrying Josh into Nicole's sleeping chamber but did not attempt to interfere.

Blackie Coleport who'd been watching all this like a spectator at a play let out a single peal of harsh laughter and then began to recite in a loud, almost inhuman voice.

*'Away in the gloomy ranges, at the foot of an iron-bark,
The bonnie, winsome laddie was lying stiff and stark.'*

Viney straightened up and went towards him.

'Shut it, Blackie,' he said with a return to his most belligerent manner. Then he added in milder tones, 'The boy'll be all right, you'll see.'

'Not your pretty little boy I'm talking about, Viney,' said Coleport, grinning. 'It was old Patsy I had in my mind.'

Viney's huge fist clenched momentarily. But it had already relaxed before Lothar interrupted.

'Please, I suggest you decide by dawn. But after that, anyone who stays stays under my command. It is essential.'

But it was just before dawn that their minds were made up for them.

Groom was standing guard. The night was full of explosions and rocket lights and on several occasions he heard the sounds of troop movement and once the neighing of horses and creaking of wheels as though a field artillery unit were on the move, but always half a mile or so distant and always fading away to the west.

Then as the day grew grey, he spotted movement on the ridge immediately above the farm. Whoever was coming down there could hardly avoid the buildings.

He returned to the barn and alerted the inmates.

'Quickly,' Lothar commanded the men. 'Out of sight. Hurry!'

They concealed themselves as best they could up in the rafters of the barn where the few boards of the old hayloft which hadn't been used for building or as fuel remained. A few minutes later they heard someone cautiously trying the door.

Lothar, stooped and using a stick like a man at least twice his age, opened it.

Immediately he was thrust back at the end of a revolver held by an exhausted-looking English lieutenant.

'Stand still. Who's in here? Come on. Speak! *Dîtes-moi, qui est ici, monsieur? Parlez vite!*'

Lothar replied, breaking from time to time into English as

execrable as the lieutenant's French, that his wife, mother, son and daughter were here, no one else, and the old lady was sick, and his young son was wounded, and he begged the lieutenant for transport and medical help to get them to hospital.

It was a nice piece of psychological judgement.

Within minutes the Englishman was apologizing for not being able to help.

'No medics, *pas de docteur, pas de transport aussi, monsieur*,' he said. And he begged permission for his men to rest in the farmyard for a while and refresh themselves with water from the pump.

They were, Lothar gathered, the remnants of a unit who had held off the advancing Germans at the expense of heavy losses and finally, giving up hope of reinforcement, they had withdrawn under cover of darkness.

'They won't be far behind us,' explained the lieutenant. 'Listen, monsieur. We are going now. *Nous départons!* You really ought to get out too. But I see your problem. Here's what I suggest. Put up a white flag. *Un drapeau blank, comprends?* Otherwise they might drop a few shells just in case. But you'll be all right with a white flag. Some of the Boche are quite decent chaps, you know. Mustn't believe all you read in *John Bull*. Sorry, don't suppose you get *John Bull* here, do you? Look, we'd better be off. *Bon chance, monsieur. Au revoir!*'

He touched his cap and left. Lothar shook his head in amused disbelief. There was something about the English middle classes, he didn't know what it was, but he feared it was going to prove a large stumbling-block to the spread of world revolution.

Shortly after the English had left, a steady bombardment of the ridge began. The artillery Groom had heard moving in the night must have relocated themselves somewhere to the south and were banging away at the high ground in an effort to inhibit the German advance. But Lothar after listening for a while knew how few guns were involved and guessed that any delay would not be for long.

'Looks like we've got to stay now, Fritz,' said Groom.

'We're right in no-man's land here. Go either way and we're likely to get shot.'

Lothar thought a while, then said, 'Yes, but we must not risk you being found as though Madame Gilbert is hiding you. Either you hide in the byre, and pretend you know nothing of the family if the Germans catch you. Or else you stay here, and when the Boche arrive, we can tell them that you are British soldiers who wish to surrender. Who knows? Perhaps you have a better chance of returning home from a prisoner-of-war camp than from a British military prison.'

'You can say that again,' said Groom feelingly. 'What say you, Nelson?'

'I say beggars can't be choosers,' said Nelson. 'Is there any grub going? Never know when we'll get our next square meal. It's all black bread and potato soup in the Boche lines, ain't it, Fritz?'

Madeleine prepared some food while Lothar and Nicole tended Josh's wound. The boy seemed stronger in himself, though his speech was still badly impaired. The old lady was still fast asleep, and it was Lothar's guess that she was slipping into a coma from which there would be no return, but it was pointless saying anything to Madeleine.

They settled down to a silent breakfast, silent as far as speech went, that is. Outside the German guns were seeking the range of the British battery and with far greater power at their disposal.

'What about this white flag, Fritz?' asked Hepworth.

'Yes, it will soon be time. Wait till the guns stop,' said Lothar. 'Once the troops are advancing, that is the time. A white flag means something to soldiers on foot. To artillery observers it is just a sighting aid. The shells are nowhere near us at the moment. Let us not do anything to attract them.'

It was well into the morning before the guns went quiet, or at least relatively so.

Madeleine had prepared a large sheet which they now nailed on to a pole which consisted of two six-foot lengths of narrow planking lashed together.

Hepworth and Groom climbed to the barn roof while some

of the others pushed the flag up from below.

Viney looked on with the indifference of one to whom all flags were meaningless. Lothar had not yet cared to probe into the Australian's intentions, not doubting that they would become clear in his own time.

Suddenly Hepworth on the roof cried, 'Hold on! Someone's coming.'

'Be quick then!' shouted Lothar anxiously.

'No! Not the Huns. Up the valley. It looks like a motorbike and sidecar.'

For a second Lothar was undecided.

Then Viney said laconically. 'If it's the Brits coming back, you'd best not have them jokers cluttering up the roof.'

He was right of course.

'Quickly, get down!' ordered Lothar. 'Everyone into the barn. Hurry!'

The two men on the roof scrambled down. The flag was resting against the side of the barn. Lothar gave it a push so that it toppled to the ground, then ushered the others through the door.

'How many?' he asked Hepworth.

'Don't know. Only saw the mo' bike,' said Hepworth. 'But there could be a whole regiment behind them.'

It sounded an unlikely formation to Lothar but, large or small, the group that was on its way spelt danger.

The only consolation was that whatever soldiers of whatever army arrived next at the farm, they would surely be far too preoccupied with large-scale military matters to have many thoughts spare for deserters.

Lothar for once was absolutely wrong.

Jack Denial had been neither displeased nor relieved when an ancient and very bellicose infantry major and a couple of platoons of riflemen had joined his irregular force in Barnecourt shortly before dawn.

The major was no more certain than anyone else of the full extent of the German breakthrough, but was clearly hopeful of grasping some late glory to rejuvenate an undistinguished and dying career.

'The next chappies who come over that ridge will be Huns,' he said with grave certainty as he made his dispositions. 'We're short on ammo and we've only got a couple of Lewis guns to back up the rifles, so we'll let them come right up to us before we let fly.'

It was as well he made this decision because, when the first approaching figures were observed a mile or more away, he would certainly have given the order to blast away had any artillery pieces been at his disposal. It was Sergeant-Major Maggs peering through Denial's binoculars who said, 'Them's some of our lads, sir.'

Denial looked, then reported this to the major who was suitably disappointed.

'Better make contact, let 'em know we're here, I suppose,' he said. 'Silly asses might take us for Boche otherwise.'

Denial, glad of an excuse to separate himself from this embodiment of all that he found most distressing among his brother officers, volunteered to go forward on his motorbike. Sergeant-Major Maggs sat in the sidecar with a white flag fluttering to reassure the approaching troops.

Fifteen minutes later he was talking with the lieutenant in charge of the retreating men.

'They'll be here soon,' forecast the young man. 'You say they've been in Barnecourt already?'

'Just an isolated unit. We were able to push them out. But they'll be back.'

'Yes,' agreed the lieutenant. 'They won't make the same mistake twice. Next time they'll blast the place flat. Any civilians left down there?'

'No, they've all packed up and gone.'

'Very wise. There are some people back at an old farmhouse up there. I tried to persuade the chappie to leave, but there's a sick old woman evidently, and she couldn't be moved. I hope they'll be OK.'

'I think I know the place,' said Denial, frowning. 'This chap, very old fellow, was he?'

'Not particularly,' said the lieutenant. 'He walks as if he were an old man, but he had quite a young face. Why, what's your interest?'

'Oh, nothing,' said Denial. But he was thinking: The boy had been killed, hadn't he? And hadn't he heard that the old man had died also? If not him, then who . . . ?

'Well, we'd better be on our way. We've rested long enough,' said the lieutenant. 'I don't want to be caught out in the open again. Coming?'

'Not yet. I think we'll scout around a little more, see what we can see,' answered Denial vaguely.

Sergeant-Major Maggs did not look too happy at the idea but said nothing.

'One thing,' said Denial as the weary soldiers rose to their feet once more. 'The major back there, he's not in touch with anyone as far as I can see. And unless he gets orders to the contrary, I think he's set on holding Barnecourt to the last drop of everyone's blood. Goodbye now.'

He sent the motorbike up the track.

Glancing back a minute later, Maggs said, 'They don't seem to be going down into the village, sir. They've moved off to the south, like.'

'Have they?' said Denial indifferently.

'And what about us, sir? Aren't we heading back to Barnecourt?'

It was a protest rather than a genuine question with the village rapidly receding behind them.

'Eventually perhaps, Mr Maggs,' said Denial. 'But first of all I want to see if I can make contact with another force. What was it that dying Welshman called them? *Viney's Volunteers!*'

8

It was a bright spring day with high clouds being swept along by a gusting wind, but it was the smaller clouds no bigger than men's fists punching away at the lower sky, that held Sergeant-Major Magg's attention. No longer did they mark a definite horizon but seemed to flourish in all directions, and their accompaniment of gasping, grunting explosions swelled and faded like birdsong in the uncertain wind.

But one thing was certain in Maggs's mind; they were heading in the one direction where there were definitely Germans. He didn't know whether to be glad or distressed when they finally reached their destination. Here they would at last be able to turn back, but not before the pair of them had checked this lonely farmhouse to see if it contained some of the most desperate criminals ever to desert from the Army. This was not how Maggs had envisaged his next encounter with Viney and his gang.

And his discomfiture was complete when Captain Denial said, 'Take care of the bike, Sergeant-Major,' and walked up to the barn door and calmly knocked.

After a while a woman answered, her narrow strong face expressing neither curiosity nor surprise, though not even her powers of control could conceal the great weariness that was eating through into the very core of her being. She did not speak.

'Madame Gilbert,' said Denial in slow, precise French. 'We have met before, madame, perhaps you recall. I am Captain Denial, Assistant Provost Marshal of this area. Would you please to tell Sergeant Viney of His Majesty's Australian Army that I am here and would like to discuss certain matters with him?'

For answer, Madeleine opened the door wide and stepped aside. Denial removed his cap and entered.

Maggs, who had wheeled the motorbike into the lee of the

byre and was standing with his rifle at the ready, shook his head in a gesture that went beyond amazement.

It came as no surprise a couple of minutes later when a burly man appeared at the barn door, looked in his direction, beckoned and said in a Yorkshire accent, 'You'd best come in too.'

'Hands up,' ordered Maggs instinctively.

'Don't be bloody daft,' said the Yorkshireman and disappeared.

Maggs waited a moment, then with a despairing shrug, followed him into the building.

As he stepped through the door, his arms were seized and his rifle removed from his unresisting hand. He was then pushed firmly but not roughly through another door and miraculously what had been a barn became a farmhouse living-room.

A fire burned in a grate. There was a smell of stewing vegetables from the pot suspended over it. In a square-shaped bed in a corner near the fire lay the waxy-pale figure of an old lady. And in front of the fire, seated by a table on a low stool, was Captain Denial.

Opposite him sat a big man whom Maggs recognized instantly.

'Hello, sport,' said Viney. 'Nice of you to pay us a visit. Take a pew.'

Maggs sat. There were other people in the room. His heart sank as he looked at them. All the stories he'd heard of subhuman cannibalistic gangs living on the detritus of ancient battlefields suddenly seemed utterly credible in the presence of this bunch of unshaven, wild-eyed, ragged-clothed madmen. No one spoke, and a silence grew inside that room that ultimately became more oppressive and intrusive than the constant rumble of artillery fire.

'What's your game, sport?' asked Viney finally.

'I just thought I'd like to be sure you really were here,' said Denial. 'It's a satisfaction to me.'

'How'd you know?'

'Small clues. A dead corporal outside HQ. Something a dying Welshman said.'

'Taff? Taff's dead?' said Hepworth in dismay.

'I'm afraid so. The Germans in Barnecourt.'

'Oh shit. He were a grand chap.'

Viney brushed aside these irrelevancies and said, 'I mean, what's your game, just walking in here? If you thought mebbe this was where you'd find me, you're taking a bit of a chance, ain't you, Cap?'

'I doubt if in present circumstances I could have got an arresting party together,' said Denial. 'The Sergeant-Major and I didn't seem to have the fire-power to take you ourselves. But I did want to be sure.'

'Well, now you're sure. What next?'

'Nothing. We sit and wait. You ran away from the war, Sergeant Viney. And your comrades here too, I presume. But you didn't run far enough. The war's coming after you. You can't avoid it now.'

'No,' said a tall blond man with bitter vehemence. 'You are wrong, Captain. None of us here ran from the war. We took the war with us. This is the dreadful thing. We took the war with us. In here. And here.'

He tapped his head and his breast.

Denial looked at him curiously.

'You, I take it, are Count von Seeberg?' he said.

Lothar was taken aback by this identification.

'My father is Graf von Seeberg, yes,' he said.

'Your father is dead, Count,' said Denial. 'I'm sorry.'

'What the hell's all this?' growled Viney, whom the new situation seemed to have restored to his old strength. 'You some sort of aristo, Fritz? And you talking like a bloody communist! Jesus!'

Lothar, pale with shock at this sudden intrusion of his old life, said, 'How do you know these things?'

'Lieutenant Cowper of the cavalry. You'll recall him, I dare say.'

'Ah, of course,' murmured Lothar. 'So. An Englishman's word? He promised not to speak of me.'

'He didn't. He hardly could, could he? It was just a note on a scrap of paper in his pocket. I checked the name.'

'Why do you say he could not speak?' asked Lothar.

'Why? Because one of your murdering bastards had stuck a knife in his ribs, hadn't he?' broke in Maggs, who had got over his first sense of intimidation.

Lothar's expression of horrified surprise convinced Denial he genuinely knew nothing of this. The German looked accusingly at Viney, who in his turn gazed steadily at Coleport, who laughed.

'You're not the only one handy with a weapon, Viney,' he mocked. 'Though me, I only use mine on the enemy!'

Viney moved towards him, this time clearly intent on knocking him down.

'I shouldn't bother,' said Denial. 'Listen.'

Out of the wallpaper of shellfire they traced a single rogue line of noise screaming towards them. The shell was a big one and even though it exploded a little distance beyond the farm, they felt its power through the trembling of the floor and walls.

'The war *is* coming,' said Denial, with some satisfaction. 'I will be curious to see if you can murder your way out of this.'

Another shell fell even closer.

Everyone was on his feet except Denial.

'They are targeting the farm,' cried Lothar. 'We have not put up the white flag.'

'Probably too late for that, sport,' cried Viney.

And as if in support of his assertion, there was a huge explosion almost on top of them, breaching one of the outer walls of the barn, ripping part of the roof away, and sending everyone crashing to the floor.

Madeleine was the first to her feet, coughing in the smoke and dust-filled air as she staggered to her mother's bed. The old lady was lying staring with alert and unpuzzled eyes at the puff-ball-filled sky which had suddenly appeared over her head. Hepworth, his head bleeding from a graze, came to join her.

'*Maman!*' cried Madeleine.

'She's all right,' assured Hepworth. 'Look, she's smiling.'

And it was true. As if in shy acknowledgement of this sudden gift of the sky, a slow smile was dawning on the old lady's face.

'The young ones!' cried Madeleine with renewed alarm. 'Heppy! What of Nicole?'

The Yorkshireman immediately made for the sleeping chamber where Josh lay, constantly tended by Nicole. But Lothar was there already.

'They're all right,' he reassured Hepworth. 'No damage. At least, not physical.'

What he meant was immediately clear when Hepworth looked at Josh. The boy had struggled upright and his bright, feverish eyes were expressing the terror brought on by the explosion even though his mouth could not articulate it.

'What's best to do, Fritz?' asked Hepworth calmly.

Lothar listened. It was a choice between staying or going. If the German artillery were bent upon razing the farm buildings to the ground, then flight was the only answer. But it seemed to him that the hit might have been almost accidental and that the guns were ranging much further down the valley. Which could only mean that ground troops were not far away. So again the question – to go or to stay?

He could see no reason for changing the former plan.

'Let's get that flag up,' he said. 'Quickly. Hepworth! Groom! Come.'

He rushed out of the barn followed by the other men. Maggs followed them as far as the door, looking bewildered at the turn of events that fate and the lunacy of Captain Denial had thrust upon him.

Hepworth cradled his hands, Groom stepped in, and with much shoving and pushing from the others, he was elevated once more to the barn roof.

'Now the flag,' cried Lothar.

Hepworth hesitated.

'What's the matter?'

'Now it comes right down to it, I'm not right sure I want to surrender to a load of bloody Jerries,' growled the Yorkshireman. 'Begging your pardon, Fritz.'

'Do you want to see the women shot down or blown up?' demanded Lothar savagely.

'No. No, of course not,' said Hepworth, and bent for the flag.

There was a sudden crackle of machine-gun fire, a short sharp scream, a thud.

Something trickled down from the roof and splashed on Lothar's shoulder.

'Jesus Christ!' cried Hepworth.

They looked up. Groom was slumped over the roof eaves, dripping blood and brains from his ruined head. Nelson who was standing next to Lothar retched with horror but did not seem able to tear his eyes away.

'The flag, we must show the flag,' cried Lothar as another burst of machine-gun fire raked the barn roof.

Hepworth and Lothar raised the makeshift flagstaff and brought it to rest an an angle against a rusty metal hook which protruded from the side of the barn, so that the flag stuck out from the corner for a distance of two or three feet.

'That'll have to do,' said Hepworth.

The move at first seemed counter-productive, attracting a hail of fire till someone among the approaching enemy identified the nature of the activity. Gradually the firing ceased.

A voice cried, 'Hey, Tommy! Let us see you.'

The men looked at each other, then Lothar said, 'Go on, Heppy. It is best for Madeleine, believe me.'

'All right, Fritz,' said Hepworth. 'If you say so. Come on.'

Lothar smiled and shook his head.

'You are surrendering as a prisoner of war, Heppy. I, you will recall, am a German deserter. I will not be welcomed so courteously. While they take the rest of you, I will try to slip away.'

'Don't be daft! You've no chance!' expostulated Hepworth. 'Give yourself up. Tell 'em you're English, then look for a chance to escape behind the lines.'

Lothar shook his head again.

'No, it is better this way.'

'Show yourself, Tommy,' commanded the voice once more. 'Or we shoot!'

'Go on,' urged Lothar. 'And good luck, Heppy.'

He turned away and found Viney at his side.

'You too, Viney,' he said. 'For Josh's sake, do this. Tell him . . . tell him, good luck.'

He moved swiftly away across the yard, keeping the barn and the byre between himself and the Germans. Hepworth, clamping his hands on his head, stepped out from behind the barn. Nelson and the three men who had returned with Nicole and Josh watched in trepidation but there was no outbreak of firing.

'Still, Tommy!' came the command. 'Your friends also. Make hurry!'

Viney said, 'Nelson, you get the others out here, sport. OK, you jokers. Move!'

'You're surrendering, Viney?' asked Nelson.

'Too bloody true.'

Reassured by this, the three Volunteers advanced from behind the protecting wall with their hands raised while Nelson went into the barn.

Maggs appeared at the door.

Viney grinned at him and said, 'You coming, Sergeant-Major, or are you one of them pommy heroes?'

Maggs withdrew without answering. Viney laughed, raised his hands and began to walk.

Behind him a voice called, 'Hey, Viney, you're forgetting something.'

It was Coleport in the doorway with Viney's Luger held awkwardly in his left hand.

Viney said, 'Chuck it away, Blackie, and step out here. You'll live to make a few sheilas too sore to sit down back in Melbourne.'

'Patsy won't be seeing Melbourne again,' said Coleport. 'But he left you a message.'

He brought the pistol up and squeezed the trigger. With his right hand he was a first-class shot, but the weapon was unsteady in his left and the bullet flew over Viney's head. Viney dropped down and rolled away across the yard till he reached the protecting wheels of the farm cart.

Coleport made no attempt to pursue him.

Laughing crazily he cried, 'All right, Viney, we'll let Jerry do the job!'

And running out of the cover of the barn, he began loosing off the pistol indiscriminately in the direction of the Germans.

The response was instant and devastating. Coleport was riddled with bullets in a couple of seconds. The trio of Volunteers, half paralysed with shock, only had time to scream *Kamerad!* and thrust their arms higher before they too were mown down.

Only Hepworth who had gone diving behind them for the shelter of the barn at Coleport's first shot managed to escape, but not without taking a bullet in his thigh.

'The fucking madman!' he cursed. 'Jesus! It feels like my leg's been chopped off.'

Viney came scrambling to join him as the fusillade died away.

'That mad mate o' thine,' gasped Hepworth. 'He's got us all killed!'

'Let's have a look at you, sport,' said Viney, hacking at the Yorkshireman's trousers with a huge jack-knife. 'They hit Blackie, did they?'

'Hit him? They cut the bugger in half. Christ, Viney! You're not trying to take my leg off with that bloody sabre of thine, are you? Aye, your mate's dead, and bad luck to him wherever he's gone. What the hell's that?'

The German fire had ceased and into the silence crept a voice, almost inaudible at first, but gathering in desperate strength.

> '. . . *he laughed as he lifted his pistol-hand,*
> *And he fired at the rifle-flash!*'

'It's Blackie,' said Viney with a slight smile. 'Take more than a few bullets to down a Digger, Heppy. It's that sodding poetry that'll be the death of him.'

The Germans must have been disconcerted too by the unexpected rhythmical chanting, but as Coleport staggered to his feet, crying '*Then out of the shadows the troopers aimed . . .*' they took this as their cue. A machine-gun chattered. The words at last came to an end.

Viney said quietly, 'Cheers, mate. Give my love to Patsy.'

'You're a cold bastard,' gasped Hepworth.

'Think so? Christ, this is a bit of a mess. But not to worry, here comes the doctor. What brings you back, Fritz?'

Lothar had dived for cover at the first outburst of firing and had begun to work his way back when the second began. Obeying an impulse stronger than that to self-preservation, he had then come at a run.

'What is happening?' he gasped. 'Heppy, you're hit! The white flag, they did not honour it?'

'They was provoked,' said Viney. 'This needs dressed. Let's get him back into the barn.'

Hepworth, lying on his back, suddenly screamed, 'Look out! Flying pigs!'

The others looked up, then flung themselves flat as they saw the mortar bomb spinning slowly overhead. It came down on the byre, exploding with a force not quite powerful enough to blow the structure apart, but with a blast strong enough to weaken the internal supports so that the walls collapsed slowly inwards, revealing the MP's motorbike standing miraculously unscathed on the far side.

Even as the men began to rise, the next bomb hit the barn. It sounded and felt like a direct hit. A piece of stone ripped at Lothar's cheek but he ignored it and sprinted to the door, terrified of what he might find inside. Sergeant-Major Maggs was slumped face down across the threshold. The second explosion had driven a fragment of metal deep into his spine. He was not dead but for the moment could move nothing except his eyeballs and these ranged slowly from left to right and back again over the few inches of floor, which in his present posture constituted his entire visible universe. Out of the smoke which filled the barn, Denial came staggering in search of the sergeant-major. Lothar pushed by him with scarcely a glance. For a second it seemed to him that a direct hit must have killed many of those within. Then he heard voices and one of the women crying and as his eyes penetrated the smoke, he realized that the damage was mainly confined to the end of the barn furthest from the interior rooms, though their drystone walls, built for privacy rather than to withstand wind and weather, had slipped crazily askew in the blast.

Nicole appeared at the door of the sleeping chamber where Josh was lying. Her eyes were wild with fear and worry but her voice was surprisingly controlled as she cried to him, 'Lott, what's happening? What shall we do?'

'How's Josh?' he demanded. But even as he spoke, behind the girl Josh himself appeared, his face almost as pale as the bandage round his throat. At least he was on his feet even though he was leaning heavily against the wall and Nicole, with a cry of alarm, turned to support him.

In the living-room he found the source of the weeping. The old lady was still lying with wide eyes peering at the sky, but she wasn't seeing anything. One shock too many, guessed Lothar, but at least her end must have been painless. Madeleine was lying across the bed, her head buried in the pillow next to her mother's, her shoulders shaking as she wept her pain and sorrow.

Nelson lay on the floor, stunned but otherwise little hurt, while Fox was absurdly trying to crawl beneath the bed in a terrified search for shelter.

Another explosion just outside the barn rocked everything, as a surface ripple moves the bottom of a pond. It was fortunate that even when a range was found, the curious floating flight of mortar bombs made then an inaccurate weapon. But it would take only another couple of near misses or one good hit to bring everything down. Viney came through the door with the wounded Yorkshireman leaning heavily on his broad shoulder. Behind him came Denial dragging the wounded sergeant-major.

'No, not in here! Do not bring them here! Out! We must all get out!' shouted Lothar.

Denial looked at him with his calm emotionless eyes.

'What happened to the surrender?' he enquired.

'We changed our minds,' said Lothar. Unexpectedly Denial grinned, the merest flash.

I made a sort of English joke, realized Lothar. But there was no time left for humour. The next bomb came, landing just short of the outer wall. The blast cracked the already weakened stones, and sent an avalanche of rubble spilling down into the bed.

'Madeleine!' screamed Hepworth, hurling himself forward, his wound forgotten in this new agony of fear.

The old lady had disappeared almost entirely and Madeleine beside her was half invisible beneath the pile. Most of it was too small to be of much consequence, but a large slab had fallen across her head and for a moment it seemed impossible to Lothar that she could have survived. But when the desperate Yorkshireman wrenched the stone clear, the crushed and bloody head moved and the face that turned up to him, though white as the hair which was all that was visible of Madame Alpert, was still informed with life and intelligence.

He tried to gather her in his arms to lift her from the bed but she pushed him away, and though the feeble thrusts of her nerveless fingers had no more force than a newborn child's, the strength of her will held him back.

'Nicole, save Nicole,' she whispered, the words coming soft and slow and almost as imperceptible as the unfolding of a flower. 'And the English boy. Take them away from here. Make them safe. Hurry.'

'Madeleine!' cried Hepworth in an agonized voice. 'I'll not leave you, I'll not . . .'

'Go!' she commanded. Their gazes locked, his striving to hold her here, hers striving to force him away. It was the woman who broke off from the struggle, but not in defeat. To Lothar, watching, it seemed as if she acknowledged that not even her magnificent will was able to drive Hepworth away while she lived, so she turned it inward upon herself, damped down her still fervent life-force, and in a few moments died.

Nicole and Josh had also come into the ruined room and the young girl stared at the still form on the bed, her face set with shock. Josh's weight was on her shoulders, holding her still, but, fearful that if she fainted they would both crash to the floor, Lothar took the boy from her. Released of her burden, she rushed forward to the bed. Hepworth caught her as she tried to fling herself on the slack still body.

'Kiss your mam goodbye,' he said sternly, his broad, open face like a marble mask of grief on a tomb.

She stooped to the still warm face and kissed the pale lips, one long kiss, then Hepworth drew her back.

'Come on,' he said. 'Time to go.'

'Go?' said Lothar, feeling Josh almost a dead weight in his arms. 'Where? How?'

And Nicole cried, 'I will not go without Josh!'

'I promised for you both,' said Hepworth with a desperate determination.

He tried to move, but his wounded leg unstressed by the despair of lost love, now gave way and he slid to the floor, cursing.

Nelson who was sitting upright now, massaging his bald head, said, 'I'd sell my old lady for a pint of beer.'

He got on to his hands and knees and crawled towards Fox who was crouched in terror against the bed, his body jerking spasmodically as a new hail of German bullets rattled against the barn wall.

'Here, Foxy, old son, calm yourself down, my boy,' he said.

He put his hand on the other's shoulder and this seemed to have a calming effect. The shaking diminished and Fox suddenly said in a strange, high-pitched voice, 'You take the patrol on, Sergeant. I'll just check that the wire's been cut for us over there.'

'That's right, my lad,' said Nelson reassuringly. 'We're all right here. Jerry's dropping back. Listen. You can hardly hear him now.'

Lothar remarked as he had often done before how the best medicine for fear was to find someone to look after, and at the same time it struck him that Nelson was right. The German onslaught had slackened off, to be replaced by the new sound of an engine. It seemed to be somehow above him. Realization came and he looked upward just in time to see the cruciform silhouette of a biplane gliding across the patch of blue exposed by the damaged roof.

'It's a plane,' cried Nelson. 'It's one of ours.'

The plane was out of sight now, but suddenly they heard the chatter of its machine-gun as it flew low over the attacking soldiers.

Viney moved out of the living area and headed for the ruined end of the barn. He returned a moment later.

'The Fly-boy's doing a good job,' he said. 'Jerry's out in the open out there and that's no place to be when someone's flying around above you. They're scattering back up the slope to where there's a bit of cover.'

It was only a temporary respite, they all knew. Shortage of fuel or ammunition would soon send the biplane back to its base.

'What strength would you say the Germans were, Sergeant Viney?' said Denial.

The Australian turned his hard gaze on the APM and said, 'I recollect telling you once before, I resigned from the Army.'

'I remember. That leaves us simply as captive and captor. Which is which is not at present clear and it's not a very helpful relationship just now in any case. Agreed?'

'Mebbe.'

'What's beyond doubt is I'm the senior officer, here. Unless you, Mr Fox, got beyond subaltern? Mr Fox?'

The man they knew as Foxy slowly raised and shook his head. Nelson said in amazement, 'Foxy? You an officer? Fuck me!'

His surprise was comic. And suddenly Lothar understood Fox's pathological hatred of officers. The poor devil! Terrified of letting his humble origins show among his brother officers; even more terrified of revealing he was an officer among the deserters! This war had shaken up society as violently as it had disrupted the earth. The countryside would show its changed contours for centuries, but with God's help this might be a step towards a new and more level society.

Viney said, 'Fox an officer? Well now. Takes a one to smell a one, Denial.'

Denial said impatiently, 'All right, Viney. Just give a simple answer to my simple question.'

'That's different. Company strength, I'd say. No more, certainly.'

'Yes, that fits in.'

'Fits in with what, for fuck's sake?'

'The Germans have broken through,' said Denial. 'They surprised everybody, not least themselves, I think. In some places units have pushed forward so rapidly that they seem to have lost contact with their main force. We mopped one of these up in Barnecourt last night.'

'You!' exploded Viney. 'All you sodding redcaps could mop up is some poor bastard's blood you've spilt on a cell floor!'

Denial ignored this.

'It's important to resist at all points,' he said. 'If the German breakthrough keeps up its momentum, they'll be in Paris next week, the French'll cave in, and we'll all be pushed into the sea. Every obstacle they meet is important; every ounce of resistance vital.'

There was a moment's silence, then Viney began to laugh. He laughed loud and long and with an unmistakably genuine amusement.

'You don't give up,' he finally spluttered through his laughter. 'You're a different race, you jokers, you really are! Can't you get it into your heads? We don't care! That's what getting out and running off means. *We don't bloody care!*'

Denial was unmoved by the outburst.

He said, 'I presume you care about each other? Not all of you are in a fit state to think of running again. Or don't you care about that either, *Sergeant*?'

'Well, sport,' said Viney reflectively, 'we're one of the younger regiments, Viney's Volunteers, and we haven't quite developed those fine old regimental traditions like you redcaps. Mind you, we don't go in much for beating poor swaddies with "Mr Dunlop" either, or making 'em run on the spot in full battle order till they drop, or keeping 'em standing outside the docks till they miss their leave-boat! No, we haven't learnt them fucking tricks yet!'

'Viney, the plane is going,' interrupted Nelson, staring

through the hole in the roof. 'The Jerries'll be back soon.'

Lothar who had been helping Nicole attend to Josh looked up from the young couple towards Viney. All the other Volunteers, including the terror-stricken Fox, were regarding the big Australian expectantly too.

He did not speak. They heard the aeroplane engine pass overhead and fade away towards the west, and still he did not speak.

'Viney!' said Lothar urgently. 'The young people . . .'

'*Viney!*' exploded the big man. 'Don't you bastards know anything else? That's all I ever hear. *Viney! Viney, make it better! Viney, pick us up! Viney, get us through! Viney, stop the fucking war!*'

He seemed to swell even bigger as he spoke, his arms reaching out menacingly, his fists clenched, his face working with a terrible rage. But Lothar realized with a cold sickness in his stomach that these were the gestures and the ragings of despair.

'It's all right, Viney,' he said softly. 'It's all right. It's not your fault.'

But the Australian's attention had been diverted.

'Where the fuck do you think you're going?' he bellowed.

Denial was stooping over Maggs and trying to raise him up.

'As you've made it clear that you've no intention of acting like soldiers,' he said contemptuously, 'and as I've no desire to stay and see what other disgusting course of action you can work out, I'm going to get the only real man here to a doctor.'

Some feeling had returned to Maggs's upper body and he could once more move his arms, but from the waist down he was still completely paralysed and his lanky form was a dead weight.

Viney watched Denial's efforts to raise him and said, 'You suit yourself, sport. It's a long bloody walk.'

Lothar said, '*Walk?*'

His gaze met Viney's.

'Jesus Christ!' cried the Australian. 'The motorbike!'

He rushed towards the door. Denial let Maggs slip back to the floor.

'Viney!' he cried.

Viney stopped and turned. Denial had his revolver out and aimed.

'You bastard. You miserable cowardly bastard!' said Denial.

He saw the big man look at him in puzzlement, saw a smile begin to pull at his lips, saw him open his mouth to reply, and suddenly he wanted to see and hear no more.

'You're not leaving anyone to die this time,' he cried.

He squeezed the trigger. The gun kicked in his hand. Viney spun away, but did not fall. He squeezed again but Lothar's diving figure bore him to the ground and the second bullet flew high.

'You fool! You madman!' cried Lothar. 'What are you doing?'

'Me mad?' gasped Denial, unresisting as Lothar dragged the pistol from his grip. 'It's you that's mad. He was running off. He was going to escape on the bike, couldn't you see that?'

'No!' cried Lothar. 'It is for the children we want the bike, for the girl here, Nicole, and for Josh.'

Something dripped on to Denial's face. He turned his head and realized it was blood. Viney was standing over him, his right arm shattered by the first revolver bullet and blood was dripping steadily from the wound.

Lothar rose and started to examine the arm. It was bad. The bullet had passed right through the biceps, shattering the bone and exiting at a downward angle just above the elbow.

'Viney,' he said. 'It is a bad wound. There's little I can do except try to stop the bleeding.'

In fact he was amazed that the big man could remain standing and show no pain.

'Do that then, Fritz,' said Viney. 'This joker's ruined my butterfly, hasn't he? Right through the middle of it.'

Quickly Lothar got a length of bandage and bound it above

the wound, twisting tight till the flow of blood slackened to a thin oozing. Viney's eyes never left Denial's face during this process.

'It wasn't because you thought I was running, sport, was it?' he said finally. 'I could see it in your face. It was because of that nurse.'

Denial nodded. It wasn't the total truth, but it was so much of the truth that argument was not worthwhile, not now, not with the end so close.

'I'm sorry about the nurse, believe me,' said Viney. 'I meant her no harm. Not that that's much help. There's millions dying out there and no one means them any harm. That'll do, Fritz. It's not to last long, for fuck's sake! Get out there and see if that bike's still in one piece.'

Lothar hurried to the door.

Hepworth said, 'You ever ridden a mo'bike with one arm, Viney.'

'Me? No. I wasn't going anyway. I'd stick out like an abo at the governor-general's garden party down there. Up here where the shit's flying, this is my kind of place, sport.'

He spoke with a kind of sad satisfaction.

'Who then, Viney?' said Hepworth. 'I'd like to oblige, but I'm tied up just now. Nelson?'

Fox seized hold of Nelson's tunic as if he would hold him back by main force and the Londoner shook his bald head.

'Not me, mate,' he said. 'Never ridden anything but the old lady in my life. Besides, there's no future back there for me. If I could talk my way into a couple of years chokey, I might think about it, but it's the firing squad certain for me. I never told anyone before, but last thing I did before slinging my hook was open this officer's gut with my bayonet. He was a right bastard, caught me sleeping on sentry, said he was going to make sure I got court-martialled. Well, they shoot you for that, don't they? So I thought: Fuck it! and slipped it into his belly neat as you like. It didn't seem any more than jointing a carcase at the time, but not a day's passed since without me wishing I hadn't done it. He was a bastard, but I wish I hadn't.'

He patted Fox's shoulder comfortingly and laughed. 'And

now I've got one of me own,' he said. 'Never knew them bastards could be scared too. Never guessed it!'

Lothar returned.

'Yes, it is a miracle, the bike is untouched except that it is covered in dust. God willing, it will start. Viney, Josh can travel in the sidecar, Nicole on the pillion.'

The girl looked up at the mention of her name and Lothar quickly translated for her, bringing a dawning hope into her eyes.

Viney said, 'But who's to drive the thing, Fritz?'

Lothar looked around the ruined barn. Nelson shook his head and turned away. Viney said, 'Looks like you're elected, Fritz, no opposition.'

'No,' said Lothar. 'You are wrong. There is someone else.'

Viney followed his gaze to Denial.

'Him?' he cried. 'You're joking!'

'No,' said Lothar. 'I might get them down to Barnecourt, but there will be English soldiers down there. A German out of uniform riding a military policeman's bike! There is no chance. It needs someone with real authority.'

'Aye, he's got that right enough,' said Hepworth. 'And what makes you think he'll not use it by locking Josh up under close arrest first chance he gets?'

'Perhaps he will. At least it will be a chance,' said Lothar.

'I can see them Jerries moving again,' said Nelson, who was peering carefully through the breached wall. 'They'll be on their way once they're certain that plane's not just hiding in a cloud.'

Viney said, 'What about it, Denial? You willing?'

The captain glanced towards Sergeant-Major Maggs, who shook his head and said, 'Don't bother about me, sir. I'm knackered, I reckon. I'll never stand my full height again and if I can't do that, I reckon I'd as lief see an end to it here.'

'Come on, man, for Christ's sake!' urged Viney. 'There ain't no time to consult a military manual.'

Denial rose and dusted himself down.

'You should understand, Sergeant,' he said steadily. 'I will not compromise on doing my duty.'

'That's up to you, mate,' said Viney. 'Here.'

He pulled something out of his pocket.

'If it'll help you any, these are the boy's dog-tags. Name of Routledge. Age about eighteen. Never harmed a soul in his life. He had a brother. They shot him for telling an officer to piss-off or some such deadly crime.'

'Routledge,' murmured Denial. Across his mind flashed an image of a slack-limbed, ashen-faced youth bound to a kitchen chair. Could it possibly be . . . ?

'And that's it,' continued Viney. 'Me, not being a greedy fellow, would reckon that one execution a family's enough, even for his gracious bloody Majesty. But I'm probably wrong. Incidentally, he can't talk. He's got papers on him saying he's Auguste Gilbert, medically discharged from the French Army. And the girl's in the family way, isn't that right, Fritz?'

Lothar, full of misgivings now that his idea was being carried out, nodded his head dumbly.

'Fritz,' said Viney. 'While the captain's making up his sodding mind, get young Josh there into that sidecar.'

Lothar helped the boy to his feet. He looked very tired and ill, but was fully conscious and he smiled at the German and tried to say, 'Lott'. Nicole steadied him from the other side and the trio moved slowly to the door. As they passed Viney, he raised his good arm and laid his hand on Josh's curly mop.

'It'll be right, son,' he said. 'You'll see. Any trouble, just call on Viney. I'll be around somewhere.'

Josh nodded, his face solemn. Then they moved on and out of the door.

Denial had watched without expression. Now he suddenly stooped down over Maggs and shook the man by the hand.

'Goodbye, Sergeant-Major,' he said.

Straightening up, he looked round the room and nodded curtly, then made for the door.

'Denial,' called Viney, coming after him. The movement obviously set pain shooting up his shattered arm and his face was grey.

'Denial,' he said in a low voice, standing close to him. 'Back in Aussie there's a kid with my name. Not mine, but that don't matter so much now, somehow. Tell my wife it was OK. Wrap it up somehow, you'll have the words. And if there's any way of squeezing any money out of the sodding army for them, that'd be a laugh. Look, here's a letter, it's got the address and everything.'

He took the much creased and tattered letter out of his pocket and pressed it into Denial's hand.

Denial nodded and said, 'I'll do what I can. I'm sorry about your arm.'

'Not to worry,' said Viney. And then with a sudden reversion to his most ferocious and terrifying manner he added, 'One thing more, Denial. Don't count on me not dropping by some day to check on just how you did your precious duty!'

'No,' said Denial. 'I'll not count on that.'

'Viney, the bastards are on the move!' cried Nelson urgently.

'OK, Captain, bugger off, quick as you can!' said Viney as Lothar returned.

'They're on the bike,' he said. 'The buildings will protect you for a little way. After that, you will be in the open for several hundred yards till you drop below the ridge.'

Denial pushed past him and disappeared from sight.

Lothar said, 'What will he do, Viney? Will Josh be all right?'

'He's getting the best deal we can give him, Fritz,' said Viney.

Behind them, Maggs laughed scornfully.

'You're dafter than you look, Digger!' he said. 'The captain'd turn in his own brother if he'd broken military law!'

'We'll see,' said Viney with surprising mildness. 'Now let's give the bastard some covering fire, shall we? Come on, Fritz.'

Under Viney's direction and with the fit helping the wounded, they took up their defensive positions along the breached wall. Maggs was propped up with his back against a pile of rubble and a rifle in his hands. Hepworth lay sprawled on his belly, also with a rifle. A thin ribbon of blood was unwinding slowly from his wounded leg. Fox had rejected the offer of a weapon with a cry of fear and he lay alongside Nelson with his head crushed against the bald man's chest.

While this disposition was taking place, they heard the kick-start of the motorbike being operated two or three times without the engine catching.

'I hope that bastard knows how to work that thing,' said Viney, sinking down beside Lothar.

The German had a rifle. Viney was armed with Denial's reloaded pistol and three grenades.

'The dust, it may have got into the engine,' said Lothar pessimistically. 'Perhaps I should go and try to help.'

As if in answer, they heard the engine suddenly burst into life. The oncoming Germans heard it too for suddenly there was a ferocious burst of fire.

'All right, sports,' yelled Viney. 'Let 'em have it.'

The defenders loosed off a series of ragged volleys and almost instantly were hit by return fire which made the previous burst seem inconsequential. They all pressed themselves as close to the ground as possible except for Sergeant-Major Maggs who in his upright position was particularly vulnerable. When the hail of bullets died away and they cautiously raised their eyes, they saw that the sergeant-major had been hit in the head. The top of his skull was almost entirely blown away.

'Now he's really got a red cap,' said Viney. 'Fritz, I didn't notice you firing.'

Lothar shook his head.

'I could not,' he said simply. 'These are my own people.'

'So they are, too.'

'Have they got away, do you think, Viney?' Lothar asked.

'God knows. Are you going to look? I'm not!'

Lothar shook his head.

'Sometimes it is best not to know things,' he said.

'You can say that again. Here, if you're not going to shoot, at least make yourself useful by reloading this thing for me.'

Lothar took the revolver and reloaded it.

'I'm sorry I cannot shoot with you,' he said. 'But it will

343

make little difference in the end, I think. Though even if I thought it would, I do not think I would change my mind.'

Viney nodded understandingly as he retrieved the pistol.

'Life's full of hard choices, Fritz, my son,' he said. 'Could you turn your head a bit to the left, mebbe?'

Puzzled, Lothar obeyed. In a short savage movement, Viney brought the pistol barrel crashing down on the exposed neck.

The German twitched violently in every limb, then became slack.

'What the fuck are you up to, Viney?' yelled Hepworth.

'No need for us all to get killed, Heppy,' said the Australian. 'Here, Nelson, give us a hand.'

Nelson slithered back, despite cries of protest from the quivering Fox. Under Viney's instructions he bound Lothar's hands and legs with lengths of bandage and rolled him into the furthest corner of the barn.

'Prisoner of war,' said Viney. 'Might work. And he's a clever cunt. He'll think of something.'

Nelson grinned and headed back to his spot. He must have momentarily exposed himself as he did so. There was a crackle of machine-gun fire. He gasped chokingly as a scarf of red wrapped itself round his neck. And then he fell dead beside Fox.

The former officer cried out in fear and rolled away.

'Nelson! Nelson, old chap!' he whispered hoarsely.

Then, as the realization dawned that there could be no reply, his face contorted with an unreadable confusion of emotions.

Suddenly he was on his feet, Nelson's rifle in his hands.

'All right, chaps, follow me! Over the plonk! Let's show the bastards!'

And he was away over the ruined wall, running drunkenly towards the enemy whose bewilderment let him cover five or six yards before they cut him down.

In the barn Hepworth said laconically, 'Looks like it's just thee and me, Viney.'

'Looks like it, Heppy,' said the Australian calmly.

'It's been a funny fucking life,' said Hepworth.

'You're not joking, sport,' said Viney in heartfelt agreement. 'It's a real pig!'

And then suddenly he smiled, a real smile a million miles away from his usual expression of savage amusement, and for a moment the smile sloughed off that impression of ageless, still, intimidating menace which was all that the world customarily saw, and he looked like the young man of twenty-four he was.

'But I'll not let the bastard beat me,' he said almost gaily. 'Nor these bastards either, if I can help it. Let's give them something to think about, Heppy!'

The two men began to fire. In a few seconds Viney had emptied the revolver and now he began to hurl the grenades. The enemy replied with machine-guns and bombs. Volleys and explosions filled the spring air with smoke and din, echoing around the broken walls and linking the ancient building with trenches and dug-outs, with ravaged woods and downbeaten townships, with shell-blasted minds and devastated lives all across the face of Europe.

Only for the riders of the motorcycle combination as it dropped out of sight below the ridge on its descent into Barnecourt, where German shells were already falling, did there seem a moment of peace and safety with nothing in the air more discordant than the creaking call of a partridge disturbed from its egg-filled nest. The wind lifted Nicole's hair and, despite everything, she felt the exhilaration of rapid movement. She looked down at Josh in the sidecar and he returned her gaze and her loving smile also, then closed his eyes and smiled again as he felt the sledge go whistling down the snowy fellside and himself warm and exhilarated in the certain arms of Wilf.

Only Denial did not smile. His enemy lay behind him, almost certainly dead. Alongside him rode a prisoner, guilty beyond any doubt. Ahead lay his duty, unambiguous and clear. But he could neither triumph nor rejoice.

EPILOGUE

THE HORSEMEN (2)

The War-to-end-Wars itself ended with the signing of an armistice on November 11th, 1918. The formal peace treaty was signed on June 28th, 1919.

Total military casualties were over 20,000,000 (twenty million) wounded, and between 8,000,000 and 9,000,000 (eight and nine million) dead. War-related civilian deaths exceeded 8,000,000 (eight million).

But at least war had been made unthinkable.

The boy heard the horses while they were still a quarter of a mile away, but he did not waken the man. They had been labouring in the fields at the golden harvest since dawn and after a mid-morning meal of bread and cheese washed down with a litre of thin beer, his father deserved a sleep.

The boy could not sleep, however, though the sun was hot and the air was drowsy. He stood up, filled with all the nervous energy of thirteen years, and peered through the heat-haze in the direction of the horsemen.

There were two of them, approaching slowly up the long sunken lane from the distant farmhouse. He screwed up his eyes to make identification certain, then shrieked in delight and pummelled his father's shoulder.

'*Papa! Papa! Les voilà! Les voilà!*' he cried.

His father awoke, stretched and yawned. Beneath the silk scarf which he always wore around his throat, the edges of a broad and unsightly scar were momentarily visible.

'*Aujourd'hui, parl' anglais,*' he growled in the low, rough tones which had nothing to do with his natural disposition. 'In honour of our guest, speak English.'

'Yes, Papa,' said the boy carefully.

Now the horsemen were near. One of them, a tall, athletic, blond-haired young man, was riding a handsome bay gelding. The other, a man of almost fifty, slightly built with greying hair, was mounted on a sturdy old chestnut, more suited to the plough than the saddle.

Pre-empting criticism, the young man cried, 'I wanted him to ride Theo, but he wouldn't.'

'I've got more sense,' grunted Jack Denial, dismounting stiffly from the chestnut. 'Well, Josh, how are you?'

'Well,' said Josh Routledge who for more than twenty years now had been Auguste Gilbert. '*Mon cousin*,' Nicole had explained to the priest, who had looked doubtfully at the pale silent young man with the bandaged throat, reprovingly at the huge swell of Nicole's stomach, and then, resignedly, married them.

The two men shook hands.

'I didn't expect you so soon,' said Josh. The gruffness of his voice after his throat wound healed had been a great help in easing him into his new life. Even now he was notorious in the countryside for his monosyllabic conversation. And his English creaked with such long disuse that his school-taught children believed they spoke it better.

'The trains were on time for once,' said Denial. 'Loti didn't let me say much more than hello to Nicole and the girls before dragging me up here!'

He smiled to show no reproach was intended.

Josh said, 'He knew how much I looked forward to seeing you. Hey, Joseph, come and shake hands with Mr Denial!'

His younger son who had been chattering excitedly to the brother he adored came shyly forward.

'How do you do?' he said carefully, offering his hand.

'I am very well, Joseph,' said Denial. 'And how are you?'

'I am very well also,' said the boy.

'Yes, you look it!'

'All right, Joseph. You and Loti go down to the house. Take the horses. Mr Denial and I will come down on the cart.'

With a yell of joy, the youngster flung his leg over the old chestnut and set it at a clumsy gallop down the slope crying, '*Je vais gagner!*'

Behind him Lothair Gilbert smiled at the two men, then set Theo in pursuit.

'Fine boys,' said Denial. 'The youngster's very like his mother, isn't he?'

'Yes,' said Josh. 'And Loti's like his dad.'

The two men exchanged glances, then Josh went to where he'd put his draught horse in the shade and began hitching him up to the haycart.

'Are you still going to tell him?' said Denial as he accepted Josh's offer of a helping hand on to the cart. As the years went by, his leg had given him increasing pain and, though he admitted it to no one, he had not been too displeased when recent promotions at Scotland Yard, had placed him pretty firmly behind a desk.

'He's twenty-one today,' said Josh. 'I always said I'd tell him at twenty-one. But we'll have the celebration first. I'm glad you could come, Jack.'

'I'd not have missed it,' said Denial.

'No. You've been a good friend to us,' said Josh.

As the cart creaked slowly down the slope towards the farm, Denial recalled that other, faster, more terrifying descent all those years ago.

There must have been a point at which he decided to enter into a conspiracy to conceal Josh's identity, but he did not know when it was. The chaos of those critical spring days had made the question of one young deserter seem unimportant. If he thought of it, it was merely as a duty temporarily postponed.

And then the chaos came under control, the Germans were first held, then pushed back, and at last it was possible to confront his responsibility.

Instantly he had recognized it was already too late. The point of decision had slipped by unnoticed after all.

But Jack Denial was not the man to shrug his shoulders and put the matter out of his mind. One responsibility ignored meant another taken on. He had betrayed his clear duty and given this man his life, and his conscience demanded that at the very least he should understand the nature of the gift.

His tutelage had necessarily been distant and intermittent but nothing that he saw of the young couple's life in the years that followed made him regret his lapse. On the Gilberts' side, suspicious gratitude had grown into true affection, and he himself knew without analysis that his relationship with this ever-growing family filled a gap in his emotional being which would otherwise have stayed a vacuum.

'Any news?' grunted Josh.

Denial was their postman, except that all the messages were one way. Through a contact in the Cumberland police he got all the news about the Routledges of Outerdale, and Josh knew of all their marriages, births and deaths. But he had never shown any desire that they should know of his existence, nor spoken any wish of making the journey back to England, even safely disguised as a French farmer.

Denial brought him up to date with a new niece and the purchase of a new pasture.

'Soon own half the valley,' said Josh. 'Owt else?'

'Nothing,' said Denial, knowing what the question referred to.

What had finally happened back in the ruined farmhouse that distant spring day had never been clear. What they did know was that early the following year a large sum of money had been transferred via a firm of German lawyers to a firm of French lawyers for the use of Nicole Gilbert.

Nicole had been terrified when the lawyers finally tracked her down, convinced it must have something to do with Josh. Instead she had found herself a comparatively rich woman, and this prosperous farm showed how wisely the young couple had invested their money and energy.

So Lothar had survived, and for a short time at least must have enjoyed his wealth and title. Denial had tried to keep him in sight but he was a moving target in a turbulent landscape. The German Revolution, the brutalities of the Freikorps, the blunderings of the Weimar Republic, the rise of the Nazis – Lothar von Seeberg had been in there somewhere, a name, a reputation, till suddenly, not long after Hitler came to power, all mention had ceased.

Denial feared the worst; though, like most Englishmen of the time, he had not yet any real concept of what that 'worst' might be.

As for Viney, there had been no trace. It was difficult to think of that huge, anarchic figure as dead, but it was just as difficult to think of him as alive and invisible. Australia had been no help. A gift of money had been accepted by his wife with no more acknowledgement than the cashing of the draft. Denial had mixed feelings. The cause he had for hating Viney would never die. Yet the eager hope in Josh's voice whenever the Australian's name was mentioned was a brake on his hatred.

They were close to the farmhouse now. Its ochre tiles seemed to throb in the late summer heat. Josh's two daughters, aged seventeen and fifteen, were filling buckets at the pump. Joseph came out of the stable where he'd been seeing to the horses. It was a scene which could have been medieval except for the sagging overhead cable by which the mixed blessing of electricity was carried to the house. Denial smiled in pleasure. In the dreadful balance sheet of that war-to-end-wars, this at least could be set on the credit side. It was a blessing to escape to this place from the pressures of his job and from the pessimism of the British newspapers, which the previous morning as he set out on his journey had been full of the news of German troops moving into Poland. At least he could forget all that in the celebration to come.

But as the cart came to a halt in the yard, Loti came rushing out of the open farm door, his face flushed with excitement.

'Papa! Papa! The news! It's on the wireless!'

'What's that then, Loti?' growled Josh.

Now Nicole had appeared in the doorway, still slight of figure and usually gay of face despite the advancing years. But now she stood there, pale and drawn, as the son she had borne twenty-one years earlier on the third of September, 1918, cried, 'The war! The war! We have declared war on the Germans! England too, Mr Denial. We are at war!'

Infected by his brother's enthusiasm, Joseph gambolled

around the yard crying, 'Hurrah! Hurrah! We're at war! We're at war!'

But Nicole and her daughters did not speak and the two middle-aged men on the cart looked at each other in sick amazement.

'They'll not let it happen again?' said Josh. 'Not again!'

The horror of it paradoxically seemed to take years off his face, turning him into a stricken young man again.

'No, they'll do something, surely,' said Denial. 'Come on Josh. Let's get into the house. I thought there was meant to be a party!'

Slowly the yard emptied. Joseph, realizing too late that his high spirits were not shared by most of the others, was the last to enter the house, subdued and resentful at what he did not understand.

The horse pulling the haycart whinnied its own resentment at this neglect, and then settled stoically to drinking water out of the abandoned buckets.

Behind it on the cart, a small butterfly which had been carried down unnoticed from the harvest field spread its brown and golden wings and took off. For a moment it fluttered outside the still open door of the farmhouse. Then it rose out of the shaded yard and let the explosion of heat from the baking tiles carry it skywards, as if it would not stop soaring till its tiny wings were shrivelled in the burning heart of the sun.

DATE DUE

MAR 0 6 1991			
4-10-91			
NOV 1 9 1993			
DEC 1 7 1993			

DEMCO 38-297

Index

Ability, 12, 14
Presbycusis, 111

Referral
 of the aphasic patient, 11
 of the dysarthric patient, 35
 of the hearing impaired, 111-112
Rehabilitation
 for aphasia, 18-21
 for dysarthria, 39-44
 for dysphonia, 99-106
 of the glossectomee, 81-86
 of the laryngectomee, 60-73
 of the maxillectomee, 89-90
Right hemisphere damage, 128-130

Sign language, 124
Spastic dysphonia, 98, 104-105
Speech production, 1-2, 55-56, 79, 86-87
Stapes, 110
Stoma, 56-57
Surgical-prosthetic voice restoration, 70-73
Swallowing. See Dysphagia

Terminal illness, 135-137
Test results, use of
 in aphasia, 16-18
 in dysarthria, 37
 in hearing impairment, 117-119
Tracheo-esophageal puncture, 71-72

Velum, 40, 86-87
Ventricular phonation, 97
Vocal abuse, 102
Vocal fold paralysis, 100-101, 132
Vocal folds, 1, 54-56, 93-97
Vocal hyperfunction, 95
Vocal nodules, 95
Vocal tract, 1
Voice disorders. See Dysphonia
Voice production, 54-56, 93

Wernicke's area, 5

Dysarthria (*continued*)
  spastic, 34
  types of, 33-34
Dysphagia
  in dysarthria, 44-48
  in glossectomy, 79-80, 82-83
  in maxillectomy, 87
Dysphonia, 93-107
  aids in identifying, 93
  causes of, 95
  defined, 93
  types of, 95

Esophageal voice, 61-62, 68-70

Family counseling
  in aphasia, 26-27
  in dysarthria, 48
  in glossectomy, 88
  in laryngectomy, 73-74

Glossectomy, 78-85
  causes of, 78
  defined, 78
  preoperative counseling for, 80-81
Glottis, 1

Hearing, 108-110
Hearing aids, 119-123
  behind-the-ear, 121
  body, 120
  eyeglass, 121-122
  in-the-ear, 121-122
  problems with, 122-123
Hearing impairment, 108-127
  acquired, 111
  aids in identifying, 111-112
  in aphasia, 20
  causes of, 110-111

  conductive, 110
  congenital, 111
  defined, 108
  mixed, 111
  sensorineural, 111
  types of, 110-111
Hearing threshold, 113
Hemianopsia, 129
Human communication
  model of, 2

Incus, 110

Laryngectomy, 54-77
  causes of, 54
  defined, 54
  preoperative counseling for, 58-59
  See also Artificial voice aids; Esophageal voice; Surgical-prosthetic voice restoration
Locked in syndrome, 134-135

Malleus, 110
Mandibulectomy, 78. *See also* Glossectomy
Maxillectomy, 86-92
  defined, 86
  preoperative counseling for, 88
Minnesota Test for Differential Diagnosis of Aphasia, 13
Motor speech evaluation, 36. *See also* Dysarthria, evaluation of
Mutational falsetto, 98, 103

Ossicles, 110

Palatal lift prosthesis, 40
Peristalsis, 45
Porch Index of Communicative

INDEX

Aphasia, 3–32
  aids in identifying, 9–11
  anterior, 5
  Broca's, 5
  case summaries, 21–24
  causes of, 6–7
  defined, 3
  evaluation of, 12–15
  fluent, 5
  global, 6
  mixed, 6
  nonfluent, 5
  posterior, 5
  spontaneous recovery from, 7–8
  types of, 4–6
Apraxia
  dressing, 129
  of speech, 6
Artificial voice aids, 62–68, 76
Audiogram, 112–117
Augmentative devices, 20, 42, 51–52, 84, 135

Blom-Singer procedure, 70–73
Boston Diagnostic Aphasia Examination, 13, 15
Broca's area, 5

Cerebrovascular accident
  in aphasia, 7
  in dysarthria, 35
Chronic brain syndrome, 130
Cochlea, 110
Communication board, 20, 42, 81, 88, 124, 135. *See also* Augmentative devices
Community resources
  for aphasia, 28
  for dysarthria, 49
  for hearing impaired, 125
  for laryngectomy, 74
Confusion, 130–132
Conversion aphonia, 98, 105–106

Deafness, 111, 123–125
  acquired, 111
  causes of, 123
  congenital, 111
  types of, 111
Duckbill prosthesis, 71–72
Dysarthria, 33–53
  aids in identifying, 35
  ataxic, 34
  causes of, 34–35
  defined, 33
  evaluation of, 36
  flaccid, 33
  hyperkinetic, 34
  hypokinetic, 34
  mixed, 34

INDEX

paraphasia replacement of one word by another that is incorrect but retains a relationship to the intended word

perseveration repetition of a response that is no longer correct

pharynx (throat) the part of the respiratory mechanism between the larynx and nasal cavities

phonation the production of voice by vibrations of the vocal folds

plosive a speech sound produced by momentarily blocking the air stream then quickly releasing it (e.g., p and t)

prosthodontist a dentist who specializes in the construction of artificial appliances designed to replace missing oral structures and parts of the face

speech formalized oral communication of ideas conveyed by the utterance of vocal sounds

speech pathology a field of study dealing with the evaluation and treatment of speech, language, voice, and allied communicative disorders

spontaneous recovery improvement in communicative, as well as other, functioning in brain-damaged patients, which occurs without apparent relationship to external influence

thrombosis blockage of an artery by a blood clot or deposit that has been formed at the point of blockage

Trach-Talk a one-way valve that can be attached to a tracheostomy tube to divert air through the larynx for phonation

Tucker valve an opening at the top of a tracheostomy tube that can be closed by a hinged mechanism during inhalation, but remains open to divert air through the larynx during exhalation

velum (soft palate) muscular portion of the posterior roof of the mouth; can be elevated to assist in closing off the nasal passages during speech and swallowing

vocal tract the part of the respiratory mechanism that includes the pharynx and oral cavity; when modified in shape and size, it is responsible for the differences in the sounds of language

deafness the absence of usable hearing; may be congenital or acquired

duckbill prosthesis also called the Blom-Singer prosthesis; a narrow tube, valved at one end, that allows lung air to be directed from the trachea to the esophagus in a laryngectomee

dysarthria a group of speech disorders caused by damage to the central or peripheral nervous system, resulting in disturbance of control over the speech musculature

dysphagia impaired ability to swallow

dysphonia impaired voice production

embolism blockage of an artery by a clot or air bubble that has been transported through the blood stream

fricative a speech sound made by forcing the air stream through a narrow opening (e.g., *f* and *th*)

glossectomee an individual who has undergone a glossectomy

glossectomy partial or total removal of the tongue

hearing impairment the amount of loss of an individual's hearing as compared to the hearing of an average normal person

hemianopsia (hemianopia) impaired or absent vision in half of the visual field

jargon unintelligible speech produced by some aphasic patients

language a system of grammatical, semantic, and phonological rules

laryngectomee an individual who has undergone a laryngectomy

laryngectomy surgical removal of the larynx

larynx (voice box) the portion of the respiratory mechanism between the upper ring of the trachea and the tip of the epiglottis; it houses the vocal folds

maxillectomee an individual who has undergone a maxillectomy

maxillectomy partial or total removal of the upper jaw

motor speech evaluation a set of speech tasks, graduated in difficulty, designed to assess the neuromotor integrity of the voice and speech musculature

nasal consonant sounds produced with resonance through the nose (e.g., *n, m, ng*)

palatal lift a backward extension on a partial or complete upper denture that assists in elevating the velum

GLOSSARY

The following terms are commonly used when dealing with communicative disorders discussed in this book.

acalculia impaired ability to solve mathematical problems as a result of brain injury
agnosia inability to recognize objects in one sensory modality, although recognition via other sensory modalities is normal
agraphia impaired ability to write
alexia impaired ability to read that cannot be accounted for by overall intellectual deficit or problems with visual acuity
anomia impaired ability to express nouns
aphasia impairment of the ability to formulate and process language symbols as a result of brain injury
apraxia of speech impairment of the ability to program the positioning and sequencing of the speech muscles and their movements as a result of brain injury
audiogram a graphic representation of a person's hearing threshold level at various frequencies
audiology a field of study dealing with the evaluation and treatment of hearing impairment; practiced by an audiologist
augmentative device one of a number of hand-made or commercially available systems designed to help nonverbal individuals communicate
cerebro-vascular accident (CVA; stroke) change in blood supply to the brain as a result of embolism, thrombosis, or hemorrhage
circumlocution using many words in an attempt to replace a single word that the aphasic patient is unable to retrieve

ber 1979, 71-78.

Lawless, C. A. Helping patients with endotracheal and tracheostomy tubes communicate. *American Journal of Nursing,* December 1975, 2151-2158.

AVAILABLE DEVICES

Olympic Trach-Talk. (Available from Olympic Medical Corp., 4400 Seventy South, Seattle, Washington 98108.)

Tracheostomy tube with Tucker valve, Pilling Co. (Available through local surgical supply house.)

even though no words are spoken and the patient exhibits problems in understanding.

As is the case with communicative disorders in general, when dealing with a dying patient, the technique or device employed is not nearly as important as is the availability of a caring, compassionate person to assist the patient in coping with and finally accepting his or her own death. Although the majority of this book has been devoted to enhancing speech production and understanding, much of human beings' most intimate and emotional communication takes place at a nonverbal level; good feelings and sensitivity for the dying patient can be understood by even the most severely involved.

SUMMARY

Communicative disorders may also be encountered in conditions not discussed in the other chapters of this volume. In most cases, the communication problem occupies a role secondary to the medical condition, but may be important in the attempt to provide good nursing care and high quality of life.

Many techniques are available to assist the nurse and others in communicating with patients exhibiting these special problems. Some techniques have been derived from working with other more common disorders, and some can be used only in special circumstances. Ultimately, however, regardless of the severity of the medical condition or communicative impairment, the attitudes and feelings of the nurse toward the patient play a far greater role than does any single technique or gadget designed to enhance communicative effectiveness.

SUGGESTED READINGS

Hawkes, C. H. "Locked-in" syndrome: Report of seven cases. *British Medical Journal,* April 1974, 379–382.

Kroner, K. Dealing with the confused patient. *Nursing '79,* Novem-

an extension of the one that has directly or indirectly caused the communication problem, such as metastatic carcinoma in a laryngectomee or glossectomee. In other cases, the communicative impairment progresses in severity as the disease process progresses, such as in amyotrophic lateral sclerosis, multiple sclerosis, and inoperable brain tumors. In still others, the communicative impairment and terminal illness are unrelated, such as severe hearing loss in a patient with advanced gastrointestinal cancer.

In all such cases, however, attempts at improving quality of life and providing supportive care are compromised to some extent by the communicative disorder, depending upon type and severity. Therefore, an already difficult situation is made worse by such patients' inability to express anger, grief, or denial; by their difficulty in making their desires known with regard to legal and financial plans; and by increased feelings of isolation and fear. The very essence of most terminal care treatment approaches is in jeopardy if patients have difficulty understanding or being understood. Unfortunately, there are no easy solutions to this perplexing problem, although there are guidelines that may be of some benefit.

Of primary importance is the need to find some method of communication that is appropriate for the individual concerned. Throughout the preceding chapters, virtually all communication techniques have been discussed; any and all should be attempted in an effort to provide the terminally ill patient with a means of expression and understanding. Gesture, writing, sign language, communication boards, artificial voice aids, and speech, no matter how distorted, can all be used to communicate fears and feelings about death. So too, the dying patient with impaired understanding may still be quite capable of discerning nonverbal indications of a nurse's feelings about death. A health care provider with uneasy feelings about his or her own death may inadvertently transmit this information,

ceptional nursing care as well as painstaking attention to establishment of a communication system.

For the locked-in patient, the first priority in communication is to obtain a visible and reliable indication that he or she can understand. This is usually accomplished via eye blinks, lateral eye movements, or raising and lowering the eyebrows. The patient should be informed as to the procedure and then instructed to respond to simple "yes/no" questions by performing the agreed-upon movement.

Once this basic form of communication has been established, more sophisticated exchanges can take place via the communication board method discussed in Chapter 2 or the game of "Twenty Questions," in which the topic is arrived at by gradually narrowing the possibilities through well-planned "yes/no" questions. Unfortunately, these methods are time-consuming and do not allow for voluntary production of thoughts and ideas by the patient; he or she is always dependent upon the availability and cooperation of a "listener."

As mentioned in Chapter 2, however, with rapid advances in technology, it is now possible in some cases, to provide adaptive switches that can be triggered by eye or forehead movements to activate an electronic communication device programmed to provide a verbal or written output. Anyone dealing with a locked-in patient should make every effort to obtain evaluation for and instruction in the use of appropriate augmentative equipment. These unfortunate individuals are totally dependent on health care providers to keep their healthy minds in contact with the world.

TERMINALLY ILL PATIENTS

It is not uncommon for the nurse to be confronted with a situation in which a patient who is receiving terminal care also exhibits one or more of the communicative disorders mentioned in the previous chapters. In some cases, the terminal illness is

with normal movement in the upper extremities, the artificial device can be used in precisely the same manner as with the laryngectomee. If the extremities are involved, however, the nurse or other listener may be required to place the instrument and operate the tone generator while the patient "mouths" the words. This takes considerably more coordination, practice, and patience, but good communication can be established and the method can be taught to spouse and family.

In very difficult cases where none of the previously discussed techniques is successful, it may be necessary to request that the patient simply "mouth" words without the benefit of vocal fold or artificial vibration. The listener is then required to read the patient's lips, and this skill is not possessed in equal degrees by all people. It may be necessary to experiment with many individuals on the ward or in the intensive care unit before one is found who can understand what the patient is trying to communicate. To assist the listener, however, the patient should be encouraged to speak in phrases and sentences, rather than individual words, to provide contextual cues.

For the patient with a tracheostomy, with or without additional respiratory assistance, consultation with the physician and speech pathologist should always be obtained prior to attempting any of the techniques mentioned here. The need to communicate is great, but an open airway free of obstructing secretions and infection must always receive top priority.

LOCKED-IN SYNDROME

The patient with so-called *locked-in syndrome* has suffered damage to the brain stem in such a way that he or she is quadriplegic and unable to perform speech movements (*anarthric*), but is conscious and has reasonably intact intellectual and language skills, in addition to some voluntary eye movements. In effect, the locked-in patient is suffering from the most severe form of dysarthria, as discussed in Chapter 2, and requires ex-

have additional involvement of their extremities and are unable to occlude the tracheal tube effectively to force the air through the larynx. If such patients do not require constant cuff inflation and have no airway obstruction above the level of the tracheostomy site, they may be candidates for a device called an Olympic Trach-Talk (see list at the end of the chapter for ordering information).

The Trach-Talk is a one-way valve that is attached to the tracheostomy tube. When the patient inhales, a spring keeps the valve open, allowing free flow of air into the lungs. Upon exhalation, the valve is closed and air is diverted through the larynx to create voicing. The device can be adapted to different tracheostomy tubes and allows for suctioning and application of oxygen or humidification. The Trach-Talk should only be used with patients who have reasonably good speech articulation and who are unable to occlude their own tracheostomy tubes during speaking attempts.

For individuals requiring tracheostomy with a cuffed tube that must be kept inflated such as for those on respirators or ventilators, it is not possible to divert the exhaled air through the larynx by occluding the tube, since the cuff seals off the trachea for these patients. It may be possible to use a "fenestrated" tracheostomy tube or one with a so-called "Tucker valve." Tubes of this nature have a hinged valve that is closed when air is taken into the lungs, but that opens when air is exhaled and the tube is occluded, thus allowing vocal fold vibration while maintaining an inflated tube cuff for adequate assisted respiration. Again, the patient should have good speech articulation and no obstruction above the tracheostomy site.

If the patient with tracheostomy is unable to benefit from any of the previously mentioned devices, but has reasonably good speech articulation, it may be possible to use either a neck-type or oral-type artificial voice aid, as described in the discussion of laryngectomy (see Chapter 3). For the patient

ing how to cope with the afflicted individual. They should be allowed to express their fears and feelings in a nonjudgmental, supportive atmosphere, while learning not to increase stress by arguing with or chiding the patient.

PATIENTS WITH TRACHEOSTOMY TUBES OR ON VENTILATORS

Tracheostomy is performed for a number of reasons other than to reroute the trachea in laryngectomy. In these other cases, the tracheostomy may be permanent or temporary and may require a cuffed or noncuffed tracheostomy tube. The effect on communication varies with the type of tracheostomy and severity of respiratory and other coexisting problems.

One of the most common situations in which a nurse encounters a tracheostomy tube is *bilateral vocal fold paralysis,* in which the vocal folds close off the airway in the larynx. In this case, the patient is unable to inhale because the vocal folds block the air on its way to the lungs, but the remainder of the respiratory mechanism functions normally.

By creating an opening in the trachea beneath the vocal folds, the surgeon bypasses the larynx, and air is inhaled and exhaled without difficulty. When the patient attempts to speak, however, the air normally used to vibrate the vocal folds rushes out of the tracheostomy tube and voicing is absent. Once the tracheostomy tube is occluded with finger or thumb, the exhaled air is forced up through the vocal folds once again, and voicing again becomes possible. The patient must learn to coordinate the occlusion of the tube for speech with inhalation and exhalation for normal breathing, but with a little practice he or she is usually able to communicate with no difficulty.

The foregoing example assumes no other problems than the bilateral vocal fold paralysis. In many cases, however, patients

communication activities. More severe disorientation requires more frequent repetition of the patient's name and details of the situation in an attempt to develop a foundation of reality upon which to build.

Confused patients require a nonstressful and extremely structured environment in which to function. Even minor changes in daily routines can be devastating to the confused patient and should be avoided as much as possible. If the patient wears a hearing aid, eyeglasses, or dentures, it is important to see that he or she is using them properly and that they are in good condition. Reduced vision and hearing create further limitations on the amount of sensory stimulation needed by the confused patient to maintain contact with reality.

When communicating with patients who exhibit confusion and disorientation, short phrases and simple directions produced in a clear (not shouted) voice, as already discussed in Chapter 1, are applicable here. To verify such patients' understanding of directions, they should be encouraged to repeat them or to show by performance that the message has been accurately received. If memory is involved, frequent reverification may be necessary.

The confused patient may lapse into rambling, disjointed speech on occasion and should be gently but firmly guided back toward normal discourse. Usually, a brief reminder as to the appropriate topic and context is sufficient to reorient the individual. If the conversation becomes extremely incoherent and attempts at correcting the patient are unsuccessful, it is probably wise to forego further discourse at that time, but communication should be resumed later.

Like individuals with other serious disorders, the individual who is confused requires patience and understanding from those entrusted with his or her care. This applies to family members as well as to nursing and rehabilitation personnel; and those family members usually require assistance in learn-

If right hemisphere damage is suspected or neurologically confirmed, an evaluation by the speech pathologist and neuropsychologist is warranted. Although treatment techniques for patients with right brain injury are nowhere near the level of sophistication as are those used with aphasic patients, test results can be used to specify areas of greatest limitation and to provide a basis for family counseling as to the best methods for dealing with such patients at home.

On the ward or in the nursing home, the nurse can use the findings of the evaluation to help patients take advantage of their strengths while working around their limitations.

THE CONFUSED PATIENT

When working with older adults, it is possible to encounter patients who are confused and disoriented as a result of diffuse brain damage, heavy medication, or abrupt transferral to unfamiliar surroundings. For patients with confusion resulting from temporary conditions, the obvious remedy is to remove the cause of the problem or modify the environment in such a way that the underlying condition is eliminated.

For the person with diffuse brain injury, sometimes referred to as *organic* or *chronic brain syndrome,* the nurse can deal with the language and behavior of confusion in several ways to minimize the disabling effects of this chronic condition. As noted in the discussion of aphasia in Chapter 1, a good rule of thumb is to deal with the patient at a level where his or her ability to function begins to break down. This same principle holds true for the disoriented or confused patient, but in the domain of total reality orientation rather than in language production and processing per se.

If, for example, the patient is mildly disoriented as to place but is aware of time and of his or her own identity, it is helpful to mention the name of the city and state frequently during

cerebral hemisphere for spatial and nonverbal activities. As such, individuals with right hemisphere damage may exhibit denial of the left side of the body or the existence of anything to their left. The patients may be disoriented to place and time, becoming easily lost even in familiar surroundings.

Whereas patients with left hemisphere damage sometimes exhibit apraxia of speech, individuals with right hemisphere involvement may have *dressing apraxia*—that is, they may be unable to place arms in shirt sleeves or legs in pants as a result of the general inability to sequence motor activities that require integrity of the relationship between body and space. Also noted in many patients with right hemisphere damage are poor judgment and failure to complete a task once it is initiated.

With specific regard to communication, the patient with right hemisphere damage may exhibit verbal overflow, or continuous talking about topics of little importance or relevance. Reading may be impaired because of the neglect of the left side mentioned earlier: A *left hemianopsia* allows the patient to see printed material on the right side only. Even with intensive treatment, it is usually difficult to teach the patient to accommodate for this disorder. Other deficits noted in patients with right hemisphere damage include inability to recognize faces, inability to draw figures, and loss of musical ability.

In some ways, patients with right hemisphere damage present more serious problems in nursing care than aphasic individuals. Even though aphasic patients may have severe communication involvement, they usually have good awareness of the environment and are cognizant of appropriate behavior, in spite of not being able to perform at the physical level. Patients with right-sided brain damage, on the other hand, superficially appear to be without serious deficit, but, in reality, require close supervision in their decision making and activities of daily living.

7

OTHER AREAS OF COMMUNICATIVE IMPAIRMENT

The foregoing chapters deal with the most common communicative disorders encountered by the nurse in an adult population. Many times, however, the nurse may be confronted with a patient for whom the communication problem does not fit precisely into one of the categories mentioned, or is secondary to the overall medical condition, but is sufficient to interfere with nursing care. This chapter is devoted to a review of these other areas of communicative impairment and the ways in which they are best handled by the nurse. Some of the techniques should be familiar, since they are equally applicable to a variety of patients already discussed. Others are applicable only to patients with unique problems. The types to be discussed include patients with right hemisphere damage; confused patients; individuals with tracheostomy tubes or on ventilators; patients with locked-in syndrome; and terminally ill patients with coincidental communicative impairment.

RIGHT HEMISPHERE DAMAGE

Unlike the aphasic patient, whose brain damage involves certain areas of the left hemisphere, the individual with right hemisphere damage usually does not have noticeable problems with language processing. Instead, the behavior exhibited by these patients reflects the underlying specialization of the right

Hull, R. H., & Traynor, R. H. Restoring communication in the hearing-impaired elderly. *Nursing Care,* June 1977, 14-32.

McNamee, C. Communicating with the hard of hearing. *The Canadian Nurse,* March 1978, 27-29.

Perron, D. M. Deprived of sound. *American Journal of Nursing,* June 1974, 1057-1059.

Rosenblum, E. *Fundamentals of hearing for health professionals.* Boston: Little, Brown, 1979.

Sabatino, L. Dos and don'ts of deaf patient care. *RN,* June 1976, 64-70.

Voke, J. Sounding out the waves. *Nursing Mirror,* December 1979, 40-41.

MATERIALS FOR PATIENTS

Alexander Graham Bell Association for the Deaf. *Statements on deafness.* (Available from the author at 3417 Volta Place, N.W., Washington, D.C. 20007.)

Council of Better Business Bureaus. *Facts about hearing aids.* (Available from the author at 1150 17th St., N.W., Washington, D.C. 20036.)

Dodds, E., & Harford, E. *Helpful hearing aid hints.* (Available from the Alexander Graham Bell Association for the Deaf, 3417 Volta Place, N.W., Washington, D.C. 20007.)

U.S. Department of Health, Education, and Welfare. *Hearing loss: Hope through research.* (Available from the Superintendent of Documents, U.S. Government Printing Office, Washington, D.C. 20402.)

SUMMARY

Hearing loss or deafness is experienced by many hospital patients and can interfere with appropriate medical care and nurse-patient communication. Although there are many types and causes of hearing loss, resulting in varying degrees of communicative handicap, nearly all can be helped to some extent with appropriate referral and rehabilitation.

The hearing evaluation performed by the audiologist helps in the determination of type and severity of hearing loss, as well as appropriate treatment planning. The nurse can also use the test results to obtain a general profile of the patient's hearing ability and potential communication problems.

Awareness of principles that enhance communicative effectiveness, as well as of the basics of hearing aid operation, can be of benefit to the nurse who deals with hearing-impaired individuals and with those patients who have been deafened prior to the development of language. Many materials, resources, and articles in the nursing literature are available to assist the nurse in this endeavor.

SUGGESTED READINGS

Carbary, L. J. Deafness: The silent world. *Nursing Care,* November 1976, 13-16.

Clark, C. C., & Mills, G. C. Communicating with hearing-impaired adults. *Journal of Gerontological Nursing,* May/June 1979, 40-44.

Farrell, D. E. Hearing loss in the geriatric patient. *Nurse Practitioner,* November/December 1978, 30-32.

Gates, N. The elderly deaf. *Nursing Mirror,* January 1977, 67-68.

Herth, K. Beyond the curtain of silence. *American Journal of Nursing,* June 1974, 1060-1061.

Holm, C. S. Deafness: Common misunderstandings. *American Journal of Nursing,* November 1978, 1910-1912.

of verbal instructions. If the deaf patient requires IVs, it is helpful to determine his or her handedness and to leave the preferred hand and arm unencumbered for writing and signing.

COMMUNITY RESOURCES

Fortunately for the nurse dealing with hearing-impaired or deaf individuals, more materials and resources are available than have been indicated for glossectomy, maxillectomy, and voice disorders. The list of suggested readings at the end of the chapter provides accessible literature written from a nursing perspective. In addition, the following community resources are available in most areas to provide information and assistance in dealing with hearing-impaired individuals.

State schools for the deaf, white pages.

State associations for the deaf and hearing-impaired, white pages.

State chapter of the National Association of the Deaf, white pages.

Local university programs, white pages.

VA medical centers, white pages.

Audiologists in private practice, yellow pages.

In most cases, hearing-impaired or deaf patients have been counseled by an audiologist and/or otologist, although it is possible for the nurse to make the initial contact. Materials that may help in educating patients about hearing loss or hearing aids are listed at the end of the chapter.

fore, the first principle, as always, is to try to impart an attitude of calm reassurance through whatever means available.

Sometimes the deaf patient has the ability to understand speech by lip reading, although this skill varies widely among individuals. To enhance lip-reading ability, speak normally without extensive exaggeration of mouth movements. Again, lighting should be on the speaker's face, and communication should always take place with speaker and receiver facing each other. Repetition of instructions and rewording can help the patient read lips more successfully, since some sounds are not as highly "visible" as others. For very important communication, it may be necessary to write the message. Most deaf patients can read and write, although some have difficulty at higher levels of complexity, since reading and writing ability is dependent to a large extent on early language development through the auditory system. If the patient is having difficulty with written material, a picture communication board should be attempted.

In many areas across the country, classes in sign language are available for normally hearing individuals who have a need to communicate with the deaf. While extensive practice and instruction is necessary to become "fluent" in the use of sign language, nurses can easily learn a vocabulary of simple signs for routine hospital needs as well as for a number of everyday phrases. Finally, the recent emphasis on the problems of handicapped persons has increased the availability of interpreters for the deaf. Every attempt should be made to gain access to such a person, especially if a deaf patient requires surgery or is encountered in an emergency situation (see "Community Resources," below).

As with the hearing-impaired person, it is helpful to identify the deaf patient for other hospital personnel, especially if he or she is being sent for diagnostic tests that require the following

ing. In these cases, the audiologist should again be notified and the hearing aid repaired.

THE DEAF

So far, this chapter has been directed toward communication with patients who exhibit acquired hearing loss. In these individuals, hearing has been reasonably intact throughout most of their early life, and they have developed essentially normal speech and language. Thus, the primary approach in dealing with such patients involves using residual hearing, amplifying sound where appropriate, creating a more conducive communication environment, and teaching the patient to make use of visual and contextual cues.

The nurse eventually encounters another type of hearing impairment that is much more serious and challenging. As mentioned earlier, some people are born with impaired hearing or lose their hearing shortly after birth before language has normally developed. Congenital deafness or hearing loss has many causes, including prenatal diseases such as rubella, postnatal problems such as anoxia, and childhood infections causing extremely high fever. In most cases where hearing is severely impaired or absent, the individual requires specialized instruction and has difficulty functioning in a world oriented to normal hearing and speech.

Congenitally deaf individuals are not exempt from other illnesses or accidents, however, and can be expected to appear in hospitals with the same regularity as their normally hearing counterparts. The remaining paragraphs are intended to suggest ways in which the nurse can better communicate with deaf patients. Throughout, it might be helpful to keep in mind that hospitals can be frightening places for many people, but that to be unable to hear and confined to the strange and unfamiliar surroundings of a medical setting can be devastating. There-

battery, and so on are arranged differently to house them in the glasses or the combined housing of the hearing aid and earmold. Precautions and problems encountered with body and behind-the-ear aids are applicable to eyeglass and in-the-ear instruments as well.

Two problem areas are commonly encountered by nurses who have contact with patients wearing hearing aids: absent or reduced volume, and feedback. If the hearing aid has been working properly and suddenly does not seem to be amplifying correctly, one of a number of easily correctable problems may be the culprit. Initially, a check should be made to determine that the "on/off" switch is in the "on" or M position and the volume control has been rotated to the position usually used by the patient. If there is still no amplification (indicated by holding the hearing aid or receiver close to the ear and listening for squealing or sound coming directly from the aid), then the battery should be checked for proper placement in the battery compartment and to determine whether it is delivering appropriate power. This can be accomplished by using a small battery tester (if available) or by replacing the old battery with a new one. If there is still no sound, the earmold and tubing should be checked for blockage by ear wax or other debris and cleaned. When all of these procedures fail to produce the desired result, the audiologist should be notified and the aid returned for repairs.

Feedback is the squealing or whistling sound that is produced when a microphone is placed in close proximity to the receiver to which it is transmitting. In a hearing aid, this phenomenon usually results from inappropriate placement of the earmold in the ear canal, but may occur if a hand or object is placed over or near the hearing aid while it is operating and being properly worn. Once assured that the earmold is properly in place and nothing is near the hearing aid, continuing feedback usually suggests an inadequately designed earmold or defective tub-

The behind-the-ear aid is probably the most common instrument worn by today's adult hearing-impaired population and is schematically represented in Figure 6-4. Its microphone, amplifier, receiver, and battery are housed in a small plastic casing designed to fit snugly behind the ear. The sound is transmitted through a tubing connecter and plastic tubing into a custom-made earmold. The instrument has a variable volume control and may have one or more tone adjustments. The volume should be adjusted for different listening environments, but the tone settings are best left to the audiologist. Some behind-the-ear aids have a switch that may be placed in the *O* (off) position, *M* (microphone, or "on") position, or *T* (telephone) position. In the *T* position, the hearing aid wearer hears the telephone conversation transmitted directly from the telephone to aid via changes in the electromagnetic field. In any hearing aid that does not have an "on/off" switch, the battery compartment should be opened and the battery removed when the aid is not in use. Otherwise, battery life is reduced and batteries need to be replaced at higher cost.

In the eyeglass and in-the-ear types of hearing aids, the foregoing principles are identical, but the microphone, circuitry,

Figure 6-4. Schematic representation of a behind-the-ear hearing aid.

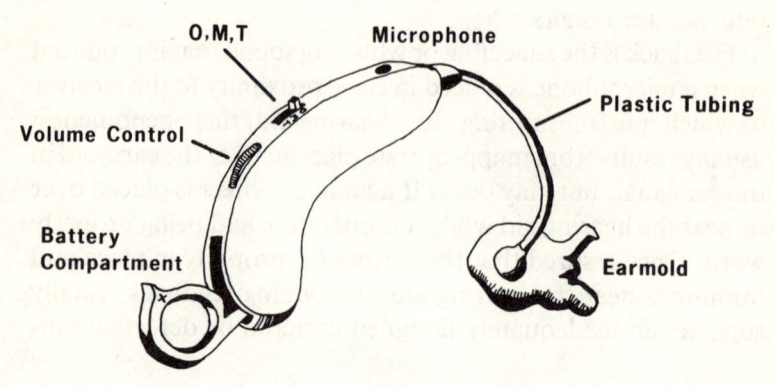

hearing aids, a basic understanding of the instruments can help during patient counseling about the use and benefits of hearing aids; it can also provide assistance when minor problems of operation are encountered. All hearing aids have certain features in common: a microphone that changes sound into electrical energy; an amplifier that increases the intensity of the electrical energy; and a receiver that changes the electrical energy back to sound. As stated earlier, a hearing aid works much the same way as a public address system or telephone does. Whereas a public address system is large and is powered by normal household current, however, a hearing aid is relatively small and powered by a battery or batteries. Its small size is an advantage in that it can be easily worn and transported, but in return it gives up a degree of fidelity. From the basic design, four general types of hearing aids have emerged: the body aid; the behind-the-ear aid; the eyeglass aid; and the in-the-ear aid. In most cases, the hearing aid transmits the amplified sound directly into the ear canal via tubing or an acrylic earmold. Sometimes, though, the sound is transmitted by a bone vibrator similar to the device mentioned in the discussion of bone conduction testing (see "The Audiogram").

The body aid is so called because the batteries, microphone, amplifier, and volume and tone controls are contained in a metal or plastic housing about the size of a pack of cigarettes, and the housing is carried in a pocket or secured close to the body with a lightweight harness. The amplified sound is transmitted to the receiver by a thin cord, and the receiver connects directly to a custom-made earmold. The body aid is usually worn by patients with severe hearing loss that requires the extra power characteristic of the instrument.

The advent of technology in microcircuitry has resulted in the ability to obtain more and more power and fidelity from smaller and smaller instruments. As such, there is a greater likelihood that a nurse working with hearing-impaired patients will encounter one of the three remaining types of hearing aids.

hearing-impaired patients' charts are identified by placement of a specially colored sticker to alert health care personnel. This is usually a hospital policy decision, however, and should be handled through appropriate administrative channels.

HEARING AIDS

Up to this point, emphasis has been placed on approaches that the nurse can take to remove some of the burden for a hearing-impaired patient, but little has been said about hearing aids. These instruments, when properly fitted and in good working order, can significantly improve the patient's ability to hear and understand speech. Like all other assistive devices, however, they do have disadvantages and limitations, the major one being that they are totally ineffective if left in the drawer of a patient's night stand or dresser. Nurses can strongly encourage hearing aid users to wear their aids at all times when hearing and listening are important. This is especially critical during the immediate weeks following fitting of a patient's first hearing aid. Many times, patients' responding positively to sounds that they have not heard for years requires a period of adjustment. The abrupt addition of long-forgotten sounds such as birds chirping, newspapers rustling, and silverware clanging can be irritating until the auditory system is given time to adapt.

Also, the patient who is a reluctant hearing aid wearer should be reminded that the instrument is an "aid" and not a cure for hearing loss. It is, in effect, a miniature public address system that makes some sounds louder but may or may not make them more understandable. A hearing aid user who accepts the instrument's limitations and learns to use its strengths in conjunction with visual and gestural cues, contextual information, and other factors conducive to good communication usually experiences significant improvement in hearing.

For nurses working with hearing-impaired patients wearing

nursing care, unless the hearing loss has not been totally alleviated by the operation. In that case, and in the case of the patient with a hearing loss of insufficient magnitude to warrant a hearing aid, further steps can be taken to enhance communicative effectiveness. Many of the principles discussed previously for improving communication with a patient exhibiting disordered speech can be used in reverse with the hearing-impaired listener. Speaking clearly and not rapidly while facing hearing-impaired persons allows them to make use of facial and gestural cues to supplement what they are hearing. To enhance this principle further, it is helpful to avoid placing hands or objects in front of the face while talking and to be positioned in such a way that lighting falls on the face of the speaker, not in the eyes of the listener.

For patients with more severe hearing problems and for whom hearing aids have been recommended, the foregoing suggestions take on even more importance. In addition, it is especially critical to eliminate background noises, such as television, radio, and competing conversation, and to avoid shouting. It should be remembered that in most cases, increases in loudness beyond the amplification effect of the hearing aid actually reduce the patient's ability to understand speech and may even cause discomfort or pain.

It is also acceptable and advisable to ask patients frankly about factors that help or hinder their ability to hear and understand. Many hearing-impaired individuals have quite sophisticated knowledge of their hearing loss and should not be disregarded as sources of information. Speaking openly about a patient's hearing loss also provides a measure of reassurance that the nurse is concerned about the patient as an individual and is interested in establishing effective communication.

Avoiding signs of impatience, rewording statements that are not understood, and getting the patient's attention before initiation of conversation are additional suggestions for communicating with hard-of-hearing listeners. In some hospitals,

has some difficulty following normal conversation. Further reductions in speech discrimination scores to 70% or 50% indicate serious problems in understanding speech.

Unfortunately, problems with speech discrimination may not always be solved by making the signal more intense and, in fact, may be made worse by increasing the loudness of speech beyond a given level for some individuals. Therefore, it is possible to encounter a patient with a relatively low SRT of 50 dB, but with very poor discrimination ability—the typical picture of an elderly patient with presbycusis. Such a patient requires different handling than does one who has reasonably good discrimination ability but needs to have the speech presented at higher levels of intensity.

USE OF TEST RESULTS

After a patient has been evaluated, the nurse and audiologist can review the audiogram and discuss the treatment plan. The results are generally categorized as follows:

1. The patient has little or no hearing loss, and no further treatment is indicated.
2. The patient has a conductive hearing loss that will probably respond to medical or surgical management by the otologist.
3. The patient has a hearing loss that cannot be remedied through medical management, but is not of sufficient magnitude to warrant a hearing aid.
4. The patient has a hearing loss that requires hearing aid amplification and a program of aural rehabilitation.

For the patient for whom surgical intervention is indicated, the postoperative role of the nurse is primarily one of routine

tion testing, the audiologist can form a reasonably accurate picture of the patient's hearing ability and has the foundation for performing further tests and initiating appropriate treatment. For the nurse dealing with a hearing-impaired patient, however, the primary needs include knowledge of how the individual's hearing loss affects his or her ability to communicate, as well as awareness of principles that maximize communicative effectiveness between the nurse and the hearing-impaired person. As such, the interpretation of results from two other basic tests performed by the audiologist are of benefit.

The speech reception threshold (SRT) is also expressed in dB and indicates the level at which a patient can just barely hear and repeat such words as "ice cream," "hot dog," "school boy," and "baseball." In most cases, the speech reception threshold is within approximately 6 dB of the average pure tone threshold obtained at the test frequencies of 500, 1000, and 2000 Hz, and is a rough indication of the amount of hearing handicap that might be expected for a given individual. For example, a person with a speech reception threshold of 10 dB falls within the normal range and should have no difficulty responding to normal speech. A person with an SRT of 40 dB, however, probably has difficulty hearing normal speech, while a patient with an SRT of 70 dB may have difficulty with even loud speech. An SRT of 80 dB requires that speech be shouted or amplified, and a person with an SRT of 100 dB may have little hearing of speech even if it is shouted or highly amplified.

Speech discrimination testing provides an indication of how well a person understands the words he or she is hearing and is only indirectly related to the SRT. Speech discrimination is represented as a percentage and is obtained by presenting a long list of words such as "toe," "owl," "she," and "earn," at a comfortable listening level. A score of 100% indicates perfect understanding, while one of 80% indicates that the patient

Figure 6-3. Sample audiogram showing results of pure tone air conduction testing for the right (O) and left (X) ears with masked bone conduction testing of the right ear ([ ). Condition A indicates normal hearing by bone conduction and suggests a conductive hearing loss. Condition B indicates a mixed hearing loss with a 10 dB conductive component and a 20 dB sensorineural component. Condition C indicates a sensorineural hearing loss, since the threshold remains at 30 dB, even though the outer and middle ears have been bypassed.

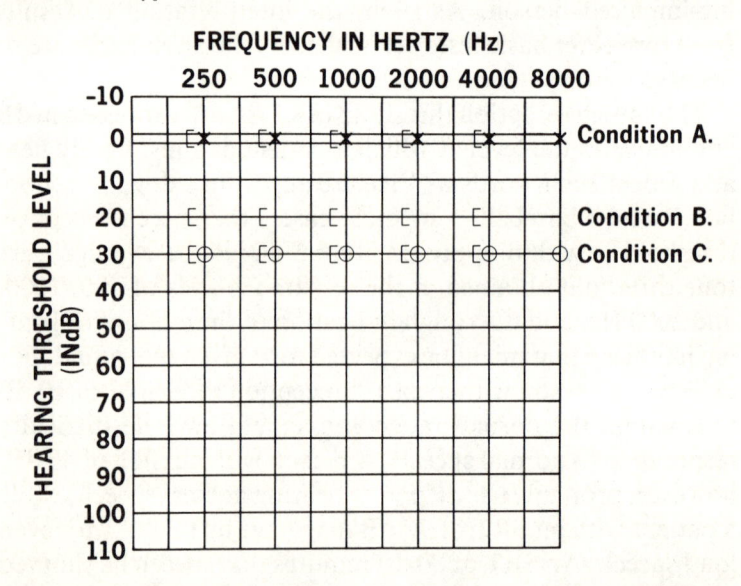

ing loss lies in the inner ear or beyond and is therefore sensorineural. Finally, if, under masking, the patient indicates that he or she can just barely hear the bone-conducted signal to the right inner ear at 20 dB, the examiner can conclude that there is a mixed loss made up of a 10 dB conductive component and a 20 dB sensorineural component. These three conditions are schematically represented in Figure 6-3.

With the additional information provided by bone conduc-

pure tone air conduction hearing test and provides a good profile of the individual's overall hearing ability. However, it tells little of what type of loss the right ear has sustained, how well the listener hears and understands speech, and whether the loss is caused by something that can be medically or surgically corrected. To obtain more detailed information, the audiologist has a large number of additional tests available, most of which are beyond the scope of this chapter. Some further information is necessary for the nurse attempting to interpret an audiogram when caring for a hearing-impaired patient, however.

In the audiogram depicted in Figure 6-2, for example, it is not possible to determine from the information given whether the hearing loss is conductive, sensorineural, or mixed. Figure 6-3 shows the same audiogram with the results of an additional test known as a *pure tone bone conduction* evaluation. This procedure is performed to determine whether there is some factor in the outer or middle ear that could account for the 30 dB hearing loss, and is accomplished by stimulating the bone behind the right ear with a small vibrator at the frequencies between 250 and 4000 Hz, thereby bypassing the outer and middle ear in question and directly testing the response of the inner ear. Unfortunately, since the bone vibration stimulates both inner ears at equal intensity, the patient could possibly respond at 0 dB because he or she is hearing the bone-conducted signal in the good (left) ear. Therefore, the audiologist must "mask" the effects of the good ear by stimulating it with specially produced noise played through an earphone. If, under masking, the patient continues to respond at 0 dB in the right ear, then the examiner can conclude that the right inner ear is functioning normally and that the problem lies in the right outer or middle ear conduction of the signal. If, on the other hand, the patient responds at 30 dB, even though the outer and middle ear have been bypassed, the examiner can conclude that the hear-

Figure 6-2. Sample audiogram showing results of pure tone air conduction testing for the right (O) and left (X) ears.

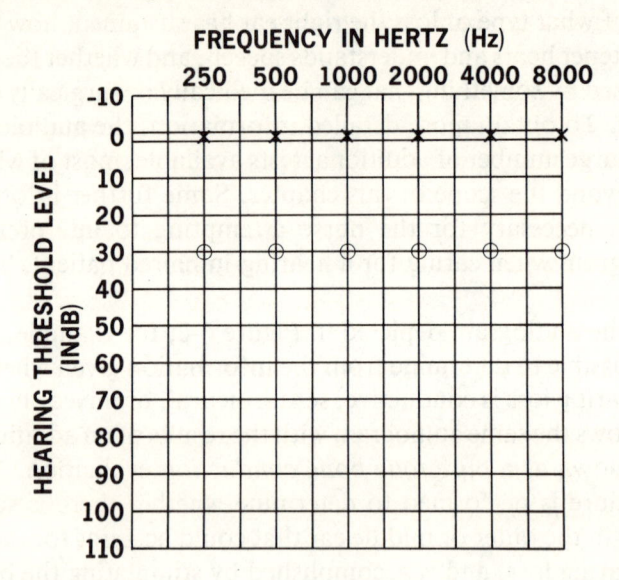

the inclusion of levels less than 0. The more dB required for a person to just barely hear a given frequency, the more severe the hearing loss compared to 0 dB.

During a hearing test, the audiologist uses the audiometer to present the various frequencies at successively lower and lower dB levels until a level is reached at which the listener can just barely hear the sound (this level is known as the *hearing threshold*). The tones are presented through earphones to each ear individually, and the results are plotted to indicate individual thresholds at each frequency for both ears. In the example shown in Figure 6-2, the left ear (indicated by X) exhibits hearing identical to the average normal ear, since each frequency is just barely heard at 0 dB. The right ear (indicated by O), however, shows a hearing loss of 30 dB across all frequencies.

The situation depicted in Figure 6-2 represents a so-called

aid technology and aural rehabilitation techniques have advanced rapidly in recent years, so that virtually everyone with a hearing loss can be helped to some extent.

In most cases, patients present easily identifiable behaviors to suggest hearing loss to the nurse. Individuals' making frequent requests for repetition, playing the radio and television at high volume, talking in abnormally loud voices, and not responding to speakers when their backs are to the speakers all constitute strong indications of hearing loss. Some patients, however, are erroneously labeled as having emotional problems because they exhibit withdrawal, frustration, and anger. Some may even appear to undergo extreme personality changes. While there are bona fide psychological problems that lead to these behaviors, it is worthwhile to consider the possibility that hearing loss is a contributing factor.

THE AUDIOGRAM

Upon referral, the patient's hearing is evaluated by an audiologist using an audiometer. The results are recorded on an audiogram similar to the one shown in Figure 6-2. In its most basic form, the *audiogram* represents the patient's hearing sensitivity for a series of tones whose frequencies are shown from low to high across the top of the grid. These frequencies correspond roughly to various pitches that are heard in everyday listening situations and are measured in Hertz (Hz) or cycles per second. Running from top to bottom on the audiogram are various intensity levels at which each frequency can be tested. The numbers represent decibels (dB), which, again, are relative measures compared to an average value indicated by "0 dB." It should be remembered that the 0 level does not mean that sound is absent, but is the amount of intensity of sound required for "an average normal ear" to just barely hear at that frequency. Some people have better than average hearing; hence

Sensorineural hearing loss is a result of pathology in the inner ear or along the nerve pathway from the inner ear and brain stem. It may be caused by such things as exposure to excessive noise or toxic drugs, Meniere's disease, tumors, and reduced blood supply. A special type of sensorineural impairment, known as *presbycusis,* is found in some elderly individuals and is characterized by an inordinate difficulty in understanding speech, compared to the overall amount of hearing loss.

Mixed hearing loss is any combination of both conductive and sensorineural components. For example, it is common to find patients with noise-induced sensorineural hearing loss that is further complicated by impacted ear wax. Of course, hearing impairment can affect one or both ears in varying types and degrees of severity. In the most severe cases, individuals are born without functional hearing and are usually referred to as being congenitally deaf. Individuals may also be born with hearing loss, but may retain a residual amount of hearing, in which case they are referred to as exhibiting congenital hearing loss. On the other hand, individuals who are born with normal hearing and eventually lose all or some of their hearing are referred to as exhibiting acquired deafness or hearing loss. The distinction between congenital and acquired dysfunction is important, since the degree of handicap and type of communication impairment is usually quite different for these groups.

IDENTIFICATION AND REFERRAL

As with the other disorders discussed in this book, the first important step in treating a hearing loss is that of early referral and appropriate evaluation. All individuals suspected of having hearing problems should be referred to an otologist and audiologist for a thorough workup. Many patients have medically treatable conditions, and even for those who do not, hearing

move in harmony with the frequency and intensity of the sound waves. Attached to the other side of the tympanic membrane is a set of three small bones (*ossicles*) in the space referred to as the middle ear. These bones—the *malleus, incus,* and *stapes*—are connected in such a way that as sound displaces the tympanic membrane and sets them in motion, they efficiently transform the pressure waves in the air into mechanical vibration in the middle ear.

At the end of the chain of bones, the portion of the stapes known as the footplate transforms the mechanical vibration into displacement of fluid in the snail-shaped *cochlea,* located in the inner ear. These fluid displacements stimulate various hair cells along the length of the cochlea, which, in turn, transform the displacements into neural activity. The neural activity is transmitted along the eighth cranial nerve to the brain stem, from where it is sent to both temporal lobes of the brain. Hearing impairment results from reduced transmission of the sound at any point along the route from the outer ear to the brain. For convenience, however, hearing loss is generally categorized as being conductive, sensorineural, or mixed.

TYPES AND CAUSES OF HEARING IMPAIRMENT

Conductive hearing impairment results from dysfunction of the outer or middle ear in the presence of a normal inner ear. Causes of conductive hearing loss include impacted wax or foreign bodies in the ear canal, a perforated or scarred tympanic membrane, fluid in the middle ear cavity, and otosclerosis. A conductive hearing loss by itself does not cause total inability to hear, since the outer and middle ear can be bypassed if sound is of sufficient intensity to vibrate the skull and thus to stimulate the inner ear directly. Conductive hearing impairment can be very serious, however, and should be treated as soon after detection as possible.

Figure 6-1. Schematic representation of the outer, middle, and inner ear.

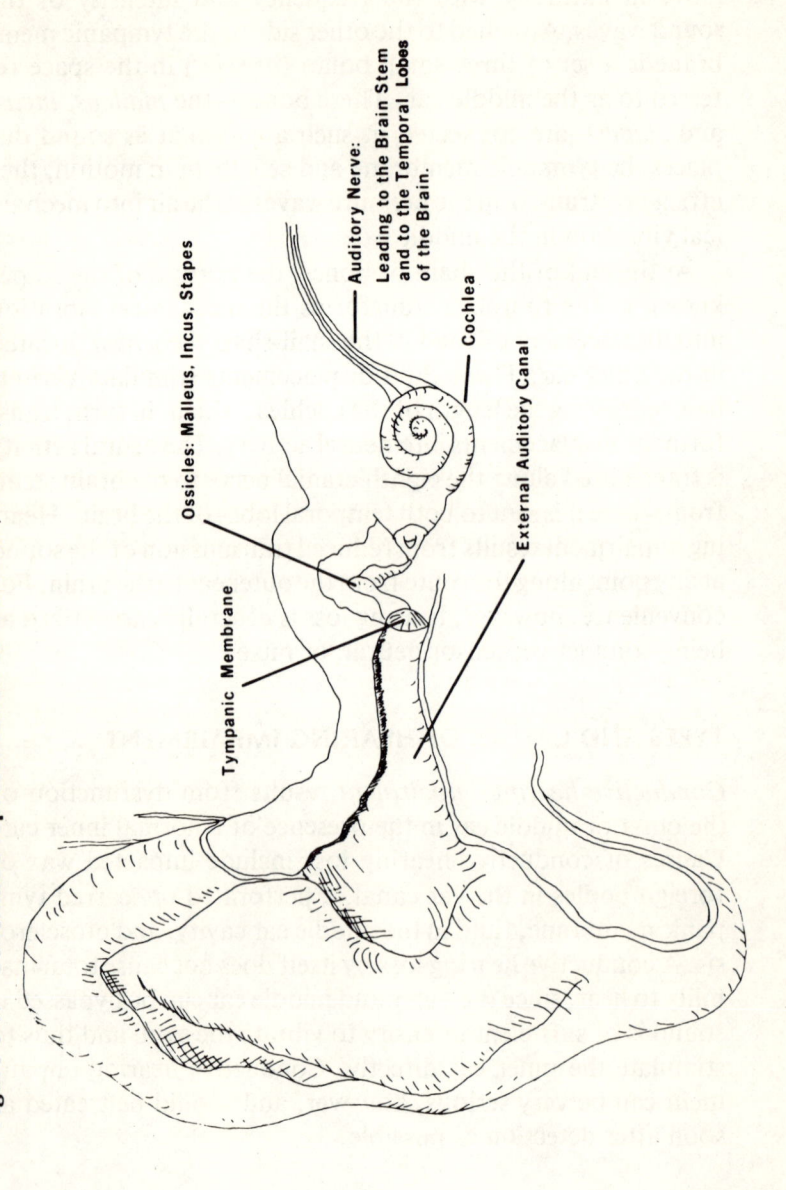

6
HEARING IMPAIRMENT

DEFINITION

In many older patients and in some younger individuals, hearing loss may create a significant problem for the patients in general communication activities, and for the nurse-patient interaction in particular. Hearing loss can be especially troublesome in patients with other communicative disorders, such as aphasia and laryngectomy. Fortunately, the fields of otology and audiology have developed a wide range of diagnostic and therapeutic techniques designed to lessen the impact of hearing loss on individuals in today's society. These techniques can only be of help, however, if the patient is properly referred for evaluation and if the results are used by everyone, including nurses, to make maximum use of current rehabilitation information. Briefly defined, a *hearing impairment* is any significant deviation in hearing sensitivity from that of an average normal ear. As such, it is not an absolute value, but a relative measure compared to what would be expected on the average in a young adult population with no known ear pathology.

NORMAL HEARING

In a normal hearing listener, sound is transmitted as a series of pressure waves through the air into the ear canal of the outer ear (see Figure 6-1). At the end of the ear canal, the pressure waves impinge upon the tympanic membrane, causing it to

and differential diagnosis are crucial for successful treatment of any dysphonic individual.

Sometimes the voice disorder has an organic basis and requires medical or surgical intervention. At other times, the dysphonia is of psychogenic origin and may reflect vocal misuse in the form of hyperfunction and abrupt initiation of voicing. A few disorders have causes yet to be determined.

Treatment of psychogenic dysphonias differs for individual patients, but generally includes elimination or resolution of causative factors, alterations of vocal symptomatology, and psychological support.

Traditionally, nurses are usually not considered in the literature dealing with dysphonia, but they may occupy a pivotal position in the identification and referral of dysphonic patients, in addition to assisting the speech pathologist during a vocal rehabilitation program.

REFERENCES

Aronson, A. E. *Clinical voice disorders: An interdisciplinary approach.* New York: Brian C. Decker, 1980.

Dedo, H. H., & Shipp, T. *Spastic dysphonia: A surgical and voice therapy treatment program.* Houston: College Hill Press, 1980.

Froeschels, E., Kastein, S., & Weiss, D. A method of therapy for paralytic conditions of the mechanisms of phonation, respiration, and glutination. *Journal of Speech and Hearing Disorders,* November 1955, 365–370.

SUGGESTED READINGS

Brodnitz, F. S. *Vocal rehabilitation.* Rochester, Minn.: American Academy of Ophthalmology and Otolaryngology, 1965.

Cooper, M. *Modern techniques of vocal rehabilitation.* Springfield, Ill.: Charles C Thomas, 1973.

ly requires many sessions with the speech pathologist, who is working simultaneously on altering the vocal symptoms and providing an outlet for the patient's resolution of the underlying conflict.

During the course of therapy, the nurse may well be placed in the role of confidant to the patient and liaison with the therapist. This relationship is quite acceptable and should be encouraged. The types of problems underlying conversion aphonias rarely represent severe psychiatric conditions, and the nurse should not be concerned with overstepping professional boundaries. On the other hand, it is helpful to work closely with the speech pathologist to ensure a free flow of information about areas of patient concern and progress that is being made.

CONCLUDING REMARKS

As with glossectomy and maxillectomy, literature directed toward the nurse dealing with the dysphonic patient is scarce. This probably results from the mistaken attitude that the nurse has little to offer in the rehabilitation program, and from the observation that voice disorders have only recently received the scientific and clinical attention of speech pathologists. It will probably take time for information pertaining to multidisciplinary detection, referral, and treatment of dysphonia to work its way into the nursing literature. Perhaps the inclusion of this chapter in the present volume may stimulate further interest in involving the nurse in the total care of the dysphonic patient.

SUMMARY

Nurses may encounter patients exhibiting voice disorders that signal underlying disease, or that result in communicative impairment or embarrassment for the speaker. Early detection

operation for similar reasons. Nevertheless, the individual awaiting such surgery should have had extensive counseling from the laryngologist and speech pathologist as to the advantages and risks of surgery. In the final analysis, the patient must decide whether the sacrifice of a nerve is too high a price to pay for potential freedom from the strain, embarrassment, and communicative impairment resulting from spastic dysphonia. The nurse may easily assume the role of an impartial observer, allowing the patient to verbalize his or her feelings about the voice disorder and the operation. It may also help to guide the patient through the decision-making process by weighing the effects of doing nothing (permanent spastic dysphonia) with proceeding with the operation (potential temporary or permanent resolution of symptoms).

If the patient decides to go through with the operation, the handling of the immediate postoperative period is important. The patient will be able to talk upon recovering from the anesthetic, but the voice may be hoarse and weak; he or she should have been prepared for this, but gentle reassurance from the nurse that the voice is normal for this stage can help to allay unnecessary fears. Also, the patient should be reminded not to push for a loud, clear voice too quickly, since the recurrence of spasm is greater in individuals who insist on rapidly achieving a loud voice. Finally, many patients awake with an overwhelming feeling of relief and release when they find that they can produce spasm-free voice. Nurses should not hesitate to join in the celebration. A normal voice is only noteworthy to someone who has experienced an abnormal one, and its recovery is like meeting an old friend.

The nurse's role in the treatment of conversion aphonia or dysphonia is not as well defined as in other disorders, but the importance of early and appropriate referral remains. Prognosis for recovery of normal voice is dependent upon the patient's willingness to accept the presence of unresolved emotional conflict and to deal with it openly. This process normal-

Unlike mutational falsetto, spastic dysphonia is one of the most frustrating voice disorders with which to work. It usually progresses in severity, beginning as hoarseness or laryngitis and gradually developing into the characteristic spasmodic interruptions during the production of voice. The progression of symptoms usually persists in spite of voice therapy, although various techniques seem to be temporarily helpful. Progressive relaxation, instruction in using a breathy voice, and production of voice at a pitch level higher than normal may result in brief periods of symptom reduction or elimination, but invariably the spasms return at their previous level of severity or worse.

Recently, Dedo and Shipp (1980) have reported a combination of surgery and voice therapy that appears to be a major breakthrough in the treatment of spastic dysphonia. In essence, the surgeon makes a small incision on the left side of the neck and severs the recurrent laryngeal nerve, thus voluntarily creating a unilateral vocal fold paralysis, as discussed earlier. When the patient attempts to produce voice postoperatively, the paralyzed cord does not move, and the preoperative spasm disappears. The typically weak, breathy voice of the individual with a vocal fold paralysis usually resolves with voice therapy, and the patient is able to produce essentially normal voice. Dedo and Shipp report high rates of success, although there is an indication that recurrence of spasm is possible if the patient does not follow the instructions of the voice therapist strictly.

The nurse may encounter the spastic dysphonic patient on a number of different levels. If the nurse is caring for a person with untreated spastic dysphonia, then referral to the laryngologist and speech pathologist should be made. If the spastic dysphonic patient is awaiting surgery to section the laryngeal nerve, he or she is probably apprehensive since it does not seem natural to destroy nerve tissue voluntarily and create a paralysis. Indeed, some surgeons do not perform this

the vocal abuser, and every effort should be made to reduce the amount of vocal output. However, total voice rest should be invoked rarely, if ever. Occasionally, the surgeon may order a brief period of voice rest following laryngeal surgery. Otherwise, the complete cessation of voice for extended time periods has not been shown to be beneficial and may, in fact, be harmful. Patients on prolonged voice rest often have difficulty "finding" the voice once they are allowed to do so, which further complicates the problem.

Finally, patients undergoing therapy for vocal hyperfunction should avoid speaking with restricted movements of their lips and jaws. Instead, they should be encouraged to increase oral activity to relax the throat and laryngeal muscles during speaking. As has been noted in Chapter 4, increased oral activity also enhances overall speech intelligibility.

Treating the patient with mutational falsetto is usually a rewarding experience for those involved. In most cases, the individual does not want to talk in a high-pitched, squeaky voice, but either is afraid to change to a lower pitch after years of using the falsetto voice or is being reinforced, sometimes subtly, to use the abnormally high voice. Usually, the speech pathologist has little difficulty eliciting a lower-pitched, normal voice. Sometimes it can be obtained merely by "talking the patient into it." More often, though, the therapist will assist the patient by having him prolong the sound that is produced by clearing the throat or coughing, or by manually lowering the patient's larynx as he produces voice. The difficulty arises in the patient's acceptance of and adjustment to the new voice. In effect, the patient must be convinced that the lower voice is, in fact, better-sounding and normal for him. A single, positive response by a nurse upon hearing the results of a therapy session can have an immeasurable influence on the success of the rehabilitation program. Continuous praise and reinforcement of the new voice should smooth the transition for using the voice with family and friends.

laryngologist may perform a "vocal fold stripping" to remove a nodule or its cousin, a polyp, the positive results of the surgery will be short-lived unless the patient learns to produce a nonabusive voice and to eliminate poor vocal habits. Contact ulcers rarely respond to surgery of any kind.

In treating the disorders caused by vocal hyperfunction, the speech pathologist employs a large number of techniques to obtain relaxed voice production, without abrupt initiation of voicing (called *hard vocal attack*) and at the appropriate pitch and loudness levels. Since the voice is sensitive to emotional conflict and psychological status, the speech pathologist does not treat the voice as a separate entity, but recognizes that it exists as part of a human being with individual strengths, weaknesses, and needs. Each voice therapy program is customdesigned for the individual, but certain principles are common to all cases of vocal abuse and misuse, and these can be reinforced by the nurse as well as the speech pathologist.

Vocal abuse patients should be frequently reminded to eliminate nonproductive coughing and throat clearing. If patients have difficulty breaking this habit, they should be instructed to swallow slowly when they feel the urge to cough or clear their throats. Swallowing can be made easier by taking sips of water, chewing gum, or sucking on sugarless candy to produce saliva. It is imperative that the patients not yell, scream, talk, or sing loudly, nor should they attack words hard or abruptly when talking or singing. On the other hand, they should avoid whispering, since this act also produces vocal fold friction and irritation.

Smoking and alcohol abuse, in addition to their overall contribution to disease and disability, are especially troublesome for the already irritated larynx. Patients should be educated as to the effects of these substances on health and vocal hygiene. Also, a person with an irritated larynx should avoid environments in which others are smoking. Inordinate amounts of talking and singing cause further vocal tension and stress for

Figure 5-4. Schematic representation of a unilateral left vocal fold paralysis with a compensating right vocal fold during voicing.

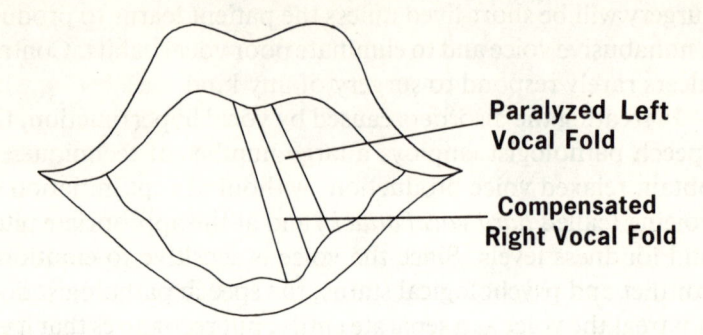

Paralyzed Left Vocal Fold

Compensated Right Vocal Fold

per day) for brief periods (approximately 10 minutes). The nurse can provide assistance by encouraging the patient to perform his or her program as scheduled, and by continued reinforcement of the idea that prognosis for recovery of a strong voice is very good in most cases of vocal fold paralysis.

Occasionally, the voice does not recover via voice therapy alone to a level that is acceptable for the patient's communicative needs. In these cases, it is still possible to achieve a voice closer to normal through the intervention of the laryngologist. In essence, the surgeon injects small amounts of Teflon in the side of the paralyzed fold away from the midline. This procedure has the effect of pushing the mass of the affected fold toward the normally functioning one, thus again creating a more efficient vocal fold vibration. Regardless of the techniques employed, the patient should be given large amounts of psychological support throughout the treatment program. Many individuals react so negatively to vocal impairment that the emotional component is sometimes worse than the actual reduction in vocal fold functioning is.

Vocal hyperfunction leading to organic changes in the vocal folds, such as vocal nodules and contact ulcers, generally falls in the domain of the speech pathologist. Even though the

In some organically based disorders, and in virtually all psychogenic dysphonias, the multidisciplinary efforts of the laryngologist, speech pathologist, and nurse take on a much more important role. In many cases, treatment extends over a period of several months, and continued support and encouragement are necessary for successful rehabilitation. The most illustrative example of an organically based voice disorder that requires multidisciplinary effort is unilateral vocal fold paralysis. In this condition, the nerve supply to one side of the larynx is interrupted by surgery (e.g., thyroidectomy), trauma, tumor pressing on the nerve, or some unknown cause. As a result, the affected vocal fold is weakened or paralyzed, making it difficult or impossible to bring both vocal folds together for normal voicing.

The voice that is produced is usually low in volume, breathy, and lacking in pitch variability. The effect can be emotionally traumatic, especially in patients whose livelihood depends upon normal voice production. Fortunately, most cases of vocal fold paralysis can be helped by the combined efforts of the laryngologist and speech pathologist. Usually, the speech pathologist teaches the patient to phonate against resistance. This simply means that the patient produces voice while pushing against a wall, attempting to lift a heavy object, or pulling up on the arms or seat of the chair in which he or she is sitting. This approach, first reported by Froeschels, Kastein, and Weiss (1955), is based on the belief that if enough tension is created while voice is being produced, the normally functioning vocal fold is pulled across the midline, where it usually stops, and meets the paralyzed vocal fold at its new position (see Figure 5-4). The so-called vocal fold compensation is possible because of the unique muscular arrangement on the posterior part of the larynx.

The patient should gradually increase the amount of phonation and resistance by practicing frequently (several times

The voice characteristic exhibited by a person with conversion aphonia is essentially a whisper during voluntary speech acts, in the presence of normal voicing during coughing, laughing, and crying. A conversion dysphonia is similar, but whispering is usually intermittent with periods of voicing interspersed during conversation.

EVALUATION AND TREATMENT

As stated earlier, the first step following the detection or suspicion of a voice disorder is immediate referral to a laryngologist. Upon examining the larynx, the laryngologist usually elects one or more of the following options:

1. Referral to other medical specialties or for further tests if the disorder appears to have an organic basis.
2. Direct intervention with medical and/or surgical treatment.
3. Referral to a speech pathologist for further evaluation, preoperative voice recording, voice therapy, or postoperative follow-up.
4. No further recommendations, such as in a case where maximum therapeutic benefit has been achieved.

In terms of the various voice disorders already discussed, those with an organic basis tend to respond best to medical intervention. For example, in dysphonias caused by endocrine disturbance or allergies, the voice may return to normal with appropriate medication and preventive measures. The role of the speech pathologist and nurse will be somewhat limited in these cases, consisting primarily of providing appropriate referral, obtaining pretreatment and posttreatment voice recordings, and administering routine nursing care.

disorder is *mutational falsetto*. This problem is typically seen in males, although females may exhibit similar problems on occasion. In essence, when the individual reaches puberty and the voice should drop to a lower pitch, the voice continues to be produced in a falsetto mode, for reasons not fully understood. It should be noted that the vocal folds are anatomically and physiologically normal in individuals with this disorder. It is unfortunate that many patients with mutational falsetto suffer undue anguish and embarrassment when, in fact, early referral to a speech pathologist usually results in successful acquisition of a normally pitched voice.

Spastic dysphonia is a term used to identify a voice disorder in which the voice is intermittently interrupted by spasms, which create such tension on the vocal folds that air from the lungs is completely impeded and no vibration results. As might be expected, this is an extremely disabling condition and, unlike most other voice disorders, does not respond well to traditional voice therapy techniques (to be discussed later). Recently, however, a surgical approach has been found to be effective in some cases and is covered in the section dealing with rehabilitation. For many years, spastic dysphonia was considered to result from deep-seated emotional and psychological conflicts. More recently, there is evidence to suggest that, in some cases, a subtle neurological disturbance may be involved.

Finally, *conversion aphonia* or *dysphonia* may be encountered on occasion in a hospital setting. According to Aronson (1980),

> A conversion voice disorder (1) exists despite normal structure and function of the vocal folds, (2) is created by anxiety, stress, depression, or interpersonal conflict, (3) has symbolic significance for that conflict, and (4) enables the patient to avoid facing the interpersonal conflict directly and extricates the person from the uncomfortable situation. (p. 142)

Figure 5-3. Schematic representation of bilateral contact ulcers.

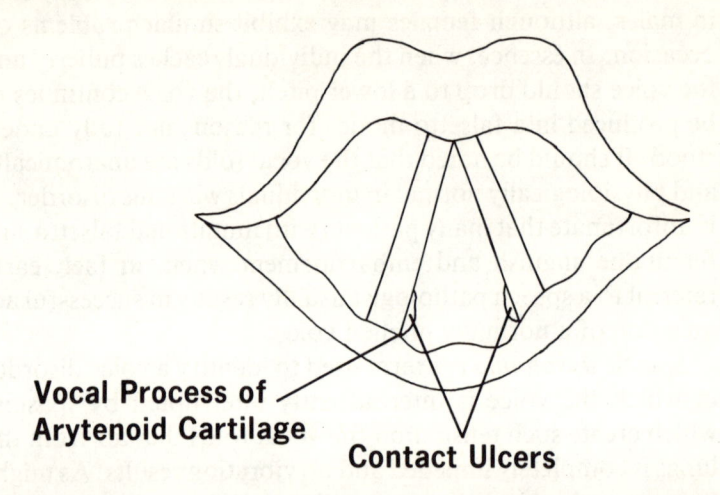

Vocal Process of Arytenoid Cartilage

Contact Ulcers

posterior point at an area referred to as the vocal process of the arytenoid cartilage. Contact ulcers are usually found in adult males who use an extremely low-pitched voice and bring the vocal folds together abruptly at the initiation of voicing. It is also common to find a history of gastric ulcers and esophageal reflux in patients with contact ulcers.

Ventricular phonation is a term used to describe a voice that is produced partially or totally with the ventricular or false vocal folds. The false vocal folds are muscles in the larynx that lie above the level of the true vocal folds and are usually only brought into contact with each other during coughing or when the individual holds his or her breath while doing heavy lifting. In some cases, however, the individual uses such extreme laryngeal tension during conversational speech that the false (ventricular) vocal folds participate in the production of voice. This activity produces a low-pitched, harsh, and strained voice, usually with accompanying visible external neck tension.

A rather interesting and relatively easily treatable voice

Figure 5-1. Schematic representation of normal vocal folds during breathing as seen on indirect laryngoscopy.

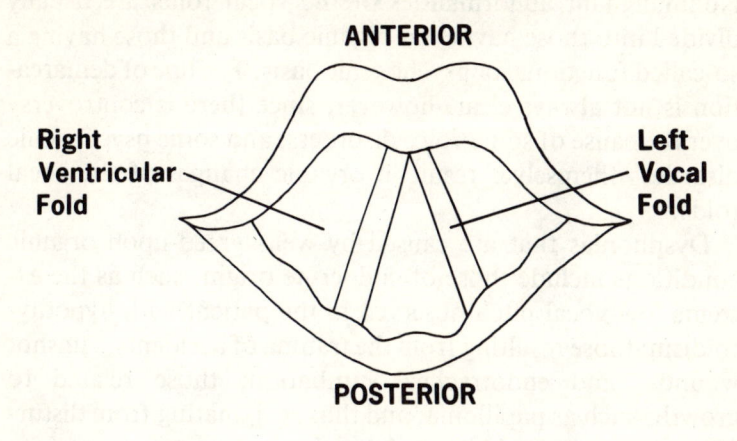

ANTERIOR

Right Ventricular Fold

Left Vocal Fold

POSTERIOR

Figure 5-2. Schematic representation of bilateral vocal nodules.

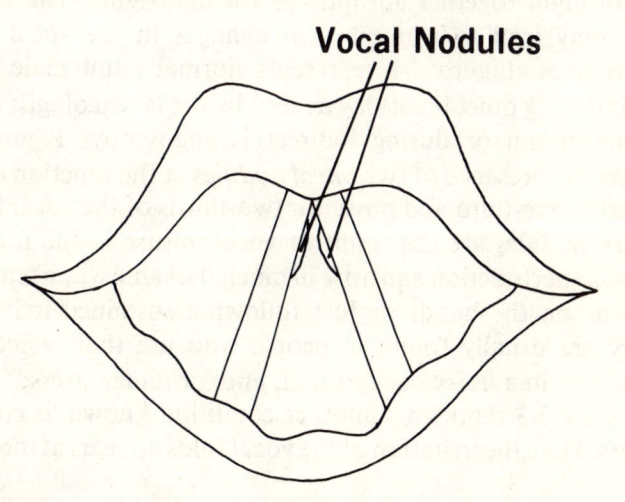

Vocal Nodules

TYPES AND CAUSES OF DYSPHONIA

Nonmalignant abnormalities of the vocal folds are usually divided into those having an organic basis and those having a so-called functional or psychogenic basis. The line of demarcation is not always clear, however, since there is controversy over the cause of some voice disorders, and some psychogenic disorders themselves result in organic changes of the vocal folds.

Dysphonias that are caused by well-agreed-upon organic conditions include those of endocrine origin, such as the extreme low vocal pitch observed in the patient with hypothyroidism; those resulting from the trauma of accidents, gunshot wounds, and endotracheal intubation; those related to growths such as papilloma; and those originating from disturbances in the central or peripheral nervous system, such as vocal fold paralysis or weakness.

Psychogenic voice disorders, on the other hand, many times involve *vocal hyperfunction*—a condition in which the vocal folds are vibrating under greater than normal tension and/or are brought together abruptly as voicing begins. This condition may lead to a number of changes in the vocal folds themselves. Figure 5-1 represents normal adult male vocal folds during quiet breathing as seen by the laryngologist in the laryngeal mirror during indirect laryngoscopy. Figure 5-2 shows the presence of two *vocal nodules* at the junction of the anterior one-third and posterior two-thirds of the vocal folds. These nodules are the result of vocal misuse in the form of vocal hyperfunction and arise in much the same way as calluses appear on the hands or feet following sustained irritation. They are usually found in people who use their voices excessively, in a noisy background, and/or under stress.

Figure 5-3 represents another condition known as *contact ulcers*. Here the irritation of the vocal folds appears at the most

them as they are closed against each other in the larynx. In order for the vocal folds to produce the smooth, pleasant sound associated with normal voice production, they must open and close with precise timing and with the correct amount of tension. Any condition that changes the size of the vocal folds, the coordination of their movement, or the amount of tension used during their vibration may cause changes in the voice that can be detected by the listener.

Terms such as hoarseness, harshness, raspiness, and breathiness are usually used to describe these changes in the voice, but they are subjective terms and rarely mean the same thing to all listeners. In addition, vastly different underlying conditions can result in similar judgments of hoarseness or raspiness. Therefore, the first rule of thumb is to refer to the laryngologist any patient who has a voice that the nurse believes is abnormal in some way.

The laryngologist then examines the larynx and makes an initial determination as to the cause of the voice disorder and recommendations for appropriate treatment. The initial examination can usually be done in a short time, so a referral should be made even if the nurse is not sure that a disorder exists. Many patients who are now laryngectomees or who have expired might have avoided such fates if they had been referred to the appropriate physician early in the course of their disease.

If the laryngologist suspects cancer, he or she usually schedules the patient for direct laryngoscopy and biopsy to verify the diagnosis. Assuming cancer is diagnosed, then radiotherapy, surgery and/or chemotherapy will probably be recommended (see Chapter 3). Most voice disorders are not caused by cancer, however, and the dysphonias discussed here are those that result from other conditions and for which the nurse can assist in the rehabilitative effort.

5

VOICE DISORDERS

DEFINITION

Disorders of the voice (*dysphonia*), with the exception of the total loss of voice in the laryngectomee, are difficult to define, since Western culture accepts a wide variety of voices as falling within the normal range. In general, a voice is considered to be abnormal if its pitch, loudness, and/or overall quality do not conform to the listener's expectations based on the speaker's age, sex, and physical stature. In some instances, a voice therapist might consider a voice to be abnormal, but the speaker is reinforced to use the voice by listeners who find the voice to be "sexy" or distinctive, as in the case of a movie star or television personality.

For the purposes of this discussion, however, the types of voice disorders that might be encountered by the nurse in a medical setting are of primary concern. As such, the remainder of the chapter is devoted to disorders of voice that may indicate an underlying disease or illness requiring therapeutic intervention, and to disorders that result in some degree of impairment of communicative ability for the patient, including voices that deviate to the extent that they draw attention to themselves or cause embarrassment for the speaker.

IDENTIFICATION AND REFERRAL

As indicated in the discussion of laryngectomy (see Chapter 3), the vocal folds vibrate as a result of air being blown through

tomy and maxillectomy, with or without additional excision of related structures.

Swallowing problems associated with glossectomy and maxillectomy usually result from the inability to transport the food from the front to the back of the oral cavity and into the esophagus. Techniques and devices are available to assist the patient in learning to compensate for the absent or immobile structures and to regain some degree of swallowing ability.

Speech intelligibility may be severely impaired in the glossectomee or maxillectomee because of the inability to produce many consonant sounds and changes in resonance characteristics of the oral and nasal cavities. Prosthetic reconstruction and speech rehabilitation can enhance communicative effectiveness by taking advantage of compensatory adjustments and increased oral activity.

The psychological impact of severe facial disfigurement in the glossectomee or maxillectomee may create the greatest challenge for the nurse, and is probably best handled with an attitude of support and acceptance of both the positive and negative aspects of such surgery.

REFERENCES

Larsen, G. L. Dysphagia. In L. J. Bradford and R. T. Wertz (Eds.), *Communicative disorders: An audio journal for continuing education.* New York: Grune & Stratton, 1981 (audiocassette).

Lauciello, F. R., Vergo, T., Schaaf, N. G., & Zimmerman, R. Prosthodontic and speech rehabilitation after partial and complete glossectomy. *Journal of Prosthetic Dentistry,* February 1980, 204-211.

Skelly, M. *Glossectomee speech rehabilitation.* Springfield, Ill.: Charles C Thomas, 1973.

SUGGESTED READING

Daly, K. M. Oral cancer: Everyday concerns. *American Journal of Nursing,* August 1979, 1415-1417.

deviations from accepted norms for physical appearance, and the disfigured patient is likely to encounter negative reactions from many observers.

Unfortunately, the availability of large numbers of materials and community resources to assist in dealing with aphasia, dysarthria, and laryngectomy is conspicuously lacking for the glossectomee or maxillectomee, and nurses are generally left to their own devices. Some help may be available through local American Cancer Society units, VA medical centers, and state and federal agencies. It is also advisable to contact a psychiatry or psychology referral service, since some counselors specialize in dealing with individuals who have undergone disfiguring surgery. In many cases, however, the nurse shares much of the responsibility for dealing with the patient and family.

Moral support and acceptance of patients on the basis of their human qualities rather than their physical appearance are two important contributions that nurses can make to glossectomees or maxillectomees. Unrealistic optimism should be avoided, since patients can easily look in a mirror and assess the extent of change in their physical appearance. Extreme pessimism, on the other hand, is also unwarranted, since many patients return to productive, fulfilling lives with a revised outlook on the meaning of physical appearance and its relationship to being alive and free of disease. Continued emphasis of the positive with gradual acceptance of the negative or unchangeable are important in dealing with any disability, but are absolutely vital in the successful rehabilitation of the glossectomee or maxillectomee.

SUMMARY

Surgical removal of structures above the level of the larynx can also affect speaking and swallowing ability. The two most commonly encountered conditions in this category are glossec-

Intelligibility of a patient's speech is related to how well the prosthodontist and/or plastic surgeon are able to reconstruct an adequate barrier between the oral and nasal cavities. Some individuals regain speech that is easily understood, while others have such large defects that speech will always be moderately or severely impaired. In most cases, however, regardless of the original level of intelligibility, understanding can be improved in several ways.

All of the techniques previously mentioned are equally applicable to the patient who has undergone maxillary resection. Especially important are reinforcing the patient to talk with as wide a mouth opening as possible and reducing the rate of speech. By increasing the amount of mouth opening, the air and sound coming from the larynx has less resistance through the oral cavity and is less likely to flow upward through the nose. If the mouth is kept relatively closed and inactive, even with a small oral-nasal defect, there is greater likelihood that the air and sound will be directed through the nasal cavities.

Reducing the rate of speaking helps to accommodate the wider mouth opening and also gives the listener added time to process the altered speech pattern and to make better use of contextual cues. When understanding becomes severely impaired—causing frustration for both speaker and listener—and rewording has been unsuccessful, it is sometimes helpful to have the patient write one or two important cue words to get the listener "tuned in," and then to proceed with speaking. This technique is also useful when dealing with dysarthric, laryngectomy, and glossectomy patients in similar situations.

THE NURSE AS COUNSELOR

Although the speech and swallowing problems encountered by the glossectomee or maxillectomee can be significant, they may be of lesser concern than the psychological adjustment to the facial disfigurement. Our society does not deal well with

REHABILITATION

As the patient's condition stabilizes, he or she is transferred to an acute care surgical ward to continue healing and begin rehabilitation. Fortunately, as has been seen with the glossectomee, it is sometimes possible for a prosthodontist to alleviate a substantial part of the problem by creating an appliance similar to a denture, but with extensions to make up for the absent muscle and bone tissues. Sometimes the plastic surgeon is able to create a palatal shelf from tissue borrowed from other sites. Rarely, however, does a prosthesis and/or surgical reconstruction result in complete recovery of function. The nurse and speech pathologist can help the patient use the prosthesis or surgical reconstruction more effectively, though, by teaching and reinforcing compensatory techniques for swallowing and speech.

Unless the tongue and larynx are involved, the problem with swallowing is usually not aspiration, but redirection of the food or liquid through the nasal cavity. There may also be difficulty in transporting the food backward from the front of the mouth if jaw mobility has been reduced. Use of the feeding syringe with tubing is advisable to overcome both of these difficulties. Gelatin and liquids can be introduced in the back of the oral cavity on the side away from the surgically involved one. This is usually sufficient to trigger a reflex swallow and to direct the material down into the pharynx and esophagus rather than up into the nasal cavity. Care should be taken to introduce small amounts initially and to increase patients' tolerance gradually while teaching them how to use the syringe. As with glossectomy, psychological obstacles may impede progress, and patients should always be "talked through" each step of the swallowing process. It is also preferable to remove the tracheostomy and nasogastric tubes if at all possible, since both are deterrents to normal swallowing behavior.

PREOPERATIVE COUNSELING

During the preoperative period, the potential *maxillectomee* (patient about to undergo some form of maxillectomy) should be dealt with in a fashion similar to that of the potential laryngectomee or glossectomee. Again, the introduction of another individual with the same type of surgery is a delicate subject, in view of the severe facial disfigurement suffered by some patients. On the other hand, questions should be answered honestly and directly, since lack of adequate information sometimes causes more apprehension than the actual knowledge of what to expect.

The nurse should be especially sensitive to the needs of the spouse and family, gradually preparing them during the preoperative period. There is probably no correct method for preparing someone for the consequences of major ablative head and neck surgery, but repeated sessions in which information is exchanged and questions are answered are usually of some benefit in leading to more rapid adjustment during the postoperative period.

POSTOPERATIVE CARE

In the immediate postoperative stage, the patient with a partial or total maxillectomy is confined to the SICU for several days until stable. A nasogastric tube is usually in place, along with a temporary tracheostomy, IV, and packing of the recently created facial cavity. Communication is most efficiently conducted via writing and gesturing, although a communication board with pictures and words of the most common needs in an intensive care unit is sometimes helpful. If the patient has had an eye removed, care should be taken to place materials in a position where vision in the remaining eye is unimpeded.

the other muscles surrounding the upper pharynx. For example, when a talker says the word "notebook," the velum and surrounding muscles stay relaxed during the production of the *n* sound, but close off the nasal cavities during the production of the remainder of the word. The typical *n* sound produced is a result of the resonating properties of the nasal cavities. If the hard and soft palates remain intact for a patient with hemimaxillectomy, nonnasal sounds should be produced normally, since they do not require resonance through the nasal cavities. The *n, m,* and *ng* sounds may change in quality, however, since they are now resonating in a much larger space caused by removal of maxillary structures. An analogous change takes place in the sound produced by a string vibrating over the body of a violin as compared to a cello. The change in sound quality can also be demonstrated by comparing the production of such words as "mom" and "none" while the patient has the surgically produced maxillary cavity packed with gauze and while it is open. Other than this change in quality, however, the speech of the patient should be intelligible and communication not a problem.

In the individual whose velum and hard palate have been removed, the effect on speech production is more pronounced. In this case, not only do the nasal sounds resonate differently in the maxillary cavity, but nonnasal sounds are difficult to produce, since air is allowed to leak through the nose and cavity produced by the surgery. Thus, it is difficult or impossible to produce sounds such as *p, t, s,* and most others, since they require the ability to force air through the oral cavity under considerable pressure. In addition, the actual articulation of some sounds may be impaired if the tongue cannot press against the palate or if the lips cannot be properly closed. In these patients, speech is likely to be severely impaired. Likewise, swallowing is difficult, since the food is pushed up into the newly created cavity rather than down into the pharynx and esophagus.

It must be emphasized that each patient presents as an individual problem and that rehabilitation efforts must be geared to that particular individual. The extent of surgical resection is not the only parameter which dictates the degree to which a patient may be rehabilitated. Factors such as the mobility of the remaining oral and paraoral tissues, neuromuscular coordination, mental competence, and motivation are equal parameters that dictate the degree to which the patient's oral functions may be rehabilitated. (p. 211)

MAXILLECTOMY: DEFINITION

Surgical removal of structures above the mouth can also have a deleterious effect on speech communication and swallowing. Partial or total maxillectomy is the most commonly observed operation to fall in this category. A *hemimaxillectomy* involves removal of half the upper jaw and may include partial or total removal of the hard and soft palate, removal of the eye and surrounding bones, and sacrifice of cheekbone and muscle.

EFFECTS ON SPEECH AND SWALLOWING

If the hard and soft palates are left intact, speech and swallowing should be essentially normal, with a few exceptions. In the normally functioning individual, most speech sounds require that the hard and soft palates close off the nasal passages to allow the sounds to be produced in the oral cavity. In the English language, however, three sounds—*n, m,* and *ng*—require that the nasal cavities be coupled to the oral cavity so that they give the characteristic nasal resonance quality to these three sounds. This coupling occurs when the soft palate (velum) is kept relaxed rather than being pulled together with

and permanent adjustments are required. Coordination between the nurse and speech pathologist should help provide a realistic, honest appraisal of each patient's current status, potential for improvement, and areas where adjustment is necessary.

With regard to specific suggestions for enhancing communication, much of what has already been discussed for dysarthria and laryngectomy also holds true for glossectomy. Creating a conducive communication environment, gaining familiarity with the speaker, and providing patient, tolerant listening all have a positive effect on communication. Especially applicable to the glossectomee is the observation that increased activity of the oral structures used during speaking results in more intelligible speech. Patients who talk with limited movements of their mouths have a more difficult time being understood and should be reminded to use as much movement as possible within the limits of their disability. On the precautionary side, it is important not to demand a level of speaking expertise that is beyond the capability of a patient. Moreover, frustration can be reduced when the patient is misunderstood by encouraging him or her to reword the message, rather than persisting in repetition of the initial phrase or sentence.

Recently, it has been reported (Lauciello, Vergo, Schaaf, & Zimmerman, 1980) that the prosthodontist can provide assistance for speech and swallowing in the glossectomee. Various prostheses have been adapted from conventional dentures in an effort to compensate for the loss of tongue and jaw tissue. This approach, although in its developmental stage, shows promise for advancing glossectomee rehabilitative techniques in a multidisciplinary environment. The following passage from Lauciello et al. (1980) was written with prosthetic rehabilitation in mind, but serves as well for all those interested in patients who have undergone glossectomy:

tion rehabilitation program as well. Again, the term *communication rehabilitation* rather than *speech* has been specifically chosen to emphasize that humans communicate in many ways other than through speech, and these may all be used individually or in combination by the glossectomee.

Writing, communication boards, electronic augmentative devices, changes in facial expression, and gesture are available to glossectomees as well as to patients discussed earlier. In addition, glossectomees can communicate a number of different moods and expressions simply through vocal inflections caused by subtle pitch and loudness changes. Nurses and other individuals dealing with glossectomees on a daily basis can enhance communication through awareness of the method or methods that work best for each patient, and by reinforcing the use of them during communication attempts.

Ultimately, the most efficient means of daily communication is through speech. In the case of the glossectomee, this usually means usable rather than normal speech and is arrived at through a lengthy therapeutic process designed to teach the patient how to use remaining structures to compensate for missing ones. Two common problems arise during glossectomy speech rehabilitation, both of which can be substantially reduced via nursing intervention: Unrealistic expectations on the part of the patient and family and loss of motivation to continue lengthy therapy are attitudes that require constant attention throughout the course of treatment.

It is important to reassure glossectomees continually that the rehabilitative effort is worthwhile, even though it may require large expenditures of time and effort. At the same time, the negative aspects of glossectomy must be kept in perspective, and patients should not be given the illusion that speech will return to normal and life will revert to the preoperative level. They must gradually accept the premise that their facial features, eating, and speaking have been permanently altered

patients that they can swallow, even though it may take time and a good deal of effort on their part. Fear of choking is as much a deterrent to a successful rehabilitation program as actual physical disability is. As with neuromuscular dysphagia, proper positioning and conscious awareness of the swallowing process are necessary. Also, patients should be routinely trained to hold their breath, swallow, and cough each time swallowing is attempted.

A feeding syringe can be used to place a flavored gelatin into the back of the mouth in preparation for swallowing. If the jaw is involved and mouth opening is restricted, it may be necessary to extend the syringe with a short length of tubing. Small amounts of gelatin and large quantities of psychological support should be used initially. In many glossectomees, thick, ropy secretions collect in the pharynx and further complicate the problems with swallowing. According to Larsen (1981), this can be alleviated by dissolving, but not swallowing, a papain tablet approximately 10 minutes prior to feeding. Cleaning the oral and pharyngeal areas with a lemon-glycerine swab dipped in meat tenderizer is also suggested as a reasonable substitute.

Once a patient is swallowing gelatin consistently and without difficulty, other foods can be tried. Usually it is necessary to blenderize bulky items, and a dietician should be consulted as to appropriate nutritional content. As they progress, patients should be gradually assuming responsibility for their own feeding by learning correct placement of the syringe and determining the amount of food that can be swallowed without difficulty. Following successful rehabilitation, many patients abandon the syringe and manage to adapt to regular spoon feedings, although they rarely return to a normal diet.

While the swallowing rehabilitation program is in progress, the speech pathologist is probably developing a communica-

glossectomee, unless, of course, the glossectomee has also had his or her larynx removed. Nevertheless, speech pathologists are often asked by nurses and others to provide an "artificial speaking machine" for a person who has just undergone a glossectomy. Perhaps in the future it will be possible to create a substitute for the tongue and jaw, but unfortunately the technology remains in its infancy, and a glossectomee must depend upon residual structure and function for communication.

Once the glossectomee's condition is stable, rehabilitation of swallowing and speech begins. The basic approach will be similar for each—that is, making maximum use of remaining capabilities and compensating for functions that have been altered or eliminated. Skelly (1979) has hypothesized that competence in swallowing is a necessary prerequisite for successful speech rehabilitation. In many clinics, however, swallowing and speech problems are addressed simultaneously.

As with dysphagia caused by neuromuscular paralysis or weakness (see Chapter 2), the two major problems resulting from swallowing dysfunction in the glossectomee are aspiration of material into the lungs and maintenance of adequate nutrition. If the sensory nerves to the pharynx and larynx have been sacrificed and the patient no longer demonstrates a reflexive cough or swallow, it is unlikely that he or she will benefit from a swallowing rehabilitation program. Esophagostomy or gastrostomy are usually the alternatives considered in such circumstances. If the reflexive cough and swallow are present but the glossectomee cannot chew and move food to the back of the mouth, then a dysphagia rehabilitation program should be initiated. Swallowing retraining for the glossectomee is similar in many ways to that outlined in Chapter 2 for neuromotor dysphagia, although notable differences exist.

In all cases, patients should be counseled extensively about normal swallowing and the changes that have occurred in their swallowing apparatus. It is important throughout to reassure

with local hospital policy on this issue. As with laryngectomy, however, the preoperative counseling of a glossectomee should be frank and complete, and should include spouse or relatives. Since the patient is usually admitted to the hospital 3 days to a week prior to surgery, the nurse is likely to become a main participant in the preoperative preparation.

Again, the amount and complexity of information given is not nearly as important as the calm reassurance that the patient is in the care of a competent team of professionals who have dealt with glossectomy and its consequences many times before. Fear and apprehension should be recognized as real and natural under the circumstances. Explaining that some temporary difficulty in swallowing can be expected should help alleviate postoperative psychological difficulties. Also, as discussed earlier, it is advisable to prepare the spouse and family if they are to visit the patient postoperatively in the intensive care unit.

POSTOPERATIVE REHABILITATION

Following surgery, the patient spends several days in the surgical intensive care unit (SICU) until stable. He or she probably has a nasogastric tube and tracheostomy, although both should be temporary if the rehabilitation program is a success. Communication can be accomplished via writing or communication board (see Chapter 2), and when the patient feels well enough, he or she may be able to make personal needs known by attempting speech while occluding the tracheal tube and redirecting the air stream through the vocal folds and vocal tract. Even if these preliminary attempts are not totally successful, many patients are given a psychological boost when they find that they can produce voice and some speech.

It should be obvious by now that an artificial voice aid such as those used by laryngectomees would be of no benefit to a

and changes in oral and pharyngeal cavity size, all of which affect resonance balance for vowel sounds and some consonants. Depending upon the individual, speech impairment resulting from glossectomy may vary anywhere between mild and virtually unintelligible.

Alterations in swallowing following glossectomy are usually limited to the first or oral phase and parts of the second or pharyngeal phase (see Chapter 2). The absent or immobile tongue tissue causes difficulty in chewing food (especially if the jaw is involved) and in directing it back toward the throat to begin the second stage of swallowing. If the surgery has been extensive and the sensory nerve supply to the pharynx and larynx is interrupted, only the reflexive third or esophageal stage may be intact. In some cases, the physical inability to swallow may be temporary, but psychologically related problems may endure.

PREOPERATIVE COUNSELING

Preoperatively, the procedures for the glossectomee are similar to those for the laryngectomee. Once the diagnosis has been made, the surgeon explains the situation to the patient. The patient should then be referred to the speech pathologist for counseling as to the effects of surgery on speech and swallowing, and for information on rehabilitation alternatives. At this time, the patient should also be audiotaped or videotaped, receive a hearing evaluation, and be referred to other services as appropriate.

Some physicians and speech pathologists advocate introducing the patient to another individual who has had a glossectomy. This is a controversial issue, however, since the physical appearance following a glossectomy/mandibulectomy/neck dissection can be much more devastating than that following laryngectomy. It is advisable to become familiar

NORMAL SPEECH PRODUCTION

The sounds of normal speech generally fall into two categories: vowels and consonants. Vowels are produced by directing the sound created by vocal fold vibrations through the pharyngeal and oral cavities (collectively referred to as the vocal tract). Changes in the size and shape of the vocal tract affect its resonating properties and account for the differences between vowel sounds such as *ee, ah,* and *oo*. If the size and shape changes occur rapidly, sometimes two vowels are combined to form a diphthong such as *ow* and *ae*.

Consonants are formed by partial or complete blockages in the vocal tract and may be voiced or unvoiced. The location and type of blockage in the vocal tract gives each consonant its distinctive properties. For example, if the air stream is momentarily trapped behind both lips and suddenly released, the *p* sound would be produced. If voicing is added, the sound would change to *b*. By the same token, if the air stream is only partially blocked by the lower lip against the upper teeth, for example, an *f* sound is formed; with voicing added it becomes a *v*. Nasalized consonants are discussed later in the chapter. With the exception of a few sounds formed exclusively with the lips or lips and teeth, the tongue is the major articulator during the production of speech. Vocal tract size and shape modifications resulting in the resonance changes characteristic of vowels and diphthongs are chiefly influenced by tongue movements. So, too, the tongue is the primary articulator in partially or completely blocking the vocal tract during consonant production.

CHANGES IN SPEECH AND SWALLOWING

In the glossectomee, communication problems are the result of interruption of normal articulation because of absence of functional tongue tissue, reduced mobility of remaining tissue,

4

GLOSSECTOMY AND MAXILLECTOMY

GLOSSECTOMY: DEFINITION

The human tongue is made up of a set of finely coordinated muscles and is a major contributor to normal chewing, swallowing, and speech. Each year, thousands of individuals develop cancer involving the tongue, much of which is treatable if detected early. Surgical procedures have been refined to the point that patients undergoing glossectomy are now a common occurrence on hospital wards across the country.

Glossectomy refers specifically to surgical removal of all or part of the tongue. In many cases, however, it is common to encounter partial removal of the lower jaw (*mandibulectomy*) and dissection of neck tissue as well. As with laryngectomy, the removal of tongue, jaw, and neck tissue is usually the result of cancer, although trauma may also be a factor in some cases. Unlike the laryngectomee, however, the *glossectomee* (the person who has undergone a glossectomy) retains voice but has reduced ability to formulate speech sounds. Also unlike the typical laryngectomee, the glossectomee accomplishes breathing in the normal manner, but chewing and swallowing are almost always impaired.

International Association of Laryngectomees. *Helping words for the laryngectomee.* (Available through local American Cancer Society unit.)

Keith, R. L. *A handbook for the laryngectomee.* (Available from Interstate Printers and Publishers, Danville, Illinois 61832.)

Lauder, E. *Self-help for the laryngectomee.* (Available from the author at 1115 Whisper Hollow, San Antonio, Texas 78238.)

Waldrop, W. F., & Gould, M. A. *Your new voice.* (Available through local American Cancer Society unit.)

Films

Diedrich, W. M. *Alaryngeal speech.* (Available from University of Kansas Medical Center, Kansas City, Kansas 66103.)

International Association of Laryngectomees. *Three critical minutes: Emergency air for neck breathers.* (Available through local American Cancer Society unit.)

International Association of Laryngectomees. *A second voice.* (Available through local American Cancer Society unit.)

Records and Tapes

Doehler, P. *Alaryngeal speech instruction.* (Available from Mrs. Doehler at 243 Charles St., Boston, Massachusetts 02114.)

Hyman, M. *Alaryngeal speech instruction.* (Available from Mr. Hyman at Bowling Green State University, Bowling Green, Ohio 43402.)

Lauder, E. *Post-laryngectomy speech instruction.* (Available from Mr. Lauder at 1115 Whisper Hollow, San Antonio, Texas 78238.)

Searcy, L. Nursing care of the laryngectomy patient. *RN,* October 1972, 35–41.

OTHER RESOURCES

Artificial Aids and Accessories

Western Electric 5A and 5B neck-type electrolarynx. (Available through local Bell Telephone representative.)
Servox neck-type speech aid. (Available through the Audiophone Company of South Texas, 712 Majestic Blvd., San Antonio, Texas 78205.)
Bart neck-type vibrator. (Available through Park Surgical Co., Inc., 5001 New Utrecht Ave., Brooklyn, New York 11219.)
Variaton neck-type speech aid. (Available through Romet, Box 102, 5733 Myrtle Ave., Ridgewood, New York 11227.)
Cooper-Rand electronic oral-type speech aid. (Available through Luminaud, P.O. Box 257, Mentor, Ohio 44060.)
Tokyo pneumatic-type speech aid. (Available through Mr. Red Woodward, 3132 Waits, Fort Worth, Texas 76109.)
Medical identification bracelet. (Available through the Medic Alert Foundation, Inc., Turlock, California 95380.)
Foam rubber stoma covers and shower collars. (Available through Medmart Inc., P.O. Box 1241, Beverly Hills, California 90213.)
Cotton stoma cover. (Available through Romet, Box 102, 5733 Myrtle Ave., Ridgewood, New York 11227.)

Materials for Patients and Families

Books and Pamphlets

American Cancer Society. *First aid for laryngectomees.* (Available through local American Cancer Society unit.)
Friedburg, Stanton A. *Information for families of newly laryngectomized patients.* (Available through American Cancer Society Chicago unit.)

able to assist the laryngectomee in recovering effective communication abilities. An understanding of these approaches and their advantages and disadvantages helps the nurse deal more effectively with the laryngectomee undergoing vocal rehabilitation.

As with many other disorders, family support and understanding are important aspects of the treatment program. Fortunately, the nurse can rely on the availability of many books, films, tapes, and community resources to assist in teaching patient and family about all aspects of life following laryngectomy.

REFERENCES

Blom, E. D., & Singer, M. I. Surgical prosthetic approaches for postlaryngectomy rehabilitation. In R. L. Keith (Ed.), *Laryngectomy rehabilitation.* Houston: College Hill Press, 1979.

Keith, R. L., Linebaugh, C., & Cox, B. Presurgical counseling needs of laryngectomees: A survey of 78 patients. *Laryngoscope,* 1978, *88,* 1660–1665.

Panje, W. R. Prosthetic vocal rehabilitation following laryngectomy: The voice button. *Annals of Otolaryngology,* 1981, *90,* 116–120.

Peterson, C. A., Stallings, J. O., & Daghestani, A. Larynx reconstruction. *Nursing '79,* March 1979, 78–81.

Ryan, W. J., Gates, G. A., & Cantu, E. *Esophageal voice acquisition: A more realistic outlook.* Paper presented at the convention of the American Speech, Language and Hearing Association, Atlanta, Georgia, November 1979.

SUGGESTED READINGS

Baker, B. M., & Cunningham, C. A. Vocal rehabilitation of the patient with a laryngectomy. *Oncology Nursing Forum,* Fall 1980, 23–26.

Minear, D., & Lucente, F. E. Current attitudes of laryngectomy patients. *Laryngoscope,* July 1979, 1061–1065.

COMMUNITY RESOURCES

One or more of the following agencies or service organizations are available in most communities to provide assistance in dealing with laryngectomized patients and their families.

Local American Cancer Society unit, white pages.

Lost Chord or Nu Voice Club (sponsored by American Cancer Society), see previous item.

International Association of Laryngectomees (sponsored by American Cancer Society), 777 Third Ave., New York, New York 10017.

VA medical center, speech pathology service, white pages.

Local university program, white pages.

Speech pathologists in private practice, yellow pages.

SUMMARY

Cancer of the larynx resulting in laryngectomy creates permanent changes in breathing and production of voice. The devastating effects of these changes require good preoperative counseling and postoperative support to insure successful rehabilitation and adjustment by the patient and family.

During the immediate postoperative period, the nurse should be concerned with providing care to enhance healing, to assist the respiratory mechanism in adapting to air intake through the stoma, and to help the patient deal with emotional changes that may accompany a laryngectomy. Following the acute phase of recovery, attention changes to rehabilitation of communication.

Restoration of voice can be accomplished in one or more ways, depending upon individual needs and capabilities. Various types of artificial aids, esophageal voice, and new techniques of surgical-prosthetic voice restoration are avail-

construction reviewed by Peterson, Stallings, and Daghestani (1979), which involves use of rib cartilage and arm or leg tendons to recreate a thyroid cartilage and vocal cords. Postoperative voice is reported to be of good quality, although large numbers of patients undergoing this procedure have not been available for study.

For the nurse dealing with laryngectomees undergoing surgical-prosthetic voice restoration, detailed consultation with the surgeon and speech pathologist is necessary. Familiarity with the entire procedure gained from following a patient through all the stages would be ideal. Postoperative problems are few, but aspiration is a concern. Once the prosthesis is in place, the patient may require a week or two of training to develop good voice production. As with other forms of alternative voice production, the nurse may have the most significant contact with the patient, providing an understanding ear and continuing encouragement. Increased neck tension and pushing are the enemies of surgical-prosthetic voice restoration as well as of esophageal voice production.

THE PATIENT'S FAMILY

Postoperative counseling of the laryngectomee's spouse and family is necessary to provide the type of continued support required for successful vocal rehabilitation and adjustment. Fortunately, numerous books, pamphlets, films, and recordings are available to assist the family with such things as first aid, care of the stoma, and types of voice retraining. In most cases, the speech pathologist provides much of this information, but in areas where speech pathology services are unavailable, much of the responsibility must be assigned to the nurse.

A sampling of currently available materials and community resources that are easily accessible and quite helpful in dealing with laryngectomees and their families is given at the end of the chapter.

taped to the outside neck. A small opening on the underside of the tube directs lung air into the esophagus when the stoma is occluded with a thumb or finger. The air passes through the minute slit in the rounded esophageal end of the tube (the duckbill) and causes the walls of the upper esophageal and lower pharyngeal area (the so-called pharyngo-esophageal segment) to vibrate, thus creating a usable voice for speech. When the patient swallows, the slit is closed, preventing aspiration. Normal breathing is unimpaired when the stoma is not occluded.

Like all other forms of alternative communication for the laryngectomee, the tracheo-esophageal puncture has advantages and disadvantages. The major advantages include its high success rate in producing voice in patients whose previous attempts have been unsuccessful, relatively little risk and expense, and reversibility if the patient is not satisfied. That is, in most cases, if the prosthesis is removed, the fistula will usually close within a day, and the patient returns to his or her preoperative status.

The disadvantages of the procedure lie in the fact that a prosthesis must remain in place at all times and must be changed every day for cleaning and retaping. Of less concern is the requirement that one hand is necessary to occlude the stoma at all times during speech. Most patients who use the prosthesis quickly develop a daily routine in which a clean prosthesis is prepared for placement and substituted for the one that has been in place. Upon removal, the used prosthesis is cleaned and stored to become the substitute prosthesis the next day. Many patients can accomplish the transfer in 5 minutes.

A modification of the Blom-Singer procedure has been reported by Panje (1981), in which a silicone device is placed lower in the trachea during an outpatient surgical procedure. The prosthesis does not require taping and apparently cannot be accidentally dislodged. Panje has reported a high success rate with the procedure, but further studies have yet to be reported. The nurse may also encounter a form of larynx re-

ing care, the method and its implications are described here in some detail.

Known as the Blom-Singer or tracheo-esophageal puncture procedure (Blom & Singer, 1979), the method has been successful in establishing voice in a large number of laryngectomees whose esophageal and/or artificial voice had been inadequate or absent. Candidates for the tracheo-esophageal puncture are carefully selected by the surgeon and speech pathologist on the basis of their communicative ability and need, understanding, cooperation, and motivation. Each patient is provided with detailed counseling and information about the procedure; in addition, the patient is usually shown videotapes of people who have successfully undergone the procedure, and/or is introduced to local individuals who have had the surgery.

The laryngectomee who elects to undergo the tracheo-esophageal puncture is taken to the operating room and administered general anesthesia via tracheal intubation. The surgeon then creates a fistula between the trachea and esophagus in the upper back portion of the trachea. A rubber catheter is inserted into the fistula, routed through pharynx and nose, and connected to itself near its entry into the fistula, thus making a continuous loop. The patient is taken to recovery and keeps the catheter in place for 2 to 3 days. The catheter is necessary to keep the fistula open while preventing aspiration during the brief healing process. Usually, the patient is able to eat and carry out normal activities.

In the next stage, the catheter is removed and immediately replaced with a small, valved device variously referred to as the "t-e," "Blom-Singer," or "duckbill" prosthesis (the latter term arises from the appearance of the end of the prosthesis that is located in the esophagus). The T-shaped prosthesis consists of a small tube of variable length that projects though the tracheostoma into the esophagus, and two flanges that are

Other problems that are commonly encountered during the course of esophageal voice training can also be minimized by an understanding nurse. In some cases, the patient makes good progress during the initial stages of esophageal voice acquisition, only to reach a plateau with no apparent progress being made for several days. In other cases, the patient may have good voice production one day, but seems to regress the next day. Both situations are experienced by many laryngectomees and are not cause for alarm. Patients should be constantly reassured that plateaus and fluctuation are part of the normal learning process and should decrease in frequency as they continue to practice and become skilled as esophageal talkers.

Surgical-Prosthetic Voice Restoration

It has long been recognized that, if a connection could be made between the trachea and esophagus in a laryngectomee, then air could be directed from the lungs and trachea to the esophagus and pharynx in such a way that the tissue would vibrate in a fashion similar to that of traditional esophageal voice production. The advantage of such an approach would be the continuous air supply from the lungs as in normal voice production, rather than the injection of air into the esophagus by mouth, which is less efficient and sometimes difficult to learn. The disadvantage of this method is that if air can be directed from trachea to esophagus, then saliva, fluids, and food can leak from the esophagus into the trachea and lungs.

Over the years, many attempts have been made to accomplish the trachea-to-esophagus air passage while preventing esophagus-to-trachea aspiration. Recently, a method has been reported that appears to have solved most problems associated with this approach and is currently being used in many hospitals in the United States and abroad. Since it requires an inpatient surgical procedure and postoperative nurs-

rehabilitation, and nearly all remain in the hospital for the critical first week or two of their esophageal voice instruction. In either or both cases, the nurse is one of the primary listeners and can assist in the patient's treatment program.

Most of the principles of communicating with artificial voice aids can be applied to esophageal voice production. Of primary importance is an understanding and patient listener who does not give the impression of hurrying the esophageal speaker. Putting pressure on the laryngectomee results in the two most serious deterrents to good esophageal voice production: increased tension in the neck, and stomal air blast. As a laryngectomee experiences greater bodily tension, especially in the neck and throat, it becomes increasingly difficult to maintain the rapid and coordinated intake and release of air. He or she then tends to attempt to force the air in and out, which further leads to increased tension and less efficient voice production.

In conjunction with increased tension, some laryngectomees experience a blast of air through the stoma, which can be quite disconcerting to the listener and can even override the esophageal voice. Stoma blast results from pushing excessive amounts of air rapidly out of the lungs in a futile attempt to produce voice or increase its loudness. This is understandable, since the normal talker habitually behaves similarly when increases in vocal loudness or emphasis are required. The habit is deep-seated and may persist in some laryngectomees, even though the air supply from the lungs is no longer connected to the voice-producing mechanism. Further evidence of the influence of lifelong habits is the persistence on the part of most laryngectomees of covering the mouth upon coughing, in spite of the disconnected airway. This is usually a minor problem, but may cause embarrassment if the stoma is not covered and mucus is forcefully expelled. Use of a stoma bib or cover is highly recommended.

activity during speaking, leaving the tone generator on for long periods of speaking without pausing for the ends of phrases and sentences, and "chopping" words by turning off the tone generator prior to completing the articulation of the final syllable. The nurse can be of assistance by being aware of these potential problem areas and discussing them as they arise with individual patients.

Esophageal Voice

As outlined earlier, esophageal voice production is used by some laryngectomees as their primary means of communication. To accomplish this, air must be trapped in the upper part of the esophagus and released quickly, repetitively, and for sufficient duration to support continued fluent speech production. In spite of numerous investigations, it is still not known why some individuals are able to learn esophageal voice production and others are not. Anatomical changes, physiological differences, psychosocial factors, and patient motivation all probably influence, in varying degrees, the success or failure of a given individual in acquiring esophageal voice.

With a patient for whom esophageal voice acquisition is a realistic expectation, the treatment program usually progresses from initial inconsistent sound production via the injection or inhalation methods described earlier to production of individual vowel sounds. As the voice develops consistency, vowel prolongation and short words are introduced, followed by connected discourse in the form of phrases and sentences. When the patient is communicating effectively with esophageal voice, nuances of pitch and loudness can be brought out to enhance the naturalness of the voice.

The course of treatment can take place over a period of weeks or months and is usually done on an outpatient basis. However, some patients are hospitalized for much of their

pneumatic-type device, however, can be used only by someone with a tracheostoma. When a person with normal voice uses an artificial voice aid, the vocal folds must be kept from vibrating or creating a whisper to get the true effect of artificial voice production. As such, it is recommended that the normal talker attempting this should hold his or her breath while the tone generator is on, mouthing the words in much the same manner as would someone trying to communicate with another who is on the other side of a thick glass partition. Using the artificial larynx will give the nurse some idea of the difficulties faced by the laryngectomee when attempting to communicate. It can also be fun and helps build rapport with the patient for future communication attempts.

Some listeners have more difficulty understanding laryngectomees than can be explained on the basis of unfamiliarity alone. It appears that these listeners focus too much attention on trying to understand every individual speech sound, rather than on listening to the overall phrase or sentence. Once a few initial sounds are misunderstood, the listener gets confused and loses the entire message—the auditory equivalent of not seeing the forest for the trees. To counteract this effect, it is helpful to listen to longer segments of speech without attempting to concentrate on individual sounds or words. At the end of the message, the listener's brain will backtrack and fill in material that has been missed. Also of benefit in this regard is limiting conversations, initially at least, to familiar topics for which both speaker and listener are aware of the context. Again, as noted in the previous chapter, phrases and sentences are better than single words, and in very difficult circumstances it is helpful for the laryngectomee to reword a comment.

Many laryngectomees inexperienced in artificially aided voice production tend to make predictable errors that interfere with communication. Common problems include placing the sound source inappropriately, failing to supply sufficient oral

66 The Nurse and the Communicatively Impaired Adult

Figure 3-5. **Schematic representation of a pneumatic-type artificial voice aid.**

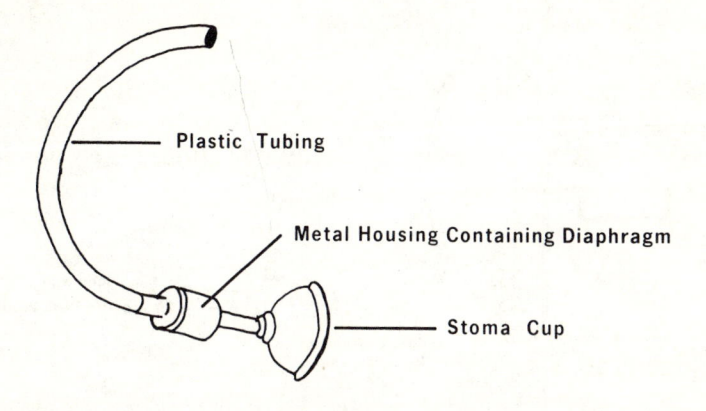

Plastic Tubing

Metal Housing Containing Diaphragm

Stoma Cup

applicable here. Background noise, competing conversation, poor lighting, and inability to see the speaker's face all reduce the level of communicative effectiveness. In addition, many listeners have difficulty adapting to the sound of speech created with an artificial device. This usually results from lack of familiarity with the artificially produced voice and the speaker's unique speech pattern. The obvious solution is to listen to the patient as much as possible, even if the first attempts result in something less than normal understanding. This approach will provide increasing familiarity on the part of the listener while also allowing opportunities for the patient to practice and improve his or her own performance. A listener should not pretend to understand if the message has not been received, nor should there be any hesitancy in asking the patient to repeat.

The nurse may also wish to try the artificial device in order to become more familiar with its operation. This is perfectly acceptable and highly recommended. Most devices have accompanying pamphlets describing their operation and maintenance, along with spare tubing for the oral-type device. The

Figure 3-4. Schematic representation of an oral-type artificial voice aid.

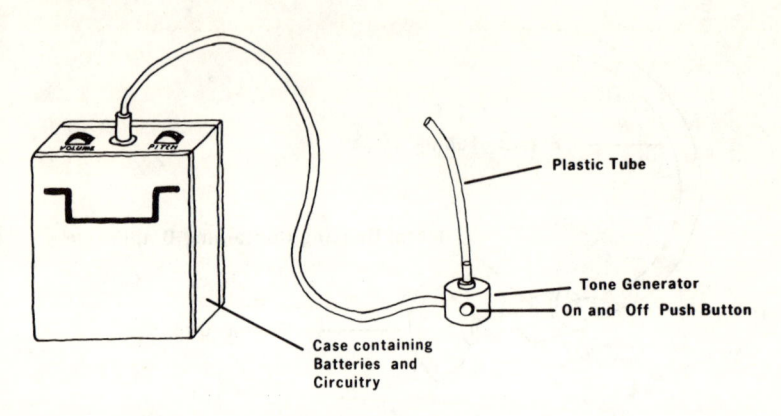

The pneumatic-type artificial aid uses air from the stoma to vibrate a reed or diaphragm located in a metal housing. Stoma air is directed into the housing through a "cup" or funnel that fits over the stoma during voice production. The sound from the vibrating reed or diaphragm is carried to the oral cavity via plastic tubing and is used for speech in a way similar to that of the oral-type device. Advantages of a pneumatic device include its relatively low price and a more natural sound. The latter is possible because air from the lungs is causing vibration much the same as it does during normal voice production. As such, changes in air flow and pressure can cause subtle changes in pitch, loudness, and quality of the resulting "voice." The primary disadvantage of the pneumatic instrument is its high visibility. Figure 3-5 represents a common pneumatic device called the Tokyo artificial larynx.

Regardless of the type of artificial device being used, communication with a laryngectomee can be enhanced in a number of ways. Of course, all of the suggestions previously mentioned with regard to creating a conducive environment are

The oral-type artificial aid introduces an electronically produced tone directly into the oral cavity via a plastic tube connected to a tone generator. The on-off button is located on the hand-held generator, which is in turn connected by wire to the battery and circuit compartment. The housing for batteries and circuitry has been designed to fit in a shirt pocket for convenience and for reduction of visibility. With the oral-type device, the patient inserts the plastic tube approximately 1 to 2 inches into the side of the mouth, presses the tone generator button, and modifies the resulting tone into meaningful speech sounds, as already discussed. Pitch and loudness controls are located on the battery and circuit housing, but once set cannot be varied from word to word during conversation. The major advantage of the oral-type instrument is that it can be introduced early in the postoperative period, since it will not interfere with healing suture lines or surgical dressing. It may also be used successfully in patients who have developed fistulas and/or who have extremely hard neck areas resulting from fibrosis. It has the disadvantage of creating a potential interference with speech intelligibility, since it introduces somewhat of an obstacle in the oral cavity. Figure 3-4 is a sketch of the Cooper-Rand oral-type electronic speech aid.

Figure 3-3. Schematic representation of a neck-type artificial voice aid.

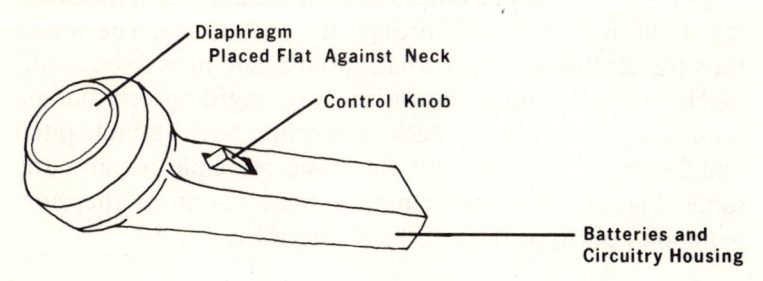

speech aid). Artificial devices have advantages and disadvantages, depending upon the type of instrument and the needs of the patient. In general, specific artificial aids can be used for effective communication very soon after the operation and long before esophageal voice therapy can be initiated. They also provide a means of communication while the patient is learning to produce esophageal voice, and, as stated earlier, may ultimately be the primary communication source if the patient is one of the many who fail to develop adequate esophageal voice.

Among the disadvantages are the monotonous or "robot" quality of many devices, the necessity for always occupying one hand, the high cost of purchase and maintenance, and the high visibility of aids. Presumably, the speech pathologist takes into consideration these and many other factors prior to recommending an artificial voice aid. In fact, many instruments should be, and probably are, tried before a final decision is made. There are many artificial aids available, a sampling of which appears at the end of the chapter (see "Artificial Aids and Devices"). In spite of the numerous variations, they all fall into three general categories: neck-type, oral-type, and pneumatic-type devices.

The neck-type device consists of a battery and circuitry compartment, which also serves as the handle, coupled to a diaphragm that emits a buzzing sound when the control switch is depressed. The portion of the instrument in which the diaphragm is housed is placed against the neck, and the buzzing sound is transmitted through the neck tissue. The sound then travels through the throat, mouth, and nose to be modified by the speech musculature into meaningful speech sounds. Some instruments in the neck-type group have variable pitch and loudness controls, but the basic principle remains the same. Figure 3-3 is a schematic representation of the most common neck-type device, the Western Electric 5A.

tion method, which involves sucking the air into the top of the esophagus by momentarily relaxing the cricopharyngeus muscle that encircles its uppermost portion. When the air has been trapped in the esophagus, it is quickly forced back out and causes the walls of the upper esophagus and lower pharynx to vibrate in a way similar to that of an audible belch. This sound is ultimately modified by the tongue, lips, and teeth in much the same way that the normal voice is modified to produce meaningful speech.

Recently, however, it has been reported (Ryan, Gates, & Cantu, 1979) that most laryngectomees do not, in fact, develop esophageal speech to a level of competence sufficient for normal communication. As such, it has become apparent that other means of communication should be explored for laryngectomees who cannot, or choose not to, produce esophageal voice. Unfortunately, many health care professionals continue to view communication other than that accomplished through esophageal voice as being inferior and not worth pursuing.

A more rational approach would be to consider esophageal voice as one alternative among many, including writing, gesturing, artificial aids, various forms of surgical voice restoration, and combinations of these. All methods should be explored with new laryngectomees to insure that they can make an informed decision as to how they wish to proceed. Nurses, as well as physicians and speech pathologists, should be careful about forcing their biases on patients whose needs may be totally different. Effective communication, rather than the method used to achieve it, is the ultimate goal.

Artificial Aids

For many laryngectomees, the first alternative form of communication after the writing stage is some type of artificial voice aid (also referred to as an artificial larynx and artificial

gastric tube, and gradually teaching the patient how to perform these tasks. This is also the time that many laryngectomees confront the reality of their situation and begin to show signs of depression—withdrawal, crying, lack of attention to self, and so forth. In most cases, the depression is temporary and gradually disappears as patients begin to feel better and see that they can do many of the things that they enjoyed preoperatively. It may also be tempered somewhat by the speech pathologist, who should be introducing an artificial larynx if it has not been introduced already; by the nurse, who is gradually releasing responsibility for care to the laryngectomee; by the family, which should be visiting on a regular basis; and by all involved, who should be acknowledging that a certain amount of depression is normal under such circumstances. A deep, long-lasting depression may require more time and care than these individuals can devote to it, however, and a psychiatrist, psychologist, or counselor experienced in working with this type of patient should be consulted as soon as possible. Much of the success or failure of a laryngectomee rehabilitation program depends upon the attitude of the individual patient at this stage of recovery.

Once the laryngectomee begins ambulating and finds that he or she can function in a comparatively normal manner, all attention usually focuses on the rehabilitation of communication. The term *communication* rather than *voice* is not used accidentally. Traditionally, postoperative laryngectomee rehabilitation has been directed almost exclusively toward learning to produce esophageal voice. This voice is produced by trapping air inside the mouth and forcing it down into the top of the esophagus. The maneuver is referred to as an injection and can be accomplished by pushing the air back during the production of certain consonant sounds such as *p, t, ch, sh,* and *st,* or by pressing the air between the tongue, hard palate, and walls of the throat in a "piston-like" action; also possible is the inhala-

POSTOPERATIVE CARE

Immediately following surgery, the laryngectomee is confined to the surgical intensive care unit (SICU) for several days, with IVs, hemovacs, nasogastric tube, and respiratory apparatus in place. The primary concern at this stage is to insure that the patient's condition is stabilized and that the healing process is progressing well. Communication with the patient is often limited to writing on a pad or magic slate, although it is possible to introduce an oral-type artificial larynx such as the Cooper-Rand electronic speech aid as early as a day postoperatively. This instrument is discussed in detail later.

Again, the primary issue is to provide a maximum amount of support for the patient at this difficult stage. The speech pathologist should be visiting regularly to introduce the artificial larynx, to check on potential complications for rehabilitation such as the occurrence of a fistula, and to develop a continuing rapport that will be important during the months of rehabilitation to follow. The nurse should be providing as much reassurance and support as possible while alerting the physician, speech pathologist, and other support personnel to changes in the patient's mental and emotional status. Also, if the spouse and family are visiting the patient in the intensive care unit, it is advisable to prepare them in advance for the array of tubes and monitors, since most lay people equate them with extremely serious illness. A preparatory explanation that this is normal apparatus for all laryngectomees should be helpful in softening the blow.

REHABILITATION

Once the patient is transferred from the intensive care unit to the acute care ward, the process of postoperative rehabilitation begins in earnest. The nurse is occupied with caring for the suture lines, suctioning the trachea, feeding through the naso-

tion the patient actually retains during the immediate postoperative period. The question has not been sufficiently studied to provide a definitive answer, but the consensus of individuals working with laryngectomees indicates that the actual information is not nearly as important as is the creation of an impression on the part of patients that they are in competent hands throughout the course of surgery and rehabilitation; that those working with them understand their apprehensions; and that everyone involved is there to assist them as much as possible. Nurses, of course, are in an excellent position to create and reinforce this impression. Nevertheless, it is probably wise to provide as much information as patients and families request and are apparently able to handle. In a study by Keith, Linebaugh, and Cox (1978), laryngectomees were surveyed about their presurgical counseling. The results indicated that "most received preoperative information only from their surgeons. Yet many indicated that laryngectomee candidates should be counseled by nurses, speech pathologists, recovered laryngectomees and others." They further indicated that the spouse should be included in preoperative counseling and that audiovisual materials would have been helpful. It should be noted here, however, that a number of potential laryngectomees have a very negative reaction after learning of their disease and pending surgery. Individual differences should be respected and needs met with common sense and rational thought.

One final aspect of the preoperative period that deserves special mention is the bed or room assignment of the patient. Every effort should be made to avoid placing the preoperative laryngectomy candidate in a room with a laryngectomee who is seriously ill, suffering from metastasis, or having complications from surgery. This seemingly self-evident rule of thumb is routinely violated in many hospitals and can have a devastating effect on the preoperative patient.

Although the change in breathing and the loss of voice present the two greatest changes for laryngectomees, other problems may exist in varying degrees, depending upon the individual patient. Nearly all laryngectomized individuals experience the temporary loss of taste, since this sense is dependent upon smell, and air can no longer pass through the nose. Swallowing may also present some difficulty, but usually returns to normal after the tissues have healed. In patients who have persistent swallowing problems, esophageal dilatation and mechanical or pureed diets may be required. Shoulder pain and decreased mobility is a common complaint of those laryngectomees who have also undergone neck dissection. Copious mucus production, stomal crusting, persistent headache, tissue fibrosis, and postoperative depression may also be encountered in varying degrees.

PREOPERATIVE COUNSELING

Good preoperative counseling of the laryngectomee is a crucial first step in the rehabilitation process. Ideally, the candidate for surgery should be referred to the speech pathologist as soon as possible. At that point, the candidate usually undergoes a voice and speech evaluation and tape recording, a counseling session in which the procedures and consequences of a laryngectomy are discussed, and a hearing evaluation. It is also at this point that referrals are made to other members of the health care team (such as the psychologist, the physical therapist, the chaplain, etc.) as appropriate. Finally, arrangements are made through the local Lost Chord Club to have a qualified laryngectomized visitor see the patient preoperatively. Sometimes this is handled through the speech pathology service, and other times it is coordinated through the nurses' station.

There is controversy over how much preoperative informa-

maining problem, of course, is the loss of ability to produce vocal fold vibration in view of the absence of vocal folds and the disconnection of the airway from the speech mechanism. It should be noted that laryngectomees have not lost the ability to "speak." That is, they still have the ability to articulate speech sounds with tongue, teeth, lips, and palate. What they have lost is the ability to produce voice.

Figure 3-2. Changes in head and neck structures after laryngectomy.

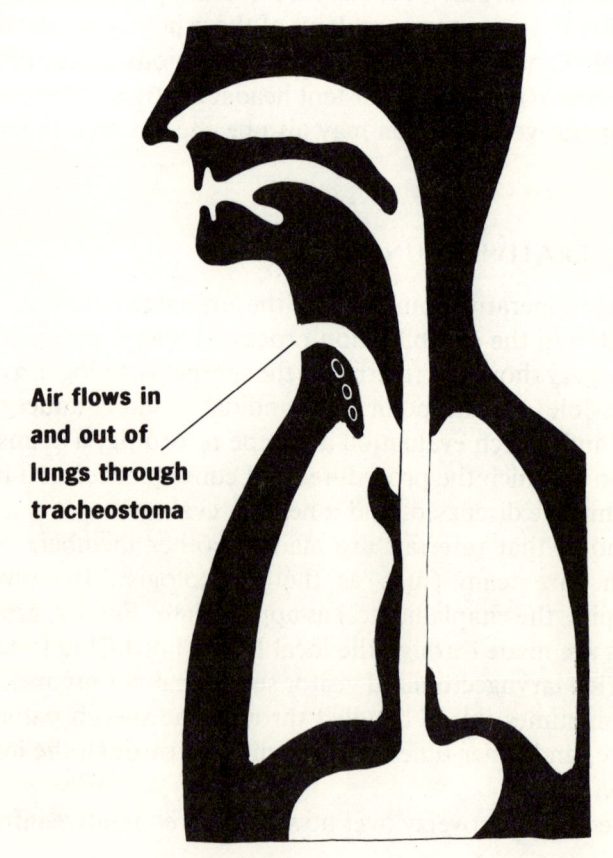

Air flows in and out of lungs through tracheostoma

wake. This drop tends to pull the vocal folds together, only to be blown apart again when air pressure increases. Cyclical activity of this nature will continue as long as air is supplied beneath the vocal folds. In males, the repeated openings and closings of the vocal folds occur approximately 125 times per second, producing a tonal quality that is subsequently modified into meaningful speech sounds by the tongue, teeth, lips, and palate. In females, the vocal folds vibrate an average of 190 times per second, resulting in characteristically higher-pitched voices than those of males.

During normal breathing and speech activities, the esophagus, lying behind the larynx and trachea, is relaxed. When the individual swallows, however, the larynx acts as a valve to close off the airway and allow saliva, food, and liquid to be passed from the mouth and throat through the esophagus to the stomach. As noted in the previous chapter, if the larynx did not perform its valve-like activity, foreign debris would enter the lungs and create a life-threatening situation.

Figure 3-2 represents the changes that occur in the relationship between breathing and swallowing in a laryngectomee (a person who has undergone a laryngectomy). Since the larynx has been removed, it can no longer provide the protective function described earlier. As a result, the airway and esophagus must be physically separated to allow for independent breathing and swallowing. To accomplish this, the surgeon rotates the top of the trachea forward, creating an opening (*stoma*) in the neck just below the area normally occupied by the larynx. A metal tracheostomy tube is inserted to keep the stoma open during healing. Air now enters the lungs directly through the tracheostoma, since the nose and mouth have had their access surgically blocked. Swallowing remains intact and functions essentially the same as before surgery, although a nasogastric feeding tube is temporarily placed to avoid swallowing and assist healing in the immediate postoperative period. The re-

Figure 3-1. Schematic representation of important head and neck structures prior to laryngectomy.

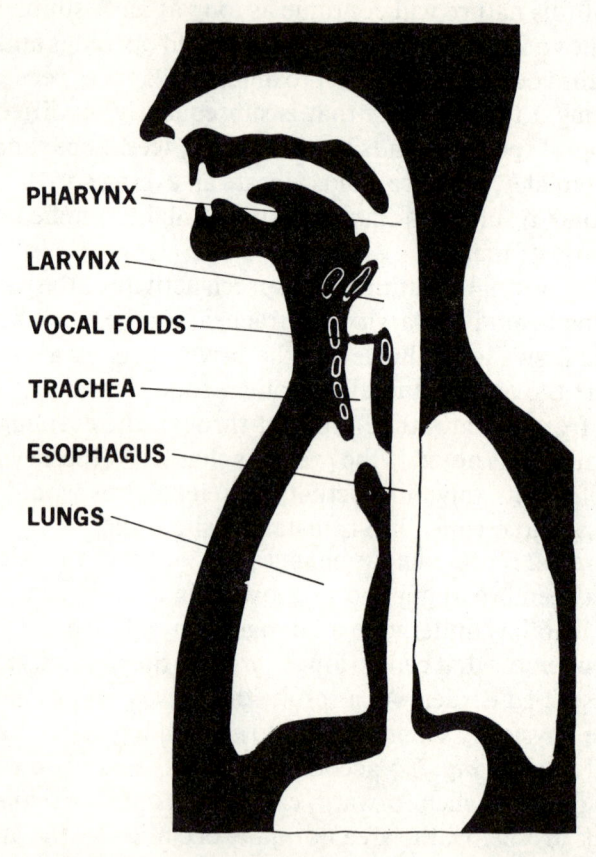

PHARYNX
LARYNX
VOCAL FOLDS
TRACHEA
ESOPHAGUS
LUNGS

out of the lungs via the same route through which it entered. If the vocal folds are held apart, the air flows out freely, thus completing one cycle of normal quiet breathing. If the vocal folds are pulled together, however, they resist the flow of air from the lungs momentarily until the air pressure is sufficient in itself to blow the vocal folds apart. A puff of air flows through the opening, leaving a slight drop in air pressure in its

3

LARYNGECTOMY

DEFINITION AND CAUSES

Nurses involved with surgical patients or patients with head and neck cancer eventually encounter communication problems that are the direct result of the presence of tumor or reflect a deficit caused by surgical excision of the tumor. The most common problem of this nature is cancer of the larynx with subsequent total laryngectomy. *Laryngectomy* is defined as the surgical removal of the larynx. This removal is almost always the result of cancer of the larynx or surrounding tissues, but is encountered occasionally as the result of severe trauma to the neck. If resulting from cancer, the surgery may be preceded or followed by irradiation and/or chemotherapy to control the tumor further. The two most immediate and significant effects of laryngectomy are the permanent changes in breathing and the loss of the ability to produce voice.

BEFORE AND AFTER LARYNGECTOMY

Figure 3-1 is a schematic representation of the interaction between breathing and voice production in the individual with a normally intact larynx. When the diaphragm contracts, air is pulled into the lungs via the irregularly shaped tube comprising the nose and mouth, pharynx, larynx, trachea, and bronchial tree. The vocal folds are spread apart to allow a free flow of air into the lungs. When the diaphragm relaxes, the air is pushed

National Health Education Committee. *What are the facts about Parkinson's?* (Available from the author at 866 United Nations Plaza, New York, New York 10017.)

Parkinson's Disease Foundation. *The Parkinson patient at home.* (Available from the New York Neurological Institute, 710 West 168th St., New York, New York 10032.)

(Available from Prentke Romich Co., R.D. 2, Box 191, Shreve, Ohio 44676.)
Array similar to the Zygo listed above.

Special adaptive equipment for speech-impaired persons. (Available from local Bell Telephone office.)
Many devices are available to assist the communicatively impaired person in using the telephone. Most local Bell Telephone offices have a representative for this purpose.

Materials for Patients and Families

ALS Society of America. *Amyotrophic lateral sclerosis: Information for patients and their families.* (Available from the author at 15300 Ventura Blvd., Suite 315, Sherman Oaks, California 91403.)

ALS Society of America. *ALLSSOAN: People helping people.* (Available from the author at 15300 Ventura Blvd., Suite 315, Sherman Oaks, California 91403.)

National Multiple Sclerosis Society. *Emotional aspects of multiple sclerosis.* (Available from the author at 205 East 42nd St., New York, New York 10017. Also available through local chapter.)

National Multiple Sclerosis Society. How your local MS chapter can help. (Available from the author at 205 East 42nd St., New York, New York 10017. Also available through local chapter.)

National Multiple Sclerosis Society. *Multiple sclerosis and the practical nurse.* (Available from the author at 205 East 42nd St., New York, New York 10017. Also available through local chapter.)

National Multiple Sclerosis Society. *M.S. Messenger.* (Available from the author at 205 East 42nd St., New York, New York 10017. Also available through local chapter.)

Myasthenia Gravis Foundation. *Help is on the way: A handbook for patients.* (Available from the author at 15 East 26th St., New York, New York 10010. Also available through local chapter.)

Myasthenia Gravis Foundation. *Myasthenia gravis: Hope through research* (DHEW Publication No. NIH 75-768). (Available from the author at 15 East 26th St., New York, New York 10010. Also available through local chapter.)

SUGGESTED READINGS

Darley, F. L., Aronson, A. E., & Brown, J. R. Differential diagnostic patterns of dysarthria. *Journal of Speech and Hearing Research,* June 1969, 246-269.

Erb, E. Improving speech in Parkinson's disease. *American Journal of Nursing,* November 1973, 1910-1911.

Goda, S. Communicating with the aphasic or dysarthric patient. *American Journal of Nursing,* July 1963, 80-84.

Griffin, K. M., Stubbert, J., & Breckenridge, K. Teaching the dysphagic patient to swallow. *RN,* September 1974, 60-63.

Valenstein, E. Nonlanguage disorders of speech reflect complex neurological apparatus. *Geriatrics,* September 1975, 117-121.

OTHER RESOURCES

Aids and Devices

Basic Communication Aids for the Nonverbal. (Available from Personalized Patient Care Co., Inc., North Salem, New York, 1978.) A set of 11 charts with pictures and words expressing common needs and problems of the nonverbal patient.

Phonic Mirror/Handivoice. (Available from H. C. Electronics, 250 Camino Alto, Mill Valley, California 94941.) An electronic device in which a message is selected by pressing one or more coded buttons and transmitted verbally by a synthesized speech signal generated by the instrument.

Zygo Eye Transfer Communication System: Zygo Model 16, Zygo Model 100, Adaptive Switches. (Available from Everest and Jennings, Inc., 1803 Pontius Ave., Los Angeles, California 90025.) An array of instruments in which various messages are scanned or coded and the patient selects the appropriate one via eye movements or other motor response.

Communicaid, Proscan, Express 1, Viewpoint, Adaptive Switches.

treatment for most dysarthric conditions is based on several common therapeutic principles, all of which can be made more successful when reinforced by an informed and understanding nurse.

Sometimes swallowing, as well as speech, is impaired in the dysarthric patient. With a thorough evaluation and swallowing rehabilitation program, however, many dysphagic patients can regain the ability to swallow and can avoid a future of liquid diet passed through a feeding tube. In some hospitals, the nurse has primary responsibility for swallowing rehabilitation, and, in others, it is a multidisciplinary effort. In either case, experience in and knowledge of dysphagia evaluation and treatment techniques is a necessary prerequisite to a successful rehabilitation program.

The process of adjusting to permanent disability or a progressively debilitating disease can be difficult for both patient and family. The nurse can assist in the reassurance of the value of rehabilitation and the gradual acceptance of disability via frank and complete patient and family counseling. Many pamphlets and community resources are available to aid the nurse in this endeavor.

REFERENCES

Darley, F. L., Aronson, A. E., & Brown, J. R. *Motor speech disorders.* Philadelphia: W. B. Saunders, 1975.

Griffin, K. M. Swallowing training for dysphagic patients. *Archives of Physical Medicine and Rehabilitation,* October 1974, 467–470.

Larsen, G. L. Rehabilitation for dysphagia paralytica. *Journal of Speech and Hearing Disorders,* May 1972, 187–194.

Schultz, A. R., Niemtzow, P., Jacobs, S. R., & Naso, F. Dysphagia associated with cricopharyngeal dysfunction. *Archives of Physical Medicine and Rehabilitation,* August 1979, 381–386.

and family. The problems are similar to the ones encountered with aphasia and can be dealt with in much the same way. Frequent, open discussions with family members allow them to verbalize their fears and frustrations in a supportive environment. These discussions also allow the nurse, physician, and therapists to educate the family gradually about the patient's condition and to help them begin to adjust to changes in physical ability and communication skills. Although not as numerous as those dealing with aphasia, books and pamphlets are available for patients and families who are the victims of conditions that may produce dysarthria. A list of readily available material for use with patients and families is located at the end of the chapter.

COMMUNITY RESOURCES

In addition to the community resources indicated in the preceding chapter, the following agencies are available in most areas to assist in dealing with dysarthric patients.

ALS Society of America, local chapter, white pages.

Myasthenia Gravis Foundation, Inc., local chapter, white pages.

National Multiple Sclerosis Society, local chapter, white pages.

Local visiting nurse association, yellow pages.

Local division of vocational rehabilitation, white pages.

SUMMARY

As in aphasia, the dysarthrias result from a variety of etiologic factors and vary widely in type and severity. Although the detailed characteristics of each dysarthria are important to the physician and speech pathologist for diagnosis and prognosis,

the preswallow breath. Once the patient is successfully and consistently swallowing ice chips, additional variety in the form of gelatin and sliced peaches should be attempted. After these are being handled without difficulty, other foods can be attempted on an experimental basis to determine which the patient likes and can tolerate.

Contrary to the practice in many hospitals, puree and mashed potatoes are usually not good foods for initial use in dysphagia rehabilitation. They are actually more difficult to swallow, may disintegrate in the pharynx (thus increasing the risk of aspiration), and may not be as effective in stimulating the sensory avenues in the mouth and throat. Other precautions include the following: (1) Avoid washing food down with liquids; (2) remove nasogastric and tracheostomy tubes, if at all possible, even to the extent of having the surgeon place a temporary esophagostomy tube during swallowing retraining; and (3) if the patient is having difficulty moving food from front to back in the mouth, introduce the food in small quantities via a feeding syringe into the rear of the mouth to stimulate the swallow reflex.

As with virtually all of the disorders discussed in this book, dysphagia requires large expenditures of time and clinical skill for successful rehabilitation. In many patients, however, a swallowing program may mean the difference between a life of having liquid nutrient passed through a gastrostomy or esophagostomy tube and one of enjoying the various tastes and textures of a regular diet modified appropriately for individual needs. Most patients who have undergone a dysphagia treatment program attest that it was worth the effort.

THE PATIENT'S FAMILY

With or without dysphagia, the dysarthric patient with moderate to severe communication problems or with progressive disease will probably have a drastic impact on the spouse

tient each step of the swallowing process while encouraging him or her to think about and visualize the structures and their movement; if the patient has difficulty with this, it sometimes helps to assist him or her manually in elevating the larynx to help stimulate a reflex swallow.

At this juncture, a determination should be made as to the strength of the reflexes and movements in the swallowing mechanism. If they are very weak, it may be necessary to stimulate them by introducing a Levin-type tube into the mouth and esophagus. The patient will get the benefit of stimulation during normal tube feeding, and the tube can be repeatedly retrieved and inserted. If the reflexes and movements are somewhat stronger, or after they have been increased via the orally passed tube, small chips of ice can then be placed in the oral cavity.

Ice chips have the advantage of possessing enough substance and temperature extreme to stimulate oral and pharyngeal sensations, but they have little or no deleterious effect should the patient exhibit some initial difficulty in swallowing them. Fear of swallowing may be present in dysphagic patients, especially if they have experienced excessive choking during previous attempts. In fact, many patients are quite capable of swallowing, but have problems in overcoming the fear of choking and not being able to breathe. Presumably, if the patient has been adequately prepared in accordance with the guidelines presented here, this fear should be minimal.

Nevertheless, when the ice chips are introduced, it is advisable to "talk" the patient through each step. First, the patient should be made aware of the presence of the ice chip in the mouth. Second, as the patient prepares to swallow, he or she should take a short breath and hold it. Third, the patient should swallow, and fourth, he or she should immediately cough. As swallowing skill develops, the patient may be able to eliminate the cough and may not have to be as conscious of

The following discussion of evaluation and management of neurogenic dysphagia is based primarily on the writings of Griffin (1974) and Larsen (1972), and on clinical experience gained at the Audie Murphy (San Antonio) and Albuquerque Veterans Administration Medical Centers. The general approaches are similar, although discrepancies exist on specific points. The principles included here reflect personal biases resulting from using various and combined approaches with a good deal of trial and error. Before attempting a swallowing rehabilitation program, the nurse should be totally familiar with the material indicated in the list of suggested readings at the end of the chapter, and should have received supervised training if possible.

Whether a dysphagic patient is a candidate for a swallowing rehabilitation program depends upon several factors. Of primary importance is the determination (1) that the patient has sufficient understanding and awareness to follow the necessary instructions; (2) that he or she can demonstrate the ability to cough, and (3) that he or she can demonstrate at least a weak reflex swallow. If all reflexes are totally absent and there is no sign of movement in the structures of the mouth and throat, the patient is not a candidate for swallowing rehabilitation and probably requires surgical intervention in the form of esophagostomy or gastrostomy.

If the patient can demonstrate a cough and a reflex swallow (as evidenced by elevation of the larynx in the neck), and can understand the instructions, he or she should be placed in sitting position with the head *flexed*. Extension of the head and neck should be avoided, since this creates a more direct route from mouth to trachea. As with the dysarthric patient, in whom the automatic act of speaking must be brought to a level of conscious awareness, so must the largely reflexive act of swallowing be brought under the voluntary control of the dysphagic patient. Therefore, it is important to explain to the pa-

swallowing mechanism. This approach is still necessary in some cases, but many individuals can be trained to swallow by undergoing a swallowing rehabilitation program. Again, however, a thorough evaluation is a necessary first step. In some hospital settings, this is carried out by the speech pathologist or occupational therapist with assistance from nursing personnel; in others, the nurse has the primary responsibility for assisting patients with dysphagia. In any case, an understanding of the mechanics of normal swallowing and techniques for treating disordered swallowing are essential for the nurse working with brain-injured patients.

Normal swallowing is usually described as consisting of three phases (Griffin, 1974; Larsen, 1972; Schultz, Niemtzow, Jacobs, & Naso, 1979). In the first phase (oral), the food or liquid is swept back with the tongue from the front of the mouth, which has been sealed by compression of the lips, along the hard palate into the pharynx. In the second phase (pharyngeal), the velum elevates to close off the passageway between the pharynx and nasal cavities, thereby preventing nasal regurgitation of the food or liquid. At the same time, the posterior movement of the tongue, along with the constriction of the muscles surrounding the pharynx, pushes the material down toward the top of the esophagus. Simultaneously, the larynx is elevated and the vocal folds are closed to protect the airway from foreign debris. The third phase (esophageal) involves the reflex activity triggered when the material enters the top of the esophagus and initiates the progressive muscular contraction and relaxation, thereby pushing the food or liquid toward the stomach (*peristalsis*). The entire cycle of swallowing takes place in 1 to 2 seconds and requires the coordination and timing of numerous muscle groups. A breakdown caused by neuromuscular paralysis, weakness, or incoordination can take place at any point in the three swallowing phases, and can, of course, affect the three phases simultaneously.

progressively deteriorating dysarthric patient. Unlike many other progressive diseases of the brain, such as Alzheimer's disease and progressive cortical atrophy, patients with progressive dysarthrias usually retain normal intellectual and mental functions until they expire. But the disease process and subsequent speech problems may last for many years. The result, then, is that persons with normal comprehension of the world around them and normal thought processes are unable to express their opinions, feelings, and needs accurately. Tolerant listeners who insure that the environment is maximally conducive to good communication provide a substantial enhancement of the patients' quality of life. In the end stage of many diseases, the nurse is such a listener.

DYSPHAGIA

So far in this chapter, only the communication problems experienced by dysarthric patients have been emphasized. Since the neuromuscular control of the speech and voice apparatus is also responsible for the acts of chewing and swallowing, however, it is common to encounter dysarthric individuals who also exhibit swallowing problems, or *dysphagia*. *Neurogenic dysphagia* may be caused by any of the etiological factors already implicated in the dysarthrias, and may vary in severity from mild drooling caused by a unilateral facial and lip weakness to a complete inability to swallow caused by total paralysis of the mouth and throat muscles. Inability to swallow or difficulty in swallowing has two important potential consequences: Patients may aspirate food or liquid into the respiratory mechanism, and they may not be able to maintain adequate nutrition.

Traditionally, patients with dysphagia have had to adjust to being fed through a nasogastric tube, or have undergone surgical esophagostomy or gastrostomy to bypass the impaired

munication devices have entered the marketplace. A sampling of the available equipment is listed at the end of the chapter (see "Aids and Devices"). It is highly recommended, however, that a speech pathologist with special training in patient evaluation, selection, and treatment be consulted when an electronic augmentative device is being contemplated. These instruments can be very expensive, but, more importantly, they may cause frustration and discouragement if used with improperly selected patients.

For dysarthric patients whose speech is impaired and can be expected to deteriorate with the progression of such diseases as amyotrophic lateral sclerosis, multiple sclerosis, and sometimes Parkinsonism, treatment is designed to maintain effective communication for as long as possible. In the early stages, the speech may be only mildly involved, and the principles of maximum use of reserve capacity and conscious attention to the act of speaking are applicable. As the disease progresses and the speech becomes more impaired, the necessity for some type of augmentation becomes apparent. In some cases, patients are reluctant to use the assistance of a communication board, typewriter, or electronic communication device, since to do so would indicate a degree of concession to the disease.

This belief should be honored, and it is never appropriate to force any communicatively impaired patient to use an augmentative device. On the other hand, if such patients can learn to use assistive devices relatively early in the course of their disease, they will be better prepared to make more effective and longer-lasting use of it when their speech becomes severely involved. Gentle encouragement from the nurse and family can be extremely helpful in making a smooth transition from unassisted communication to speech that is totally unintelligible without augmentation.

Finally, nowhere is it more important for the listener to take on increasing responsibility for communication than with the

Figure 2-1. Example of a communication board for use with nonverbal patients.

A B C D E F G H I J
K L M N O P Q R S T
U V W X Y Z

I	WANT	FOOD	NURSE
	FEEL	DRINK	DOCTOR
	AM	PAIN	WIFE
	HAVE	TIRED	CHAPLAIN
		THIRSTY	
		TOILET	
		HUNGRY	

| NO | | YES |

with a pointer or finger until a "yes" response is given. Then the process is repeated until the message has been spelled out letter by letter. A more efficient method using a similar principle is to point to each *row* of letters until a "yes" response is given, then scan the letters in the row until the appropriate letter is arrived at. Use of this type of device, of course, is successful only with literate individuals who have the ability to understand the principle and to construct meaningful phrases and sentences. For individuals who are illiterate or who have language problems in addition to their dysarthria, there are picture boards available (see "Aids and Devices" at the end of this chapter), or one can be easily constructed for the individual patient's needs.

With the rapid advancement in computer and electronic technology, a number of sophisticated, electromechanical com-

should encourage him or her to keep it in place during all speech activities, and during eating as well if nasal regurgitation is a problem. This is especially critical in the early stages, since the patient may require time to adapt to the device and may not give it a fair trial before discontinuing its use. Most patients with a palatal lift prosthesis are followed closely by the speech pathologist and prosthodontist. The nurse, however, should become familiar with the device, report areas of irritation in the oral cavity, and insure that good oral hygiene is being maintained.

As stated earlier, some patients' speech is so severely and permanently impaired that it can no longer be used as an effective means of communication. In these cases it is necessary to devise alternative means by which the individuals can communicate with others. Writing or printing is usually the first alternative to be explored in severely dysarthric patients. Unfortunately, many of the causes of dysarthria are not so selective as to impair the speech and voice mechanism while sparing the hands and fingers for writing.

If writing is impaired, but gross movements in the arm and hand remain, it is sometimes possible to teach the patient to type messages on a conventional typewriter. Even without controlled hand movements, some patients have sufficient head movement to allow a pointer to be attached to a headband for access to a typewriter keyboard or a communication board similar to the one shown in Figure 2-1. A communication board of this type can be easily constructed by printing or writing the alphabet and common words used to formulate everyday requests on heavy white paper or cardboard. To communicate, the patient indicates, by whatever means available, the appropriate letters and words to spell out a message. If the patient is unable to indicate the words or letters motorically but can indicate "yes/no" via vocalization, eye blinks, or some other type of movement, the nurse can scan the letters

responses from the patient, since he or she may have a tendency to tire more rapidly. On the other hand, it is probably not advisable to encourage extremely short or one-word responses, since the listener loses the benefit of contextual cues. For example, a patient with severe dysarthria who says "water" may be totally unintelligible even with numerous repetitions. If the patient says, "I'm thirsty, I want a drink of water," however, enough information may get through to the listener so that the message is accurately interpreted.

The loudness of the voice is often reduced in dysarthric patients and can further interfere with understanding if coupled with imprecision of speech sounds and a rapid speech rate (as observed in Parkinsonian patients, for example). Some therapeutic techniques designed to slow the rate or increase articulatory precision also have the secondary benefit of increasing vocal loudness. If the patient's voice remains too soft for effective communication, however, the nurse or other listeners can provide a degree of compensation. Such things as reducing background noise, allowing only one person to speak at a time, and having speaker and listener directly face each other should enhance communication and reduce frustration for all parties involved. It should become clear by the end of this volume that these principles apply equally as well to most other communicative disorders encountered in the adult population.

Finally, for the person who exhibits hypernasality and inappropriate loss of air through the nose when speaking, it may be possible to have a prosthodontist construct a device known as a *palatal lift prosthesis*. It is similar to a partial denture but has a backward extension designed to assist a weakened or paralyzed *velum* (soft palate) in blocking the sound or air from entering the nasal cavities during the production of certain speech sounds. (See Chapter 4 for a more detailed account of this aspect of speech production.)

If a patient is wearing a palatal lift prosthesis, the nurse

Since this is contrary to the way most people speak under normal circumstances, intensive and frequent therapy sessions are usually required to develop the appropriate habits of self-monitoring and self-correction.

A crucial issue in the success of this approach is the degree of patient motivation and cooperation. The nurse and the speech pathologist, working together, have an excellent opportunity to provide support and encouragement during the course of treatment. It should be understood by everyone involved, however, that there are no magic cures and that the rehabilitative process is sometimes a long and arduous one. Many times, a patient is highly motivated at the beginning of the treatment program, only to lose interest after a week or two when progress has not been as fast as anticipated. The nurse must be sensitive to this change in attitude and must reassure the patient that the rehabilitative effort is worthwhile. By the same token, realistic expectations must be shared among therapists, nurses, patients, and families to avoid giving false hope for total recovery when the circumstances indicate that something less can be expected. It is necessary for patients and those around them to work diligently toward rehabilitation, but also to accept certain limitations and begin to adjust to these in the patients' everyday lives.

With regard to specific therapy goals, the speech pathologist most often concentrates on increasing articulatory precision, since it is the accuracy with which the sounds of language are produced that has the most influence on how well the listener understands the message. One approach commonly used to achieve articulatory precision is to slow the rate of speaking to allow the affected speech muscles sufficient time to make the appropriate adjustments. The nurse, as a listener, should be patient and tolerant of the reduced speaking rate, and in fact should remind the patient to monitor his or her rate of speaking continually if the message is not being understood.

It is also helpful to reduce the demand for long and complex

REHABILITATION

In milder cases, even when a formal treatment program is not recommended, there may be times or circumstances in which it is difficult to understand the patient. A good rule of thumb is to recognize that most talkers, normal or impaired, do not make maximum use of their abilities at all times. As such, most individuals possess a certain amount of reserve upon which they can call when greater precision and accuracy of speech production is required. Therefore, it is perfectly acceptable to indicate to the patient via word or gesture that the message has been distorted and that greater effort on his or her part is necessary. Most speech pathologists will have prepared their dysarthric patients for this eventuality, and the nurse should not hesitate in reinforcing it. Furthermore, it is a disservice to both talker and listener to feign understanding when, in fact, the message has not been accurately transmitted or received.

In any type of communication breakdown caused by organic pathology, the main objective of rehabilitation is to return the individual to a level of functioning as close to his or her premorbid level as possible, but within the limits imposed by permanent physical disabilities. Most patients exhibiting dysarthria do not return to their premorbid level, although they may recover to a level where effective speech communication is possible in spite of persisting deviations from accepted norms. Where a formal treatment program is initiated, the speech pathologist has a number of techniques available, all of which can be enhanced in their usefulness if understood and reinforced by nursing personnel.

In addition to the principle of making maximum use of residual capabilities already mentioned, the clinician working with most dysarthric patients attempts to heighten awareness of the speech act on the part of the patients and attempts to bring the process of speaking more under the voluntary, conscious control of the individuals. In this way, the patients not only attend to what they are saying, but also to how they are saying it.

speech disorders. Such factors as pitch level, hypernasality, rate, imprecise consonants, and many more are studied to indicate the type and severity of dysarthria, and to obtain areas of difficulty that may be amenable to treatment. In spite of the different types of dysarthria, however, the findings will generally fall into one or more of the following categories:

1. The speech is normal or so mildly impaired that treatment is not warranted.
2. The speech is impaired, but is still the most efficient means of communication. Treatment is indicated to enhance communicative effectiveness.
3. The speech is so severely and permanently impaired that it can no longer be used as an effective means of communication without some type of augmentation.
4. The speech is impaired and can be expected to deteriorate with the progression of disease. Treatment is indicated to maintain effective communication for as long as possible.
5. Swallowing, as well as speech, is impaired and requires treatment.
6. The patient is so severely impaired and medically unstable that treatment is not indicated (to be discussed in Chapter 7).

By discussing the results with the speech pathologist, the nurse should be in a better position to deal effectively with the patient and family. Having knowledge of the type of dysarthria and its underlying cause(s) is also helpful in establishing prognosis for recovery, realistic expectations, and techniques for enhancing communication.

capitalize on residual neuromuscular function and compensatory adjustments to improve communicative effectiveness.

EVALUATION

Once a referral to the speech pathologist has been made, the nurse can assist in preparing the patient by explaining the necessity for an evaluation so that an appropriate treatment program can be initiated. Unlike the aphasic patient, the dysarthric patient usually has normal understanding of speech and can be talked to in a manner commensurate with his or her intelligence and vocabulary. This should also be verified, however, because any time the brain is damaged, it is possible to have subtle comprehension problems that interfere with understanding of instructions.

Upon arrival in the speech pathology clinic, the patient is usually evaluated with a test battery similar to the one discussed in the preceding chapter if aphasia is suspected. In addition, a systematic structured protocol known as a *motor speech evaluation* is usually followed. Developed primarily at the Mayo Clinic (Darley et al., 1975), the motor speech evaluation includes the following: maximum prolonging of vowel sounds; rapid and sustained repeating of nonsense syllables; producing selected multisyllabic words; producing words of increasing length, such as "zip, zipper, zippering"; counting; repeating sentences; reading a standard paragraph; and carrying on approximately 30 seconds of everyday conversation.

USE OF TEST RESULTS

The motor speech sample is usually videotaped or audiotaped and analyzed on the basis of a large number of dimensions that have been found to exist in patients with neurogenic motor

Loss of blood supply (CVA) by blockage of the various arteries of the brain is a leading cause of dysarthria as well as of aphasia. Tumors and trauma also are found as contributing factors. Diseases that have been implicated in various dysarthrias include myasthenia gravis, Parkinson's, multiple sclerosis, amyotrophic lateral sclerosis, Wilson's disease, encephalitis, meningitis, and a number of metabolic and systemic disorders.

IDENTIFICATION AND REFERRAL

As with many other disorders, early diagnosis of the type and severity of dysarthria not only is important for treatment planning and prognosis, but may assist the neurologist in establishing overall diagnosis of the neurologic disorder and the site(s) of lesion in the nervous system. Various behaviors may assist the nurse in identifying dysarthria for proper referral to the physician. If a patient exhibits a speech pattern that appears to be extremely slow or very fast in rate, he or she may have an underlying neuromotor dysfunction. Speech that may be described as "slurred," "sounding drunk," or "thick-tongued" may also indicate dysarthria. Excessively loud or soft voice and the impression that the patient is talking though the nose (hypernasality) are sometimes indicators of dysarthria, as are voices that sound hoarse or wet.

In many cases, referral to a neurologist and subsequent diagnosis has already been made. As with aphasia, the nurse may be in a position to refer directly to the speech pathologist or to suggest such a referral to the physician. Even in progressively deteriorating disorders such as amyotrophic lateral sclerosis, the speech pathologist can assist the patient and nurse in maintaining effective communication throughout the course of the disease. In less severe or stable disorders, the speech pathologist can develop a treatment program to

acteristics of flaccid dysarthria include breathiness in the voice, reduced loudness of the voice, hypernasality, nasal emission of air, and imprecision of consonant sounds. *Spastic dysarthria* occurs when there is bilateral damage to nerve fibers as they travel through the brain stem and midbrain structures. Characteristics of spastic dysarthria include slow rate of speaking, harsh voice quality, imprecision of consonant sounds, low pitch, and hypernasality. *Ataxic dysarthria* results from damage to the cerebellar system. Since this system is primarily responsible for coordinating movements, the following are characteristic of ataxic dysarthria: irregular breakdown in articulatory precision, distortion of vowels, prolonged sounds, and reduced rate. *Hypokinetic dysarthria* is a disorder commonly seen in patients with Parkinsonism and those with damage in special areas of the midbrain structures. Hypokinetic dysarthria is characterized by short rushes of speech, rapid rate of speech, reduced loudness, monotonous voice, and imprecision of consonants. *Hyperkinetic dysarthria* also results from damage to specific areas of the midbrain region and nerve cells deep within the brain, such as in Huntington's chorea. Characteristic features include variable speech rate, inappropriate silences, distorted vowels, and smacking noises. *Mixed dysarthria* may have components of ataxic, spastic, and flaccid dysarthria. It is usually found in disorders such as multiple sclerosis, amyotrophic lateral sclerosis, and other conditions that affect different areas of the nervous system simultaneously (Darley et al., 1975).

CAUSES OF DYSARTHRIA

As with aphasia, any factor that leads to damage of neural tissue may result in dysarthria. The type of dysarthria, however, will be dependent upon the location of the damage in the nervous system, rather than upon the causative agent itself.

2

DYSARTHRIA

DEFINITION

Brain damage does not always result in aphasia. On some occasions, the damage may spare the brain areas associated with language processing, but may affect groups of nerve cells or fibers responsible for transmitting information to the muscles used for speech. A breakdown in control over muscular speech movements as a consequence of central or peripheral nervous system damage is called *dysarthria*. A dysarthria may exist with aphasia or by itself. Unlike aphasia, however, dysarthria only affects neuromuscular control, with understanding of speech and language remaining essentially normal. In terms of the diagram in Figure I-1, damage occurring in the area between "Language Concepts and Motor Speech Patterns" and "Breathing, Voice, and Speech Musculature" in the speaker is likely to cause some type of dysarthria.

TYPES OF DYSARTHRIA

Six general categories of dysarthria have been recognized, although there is frequently a good deal of overlap in the individual patients (Darley, Aronson, & Brown, 1975). As in aphasia, however, one type of dysarthria usually predominates. *Flaccid dysarthria* results from damage to the lower brain stem and/or nerve fibers leading away from the brain stem to the muscles of breathing, voice, and speech. Char-

Cohen, L. K. *Communication problems after stroke.* Minneapolis: Kenny Rehabilitation Institute, 1971, 26 pp.

Keenan, J. S. *First aid for aphasics.* Chicago: National Easter Seal Society for Crippled Children and Adults, 1974, 17 pp.

Knox, D. *Portrait of aphasia.* Detroit: Wayne State University Press, 1971, 120 pp.

Ritchie, D. *Stroke.* Garden City, N.J.: Doubleday, 1961, 192 pp.

Tarzana Speech and Language Services. *Stroke education: Speech recovery.* Audiocassette, 1980. (Available from the author at 18455 Burbank Blvd. #412, Tarzana, California 91356.)

Taylor, M. *Understanding aphasia: A guide for family and friends.* New York: Institute of Rehabilitation Medicine, New York University Medical Center, 1958, 48 pp.

Wulf, H. *Aphasia: My world alone.* Detroit: Wayne State University Press, 1973, 179 pp.

adian Nurse, June 1974, 23-27.
Weinhouse, I. Speaking to the needs of your aphasic patients. *Nursing '81,* March 1981, 34-35.

OTHER RESOURCES

Tests Commonly Used with Aphasic Patients

Eisenson, J. *Examining for aphasia.* New York: Psychological Corporation, 1954.

Di Renzi, E., & Vignolo, L. A. The token test. *Brain,* 1962, *85,* 665-678.

Goodglass, H., & Kaplan, E. *The assessment of aphasia and related disorders.* Philadelphia: Lea & Febiger, 1972.

Porch, B. E. *The Porch Index of Communicative Ability.* Palo Alto, Calif.: Consulting Psychologists Press, 1967.

Raven, J. E. *Progressive Matrices.* New York: Psychological Corporation, 1958

Sarno, M. T. *The Functional Communication Profile* (Rehabilitation Monograph 42). New York: New York University Medical Center, 1969.

Schuell, H. *Minnesota Test for Differential Diagnosis of Aphasia.* Minneapolis: University of Minnesota Press, 1965.

Wechsler, D., & Stone, C. P. *Wechsler Memory Scale.* New York: The Psychological Corporation, 1948.

Wechsler, D. *The Wechsler Adult Intelligence Scale.* New York: The Psychological Corporation, 1955.

Wepman, J. M. *Auditory Discrimination Test.* Chicago: Author, 1958.

Materials and Pamphlets for Families

American Heart Association. *Aphasia and the family.* New York: Author, 1965, 26 pp.

Boone, D. R. *An adult has aphasia.* Danville, Ill.: Interstate Printers and Publishers, 1965, 24 pp.

Belt, L. H. Working with dysphasic patients. *American Journal of Nursing,* July 1974, 1320–1322.

Carbary, L. J. Aiding the patient with aphasia. *Nursing Care,* January 1976, 22–24.

Dreher, B. Overcoming speech and language disorders. *Geriatric Nursing,* September/October 1981, 345–349.

Drew, N. How to cope with speech defects in stroke patients. *Nursing '74,* February 1974, 20–21.

Ford, M. H. Communicating with the aphasic patient. *Journal of Practical Nursing,* April 1978, 20–21.

Fox, M. J. Talking with patients who can't answer. *American Journal of Nursing,* June 1971, 1146–1149.

Fox, M. J. Patients with receptive aphasia: They really don't understand. *American Journal of Nursing,* October 1976, 1596–1598.

Goda, S. Communicating with the aphasic or dysarthric patient. *American Journal of Nursing,* July 1963, 80–84.

Gough, D., & Powell, C. Aphasia and the nurse. *Nursing Mirror,* August 1978, 28–31.

Graham, L. Stroke rehabilitation—a creative process. *Canadian Nurse,* February 1976, 22–25.

Grey, H. A. The aphasic patient. *RN,* July 1970, 46–48.

Kester, J. A speech pathologist looks at care of the aging: The stroke patient. *Australian Nursing Journal,* August 1977, 44–46.

Louis, M. D., & Pouse, S. Aphasia and endurance: Considerations in the assessment and care of the stroke patient. *Nursing Clinics of North America,* June 1980, 265–282.

Miller, B. E. Assisting aphasic patients with speech rehabilitation. *American Journal of Nursing,* May 1969, 983–985.

Nelson, J. You and your aphasic patients. *RN,* March 1974, 42–43.

Piotrowski, M. M. Aphasia: providing better nursing care. *Nursing Clinics of North America,* September 1978, 543–554.

Schultz, L. D. M. Nursing care of the stroke patient. *Nursing Clinics of North America,* December 1973, 633–642.

Schwartzman, S. T. Anxiety and depression in the stroke patient: A nursing challenge. *Journal of Psychiatric Nursing and Mental Health Services,* July 1976, 13–18.

Stoicheff, M. D. Communicating with the aphasic patient. *Can-*

All of the qualities that make a good nurse in any setting—compassion, patience, tolerance, and understanding—are doubly required when dealing with aphasic patients. These qualities and the concept of communicating with each patient at his or her individual level of functioning are the major prerequisites for establishing good rapport with aphasic individuals.

An extensive group of resources exists to help the nurse in dealing with aphasic patients. Speech pathologists, journal articles, books, pamphlets, and community agencies are all available to aid the nurse in dealing effectively with those whose lives have been disrupted by the onset of aphasia.

REFERENCES

Baratz, R., & Norman, S. *The management of patients with aphasia: A guide for nurses.* Newton, Mass.: Robin Baratz, 1979.

Darley, F. The efficacy of language rehabilitation in aphasia. *Journal of Speech and Hearing Disorders,* 1972, *37,* 3-21.

Goodglass, H., & Kaplan, E. *The assessment of aphasia and related disorders.* Philadelphia: Lea & Febiger, 1972.

Porch, B. E. *The Porch Index of Communicative Ability.* Palo Alto, Calif.: Consulting Psychologists Press, 1967.

Porch, B. E. *PICA interpretation: Differential diagnosis.* Albuquerque: (videotapes), 1971.

Porch, B. E. Personal communication, 1981.

Schuell, H. *Minnesota Test for Differential Diagnosis of Aphasia.* Minneapolis: University of Minnesota Press, 1965.

Skelly, M. Aphasic patients talk back. *American Journal of Nursing,* July 1975, 1140-1142.

SUGGESTED READINGS

Amacher, N. J. Touch is a way of caring. *American Journal of Nursing,* May 1973, 853-854.

Behringer, S. M. You and your aphasic patients. *RN,* March 1974, 43-53.

and other health professionals to assess the type and degree of aphasic involvement, as well as to develop goals for patient improvement.

Finally, awareness of available community resources can be of assistance for the nurse dealing with aphasic patients. The following list includes those agencies and groups that are represented in a large number of communities.

American Speech-Language-Hearing Association, 10801 Rockville Pike, Rockville, Maryland 20852.

State speech-language-hearing association.

Local American Heart Association chapter, yellow pages.

Local Stroke Club or Aphasia Society, yellow pages.

Local Veterans Administration medical center, white pages.

Local university program, white pages.

Speech-language pathologists in private practice, yellow pages.

Local office of senior affairs, white pages.

SUMMARY

Aphasia is a disorder of language processing suffered by large numbers of brain-damaged patients each year. While the speech pathologist has primary responsibility for assessing and treating the aphasic patient's communication impairment, the nurse plays a broader role in the aphasic individual's medical care, rehabilitation program, and adjustment to disability. A good understanding of aphasia and its consequences should assist the nurse in reinforcing modalities under treatment, as well as in dealing with the psychological impact of aphasia on the patient and family.

and should not hesitate to do so, although coordination with the other members of the rehabilitation team is recommended. Almost as bad as no information is conflicting information from various professionals treating an individual patient.

A partial listing of materials and pamphlets that have been specifically prepared for families dealing with strokes and aphasia is found at the end of this chapter. In many cases it is helpful to provide the family with some general initial counseling about aphasia, to provide them with one or two of the pamphlets listed, and then to respond to their specific questions after they have had sufficient time to read the material.

WHEN A SPEECH PATHOLOGIST IS NOT AVAILABLE

Most of the foregoing has been written from the perspective that the aphasic patient is in a medical setting where multidisciplinary rehabilitation services are readily available. Unfortunately, this is not always the case, and the nurse may be confronted with an aphasic patient without benefit of support personnel trained to deal with this complex and frustrating problem. Although this is not the best situation, neither is it hopeless. Many of the ideas already expressed to help the nurse in dealing with communication impairment in aphasia are equally applicable with or without extensive formal testing carried out by a speech pathologist. The attributes of reassurance, patience, tolerance, and support apply in any situation.

In addition, a number of articles and materials have been developed to assist the nurse in dealing with aphasic patients. The list of suggested readings at the end of the chapter provides a selection of readily accessible sources of information. Also of benefit is a booklet entitled *The Management of Patients with Aphasia: A Guide for Nurses*, by Baratz and Norman (1979). This relatively inexpensive resource can be used by the nurse

by patients as having a negative influence on their communication and overall attitude. Television sets and loud voices of ward personnel were singled out as being especially irritating. Annoying too, was the feeling that the patients were on display for large groups of unidentified personnel without obvious relationship to their medical treatment or rehabilitation. It is apparent that health care providers in general and nurses in particular must be constantly attentive to the ways in which they deal with aphasic patients and to the environment in which aphasic patients find themselves following brain injury.

THE PATIENT'S FAMILY

It is the opinion of many clinicians working with aphasic patients on a daily basis that success or failure of a rehabilitation program is directly linked with family support and understanding, or lack of it. Everything that can be done should be done to provide families with information, guidance, and involvement in the treatment program for the affected member. Fears and frustrations are normal in families that have recently experienced a stroke in their midst, and these should be allowed to be expressed in a supportive and understanding atmosphere.

Once again the nurse is in an excellent position to provide for the needs of the family and at the same time to alert physicians, therapists, social workers, and counselors to potential problem areas in the family or patient-family interaction. Oversolicitousness or avoidance are common reactions of family members to someone with aphasia. Indeed, it is not uncommon to find both reactions in the same family. They usually result from guilt and lack of understanding of what aphasia means, what the patient can and cannot do, and what the future holds. The nurse can answer many of these questions

PRECAUTIONS

It is advisable to discuss here some of the pitfalls encountered by nurses and other health professionals when dealing with aphasic patients. In the final analysis, it is the attitude toward the aphasic individual and the dignity with which he or she is treated that assumes infinitely more importance than the actual treatment techniques employed. This point was painfully illustrated in a study by Skelly (1975), in which a large number of aphasic patients were interviewed about their poststroke hospital experiences after they had recovered a useful level of communication. The entire article should be required reading for anyone dealing with aphasic patients, and only the salient points are reviewed here.

In general, patients with aphasia were treated as though they were children or as though they did not exist, perhaps reflecting the often observed belief that people who cannot express themselves also cannot understand. As a result, expressions of reassurance and encouragement were notably absent at a time when they were needed the most. Equally frustrating were the rapidity and amount of talking directed toward the patient. A slower rate of speech with ample time allowed for introduction of new information would have substantially enhanced the understanding of the patients interviewed.

The patient's own responses were slower, but few listeners allowed sufficient response time. This created the double problem of interfering with the process of language formulation and giving the impression that what the patient had to say was of so little importance that no one wanted to take time to listen. Patients also reported that their morale and progress were affected by subtle, nonverbal indications of annoyance and impatience on the part of the hospital personnel, even though these feelings were not overtly expressed to the patients.

As discussed earlier, noise and distractions were often noted

his stroke. His speech and understanding are functional for expressing his wants and needs and following one-to-one conversation. He has little difficulty carrying out one- and two-step simple commands. More complex commands may require repetition in order for him to complete the task. He has only occasional delays and self-corrections on naming, sentence completion, and imitation tasks.

In Case Five, the patient has less difficulty communicating than any of the patients previously discussed, and may also improve further in view of his relatively recent CVA. To enhance communication, the listener should allow sufficient response time and repeat when necessary; group conversation should be avoided. Obviously, there should be no demand for reading and writing performance because of the patient's premorbid illiteracy.

Case Six

Mr. E. O. is a 49-year-old male who suffered a left CVA 4 months ago. Test results suggest a mild communicative deficit. His speech is functional for expressing wants, needs, and ideas in most communicative environments. Occasional mild word-finding difficulties are present, but he compensates by rewording his ideas. Understanding is functional for complex two- and three-step instructions, with occasional additional processing time needed for three-step instructions. Reading and writing are functional at the paragraph level, with additional time required to process complex technical information.

Case Six is the least involved patient presented and has potential for continuing gradual recovery. Even though his communication is only mildly impaired, he may have difficulty in highly complex, stressful, and/or distracting circumstances. The patient is already compensating for this to some degree and should be reinforced in his attempts to do so.

suggest that his prognosis for further significant recovery would not be as good. People communicating with this patient can increase communicative effectiveness by using "yes/no" questions to determine basic needs and allowing extra response time for simple instructions.

Case Four

Mr. F. A. is a 43-year-old male who suffered a left cerebral hemorrhage and subsequent surgical evacuation 5 months ago. His test results suggest a moderate communicative deficit. His speech consists primarily of fluent automatic speech with occasional repetitive phrases such as "Oh, I just don't know." He does much better in general conversation when he is controlling the context and when the information requested of him is of a general rather than a specific nature. He has difficulty with questions such as "What is your oldest son's name?" It is much easier for him if the question is more open-ended, such as "Could you tell me about your family?" His understanding is better for longer, more redundant material than for shorter stimuli.

The patient in Case Four is functioning at a slightly higher level than the one in Case Three and has a better prognosis for further recovery, based on the cause of his aphasia (hemorrhage) and the relatively short time since onset. The listener can maximize communicative efficiency by providing a relaxed, nonstressful atmosphere and encouraging the patient to converse; asking simple, redundant "yes/no" or "either/or" questions; and repeating questions or instructions as necessary.

Case Five

Mr. E. B. is a 53-year-old male who suffered a left CVA 4 months ago. His test results suggest a moderate communicative deficit in the verbal and gestural modalities; he was illiterate prior to the onset of

tasks and did not vocalize during the entire evaluation. His only utterance was a whispered "no" after he became very agitated. His auditory comprehension is markedly involved. He is unable to understand commands involving object function, such as "Point to the one used for writing and erasing," and he shows significant delays in responding to commands involving nouns, such as "Point to the pencil."

In Case Two, the picture is somewhat brighter than in Case One. Although the patient is quite impaired, he is only at 1 month after onset and already shows slightly higher levels of communicative ability than those indicated in Case One. The observation that the patient refused all speech tasks does not necessarily imply that he cannot perform speech tasks. He may in fact have speech problems that cause such a high level of frustration that he will not attempt any form of speech response. He is also able to respond to simple verbal commands, albeit delayed; people communicating with him should speak in simple sentences, allowing ample time for response and providing repetition as needed.

Case Three

Mr. R. G. is a 52-year-old male who suffered a left CVA 18 months ago and another 4 months ago. His test results suggest a marked communicative deficit across all modalities. His speech is characterized by minimal verbal output that is perseverative in nature. He can express basic wants and needs by correctly answering "yes/no" questions directed to him. Understanding is limited to short phrases, with greater comprehension exhibited when these are repeated and accompanied by gesture.

Although the patient in Case Three is currently functioning at a slightly higher level than the one in Case Two, the observation that he is at 4 months after onset of his second CVA would

CASE SUMMARIES

The following case summaries are presented to provide a general picture of types and severity of aphasia encountered on a day-to-day basis. While reading the summaries, it might be helpful to imagine what techniques and approaches a nurse could use in dealing with each patient.

Case One

Mr. C. H. is a 59-year-old male who suffered a massive left CVA 1 year ago. His current test results suggest only minimal improvement, compared to results of previous tests, and indicate a severe communicative deficit across all modalities (speaking, understanding, gesturing, reading, and writing). His speech remains severely impaired and nonfunctional for expressing basic wants and needs. His understanding remains nonfunctional for even simple commands; responses to very simple "yes/no" questions are inconsistent.

In Case One, communication can take place only at the most primitive level. All instructions or conversation should be simple and concise, accompanied by gestural demonstration and repeated if necessary. Responses to "yes/no" questions should always be verified. The observation that the patient is at a low level of functioning at 1 year following a massive CVA would suggest that further significant recovery cannot be expected. Extensive family support and counseling are probably necessary to assist members in dealing with the needs of this patient and in their adjustment to this very difficult situation.

Case Two

Mr. F. D. is a 56-year-old male who suffered a left CVA 1 month ago. His test results indicate a marked to severe communicative deficit across all modalities. During testing, he refused all speech

patient is facing the talker and that the vocabulary is adjusted to the patient's level of understanding with each word clearly pronounced. If hearing loss is involved, it is imperative that the patient be referred for an audiometric evaluation as soon as possible (see Chapter 6 on hearing loss).

Abstract ideas and highly emotional topics should be avoided with most aphasic patients, although frank discussions about the patient's condition are warranted if handled at the appropriate level of complexity. Repetition of questions or instructions is often necessary for individuals with aphasia. If verbatim repetition of the comment does not result in accurate comprehension, it may be possible to reword the question or instruction in a way that will enhance communication.

Sometimes writing a word or question while also saying it results in better comprehension by the patient. This approach should be attempted with caution, however, since a number of aphasic individuals become confused with too much information. By the same token, if a patient is attempting to communicate something to the nurse, the patient may be able to point to a picture or write a few letters to get his or her point across. So-called communication boards or language boards are available for this purpose and are discussed in detail in Chapter 2. It should be noted here, however, that communication boards may not be useful to the aphasic patient, since such a patient's problem involves the ability to process language and may include the recognition of pictures and words. Thus, the very deficit that prevents normal speech communication also interferes with the underlying processing necessary for the successful use of alternate communication devices or systems. Prior to using a communication board, it is advisable to consult with a speech pathologist, if possible. In any case, all attempts at communication by an aphasic patient via any means available should be strongly encouraged and reinforced.

tients, communication may break down at higher levels of complexity, such as that used in the previous example. As stated, in that case, limiting communications to short, simple exchanges usually results in successful transfer of information. Again, the main principle to keep in mind is to refrain from overtaxing the patient's communication system, but also to avoid dealing with the patient at a level far below his or her capability.

Aphasic patients, even those with comparatively mild communication problems, are especially vulnerable to distraction by background noise and activity. Therefore, every attempt should be made to talk with these individuals in a quiet, relaxed atmosphere. Radios and televisions should be turned off and doors to corridors should be closed prior to attempting communication with an aphasic patient. It is also helpful to "warm up" a patient prior to initiating important communication. For example, upon entering a patient's room, it is recommended that the nurse spend several minutes chatting (but not chattering) with the patient at a level that does not tax his or her system. Then the business at hand can be introduced in short, simple sentences accompanied by frequent gestures. Instead of hurrying into a room and telling a patient that it is time to take his or her medication, the nurse might enter the room, close the door, greet the patient, and comment on the weather, all the time pausing and giving the patient time to respond and to initiate communication if it is apparent that he or she is attempting to do so. The nurse can then explain that the doctor would like to have the patient take some pills, at the same time gesturing with the water glass and pills. If the patient is at a sufficient level of understanding, the nurse can show on a wristwatch the times that the medication will be administered, thus enhancing the patient's understanding of the medication while involving him or her in the details of day-to-day care. It is not necessary to yell at the patient. It is much more effective to insure that the

results and treatment plan with the speech pathologist and other members of the rehabilitation team. Knowledge of the relative strengths and weaknesses in a patient's communication system can be used to reinforce areas of concentration in therapy and to avoid pushing the patient beyond the limits of his or her capability. For example, if the test results indicate that a patient can accurately follow one-step commands such as "Point to the comb," but fails at following more complicated commands such as "Point to the pen and pick up the cup," then the nurse should have an easier time communicating with this patient if questions or instructions are kept brief and limited to a single topic. Long and complicated instructions or questions are frustrating for such patients, since their communication systems are overloaded and they are unable to comply.

REHABILITATION

It is difficult to provide detailed information about how to deal with specific patients, since there is such a wide variety of type and severity of disorders within the population of aphasic patients. As a general statement, however, it is usually acceptable to communicate with the aphasic individual at the level where communication just begins to break down. For some, this may mean limiting conversation to simple questions that can be answered with "yes" or "no." Even here, though, caution should be applied, since many patients may respond "yes" to questions because of the tone of voice used by the talker, but may not understand the content of the question. The ability of the patient to respond accurately to "yes/no" questions should be verified either by consulting with the speech pathologist or by interspersing questions for which "no" is the correct answer, rather than simply asking two or three questions, all of which can be answered in the affirmative. For other pa-

promise of recovery as do those experiencing more limited damage resulting from a single episode. Obvious factors such as coexisting medical or emotional problems, patient motivation, and family support all have varying degrees of influence on the eventual level of recovery in any individual aphasic patient.

Test data, clinical observation, and personal experience are used by the speech pathologist in arriving at a decision as to whether to treat a patient. It must be understood that all patients are not candidates for a treatment program, and that a recommendation "not to treat" does not indicate disinterest or lack of caring on the part of the speech pathologist. On the contrary, many speech pathologists err on the side of caution and see patients for a trial period of therapy, even though all indicators argue against the expectation of improvement. Budgetary restrictions, staff shortages, and space limitations, however, necessitate the establishment of a priority system in most speech pathology clinics. As such, patients with the greatest potential for recovery or most significant need for communication usually receive intensive treatment.

A decision not to treat a patient does not suggest that nothing can be done. In many cases, the individual may benefit from the socialization and general language stimulation of a structured group environment. Also, some patients do not follow predictable patterns of recovery and may benefit from treatment at a later date. Nurses should be alert to changes in patients' condition, attitude, and day-to-day communication that warrant a re-referral for an updated speech and language evaluation. Even in severe cases, most aphasic patients are aware of their surroundings, can handle bodily functions, and have some residual minimal communication skills.

If the speech pathologist recommends that a formal treatment program should be initiated, the nurse can be instrumental in the rehabilitation process by discussing the test

USE OF TEST RESULTS

It should be noted that no single test instrument will be suitable for all patients at all times. As a result, numerous supplementary tests have been developed and are variously employed by the speech pathologist, depending upon the needs of the patient. Regardless of the test or tests used, the speech pathologist analyzes the results with the following questions in mind:

1. What is the nature of the deficit? That is, does the patient have difficulty in understanding and following commands? Is there difficulty in naming common objects? and so on.
2. How severely impaired is the patient's ability to perform in the various communicative modalities?
3. Is the patient a candidate for treatment?
4. What is the prognosis for recovery of communicative functioning?
5. If treatment is indicated, how should it proceed?

These are difficult questions to answer even with the best test materials available, since individuals vary considerably in terms of their reaction to stroke, their premorbid communication abilities and needs, their overall health, and other factors. Nevertheless, there are guidelines.

In general, younger patients have a better prognosis for recovery than do those of more advanced age. The amount of time since onset of the stroke or accident is also an indicator of potential for recovery. A patient at 2 years after onset would not have as good a chance for significant recovery as would a patient with similar test performance at 1 month after onset. Individuals who have suffered severe and extensive brain injury, or who have had multiple CVAs, do not show as much

Figure 1-4. Typical BDAE scoresheet showing patient's level of communicative functioning for various tasks.

Patient's Name ___D. M.___ Date of rating ___9-16-80___
                                  Rated by ___W. J. R.___

APHASIA SEVERITY RATING SCALE

0. No usable speech or auditory comprehension.

1. All communication is through fragmentary expression; great need for inference, questioning and guessing by the listener. The range of information which can be exchanged is limited, and the listener carries the burden of communication.

2. Conversation about familiar subjects is possible with help from the listener. There are frequent failures to convey the idea, but patient shares the burden of communication with the examiner.

3. The patient can discuss almost all everyday problems with little or no assistance. However, reduction of speech and/or comprehension make conversation about certain material difficult or impossible.

4. Some obvious loss of fluency in speech or facility of comprehension, without significant limitation on ideas expressed or form of expression.

5. Minimal discernible speech handicaps; patient may have subjective difficulties which are not apparent to listener.

RATING SCALE PROFILE OF SPEECH CHARACTERISTICS

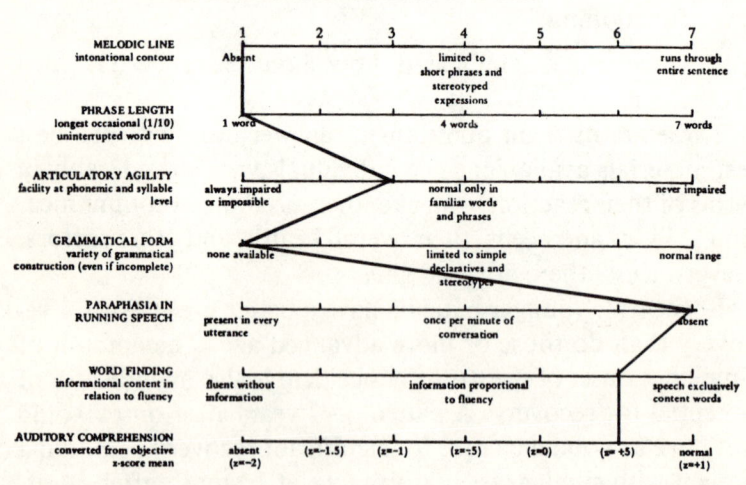

Characteristic	1	4	7
MELODIC LINE — intonational contour	Absent	limited to short phrases and stereotyped expressions	runs through entire sentence
PHRASE LENGTH — longest occasional (1/10) uninterrupted word runs	1 word	4 words	7 words
ARTICULATORY AGILITY — facility at phonemic and syllable level	always impaired or impossible	normal only in familiar words and phrases	never impaired
GRAMMATICAL FORM — variety of grammatical construction (even if incomplete)	none available	limited to simple declaratives and stereotypes	normal range
PARAPHASIA IN RUNNING SPEECH	present in every utterance	once per minute of conversation	absent
WORD FINDING — informational content in relation to fluency	fluent without information	information proportional to fluency	speech exclusively content words
AUDITORY COMPREHENSION — converted from objective z-score mean	absent (z=-2)	(z=-1.5) / (z=-1) / (z=-.5) / (z=0) / (z=+.5)	normal (z=+1)

Source: Reproduced by special permission from *The Assessment of Aphasia and Related Disorders* by H. Goodglass and E. Kaplan; copyright 1972, published by Lea & Febiger, Philadelphia.

Figure 1-3. Typical PICA scoresheet showing patient's level of communicative functioning for various tasks.

Porch Index of Communicative Ability
RATING OF COMMUNICATION DEFICIT

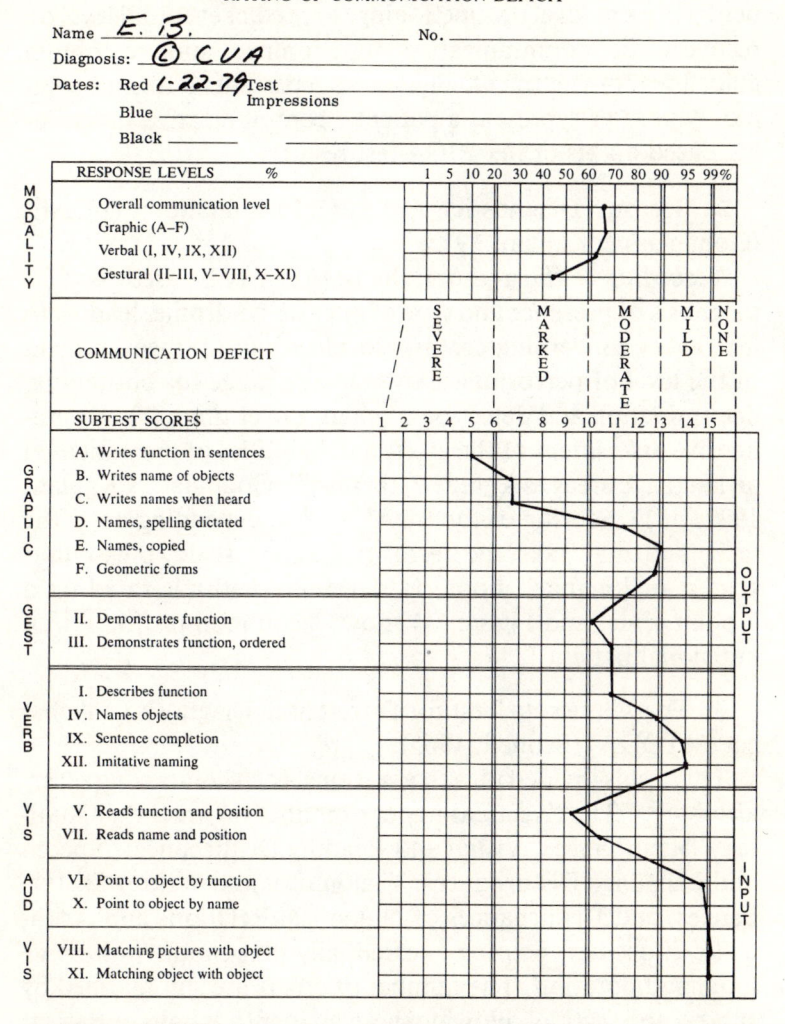

Name _E. B._ No. _____
Diagnosis: _© CVA_
Dates: Red _1-22-79_ Test _____
　　　　　　　　　Impressions
　　　Blue _____
　　　Black _____

MODALITY	RESPONSE LEVELS	%	1 5 10 20 30 40 50 60 70 80 90 95 99%
	Overall communication level		
	Graphic (A–F)		
	Verbal (I, IV, IX, XII)		
	Gestural (II–III, V–VIII, X–XI)		

	COMMUNICATION DEFICIT	SEVERE	MARKED	MODERATE	MILD	NONE

	SUBTEST SCORES	1 2 3 4 5 6 7 8 9 10 11 12 13 14 15
GRAPHIC	A. Writes function in sentences	
	B. Writes name of objects	
	C. Writes names when heard	
	D. Names, spelling dictated	
	E. Names, copied	
	F. Geometric forms	
GEST	II. Demonstrates function	
	III. Demonstrates function, ordered	
VERB	I. Describes function	
	IV. Names objects	
	IX. Sentence completion	
	XII. Imitative naming	
VIS	V. Reads function and position	
	VII. Reads name and position	
AUD	VI. Point to object by function	
	X. Point to object by name	
VIS	VIII. Matching pictures with object	
	XI. Matching object with object	

(OUTPUT / INPUT)

Source: Reproduced by special permission from *The Porch Index of Communicative Ability* by B. E. Porch; copyright 1967, published by Consulting Psychologists Press, Palo Alto, California.

Responses are scored on a 1-15 point scale on the basis of accuracy, responsiveness, completeness, promptness, and efficiency. The examiner can use these scores to describe the patient's present level of functioning, to predict eventual level of recovery of communicative functioning, and to obtain guidelines for treatment. Figure 1-3 represents a format commonly used to represent a patient's communicative functioning based on his or her PICA test results.

2. Boston Diagnostic Aphasia Examination (BDAE) (Goodglass & Kaplan, 1972).

According to the authors, the BDAE may be used for "(1) diagnosis of presence and type of aphasic syndrome, leading to inferences concerning cerebral localization; (2) measurement of the level of performance over a wide range for both initial determination and detection of change over time; (3) comprehensive assessment of the assets and liabilities of the patient in all language areas as a guide to therapy" (Goodglass & Kaplan, 1972, p. 1). Scoring of the BDAE varies, depending upon the task, and may take the form of a rating scale, plus-minus scores, or longhand notation. Overall severity is rated on a 7-point scale, and Figure 1-4 shows a common profile for an individual patient.

3. The Minnesota Test for Differential Diagnosis of Aphasia (MTDDA) (Schuell, 1965).

This test is divided into five sections, each comprising several subtests. The five areas under examination are these: Auditory Disturbances; Visual and Reading Disturbances; Speech and Language Disturbances; Visuomotor and Writing Disturbances; and Disturbances of Numerical Relations and Arithmetic Processes. Scoring is essentially plus-minus with some longhand notation. The number of errors are summarized by subtest and section, allowing the examiner to assign the patient to one of five major and two minor categories of aphasic impairment.

the patient for evaluation. Most people experience anxiety before and during test taking. In many cases, this anxiety is magnified in aphasic patients, who are already aware that they are having difficulty communicating and are hesitant about revealing the extent of the problem, or perhaps do not want to accept the notion that a deficit exists. At any rate, much of the fear and anxiety can be eliminated by providing reassurance that the evaluation is not the type that can be passed or failed and that there are no right or wrong answers. Rather, the testing is designed to assess strengths as well as weaknesses in communicative functioning so that appropriate treatment can be planned to assist patients during their recovery. It might also be pointed out to patients that the evaluation will not be painful or cause discomfort, and that it will take from 1 to 2 hours, depending upon the tests used and individual clinic procedures and schedules.

APHASIA EVALUATION

When the patient arrives at the speech pathology clinic, he or she usually undergoes a battery of tests, including some of those mentioned in the bibliography for this chapter (see "Tests Commonly Used with Aphasic Patients"). In an attempt to provide a general overview of the kinds of tests available and to provide familiarization with the tests most commonly used with aphasic patients, only three instruments are reviewed here.

1. Porch Index of Communicative Ability (PICA) (Porch, 1967).

This widely used test is designed to elicit verbal, gestural, and graphic responses pertaining to 10 common objects. The 18 subtests sample communicative functioning in the three modalities across a series of tasks of varying complexity.

duals when involved in language processing. The person exhibiting slow rise time tends to miss the initial parts of incoming linguistic material or miss the entire message if it consists of one or two short words. For example, given the command "Touch the cup," the patient may respond by pointing to the cup or picking up the cup, since he or she has processed that portion of the message dealing with the cup, but has not handled the initial part of the command. By the same token, this patient might have difficulty identifying the object, given the command "cup," but has no difficulty if his or her language-processing system has been prepared by a lead-in sentence such as "Show me the cup."

Noise buildup is a deterioration in performance over time, even though the complexity and length of the stimulus and response remain unchanged. The patient appears to be unable to clear his or her system of what came in previously in order to process more recent information. An example of this type of behavior is seen in the patient who may respond correctly by pointing to the initial portions of a list of common objects, only to deteriorate progressively in performance toward the end of the list.

REFERRING THE PATIENT

If the patient exhibits these or other behaviors that suggest a problem in communicative functioning, a referral to a speech pathologist should be initiated. In many cases, the speech pathologist visits the patient at bedside or on the ward to establish rapport and to get an idea of the patient's general level of functioning. Also, the speech pathologist usually uses this opportunity to advise the patient that a formal evaluation is being planned and will take place sometime during the next several days.

Here, again, the nurse can play a critical role in preparing

follow commands such as "Point to the chair." Aphasic individuals also have been observed to have difficulty processing and retaining long, complex, or unfamiliar material. For example, the patient may perform very well if asked to "Point to the chair" (response), "Stand up" (response), "Open the door" (response); but may have considerable difficulty in responding to the command "Point to the chair, stand up, and open the door."

Changing from one task to another may be extremely difficult for some aphasic patients and manifests itself as an abnormally high number of errors initially, with a gradual increase in correct performance over a period of time when change occurs. For example, an individual may be participating in a conversation about baseball, but may have difficulty if suddenly the topic changes to politics. After a period of time, however, he or she may again begin to participate effectively in this conversation.

The possibility of aphasia should be suspected in the patient who exhibits a large number of self-corrected errors in the process of speaking or writing. For example, the individual who says "I wite . . . wride . . . write with a pen," or "I smoke a match . . . no . . . cigarette," may be having difficulty with internal language-monitoring systems that can be disrupted by brain damage. Intermittent problems are also characteristic of aphasic patients. The individual may perform a task requiring language processing for a period of time and then experience a period of decreased performance followed once again by appropriate responses. An example would be the patient who appears to "tune out" during a conversation or who responds appropriately to a question but later responds inappropriately to the same question or one of similar content and complexity.

Two terms borrowed by Porch (1971) and others from the field of electronic systems—*slow rise time* and *noise buildup*—also characterize the behavior of many aphasic indivi-

beyond his or her residual level of deficit, since the brain tissue has been permanently damaged at that point. Therefore, referral to the speech pathologist for testing and treatment would be useless.

Recently reported experimental and clinical evidence (Darley, 1972), however, supports the contention that therapeutic intervention can bring about improvement beyond that attributable to spontaneous recovery. It should be noted, though, that not all aphasic patients can be aided by a formal treatment program. As such, early referral to the speech pathologist for accurate and detailed testing of each patient is recognized as an essential first step in determining rehabilitation potential and in designing a rehabilitation program for those who show promise of benefiting from therapeutic intervention. For those whose prognosis is poor, testing may illuminate modalities that can be used by nursing personnel and families for establishing some level of communication with the patient. A detailed evaluation can also be of help in planning for long-term placement and developing realistic expectations for family members and health care providers. Depending upon the setting, a neurologist or other physician makes decisions regarding referral of a brain-damaged patient for aphasia testing. In many cases, however, the nurse either may be the primary referral source or can be extremely influential in bringing to the attention of the physician the need for referral of an aphasic patient.

IDENTIFYING THE APHASIC PATIENT

Various behaviors may assist the nurse in identifying aphasic patients for referral to the speech pathologist. Increased latency of response, indicated by long delays in processing incoming language, is frequently observed in individuals with aphasia. The patient may require longer than normal periods of time to answer simple questions such as "What is your name?" or to

Figure 1-2. Patterns of recovery of communicative function for aphasia resulting from various causes.

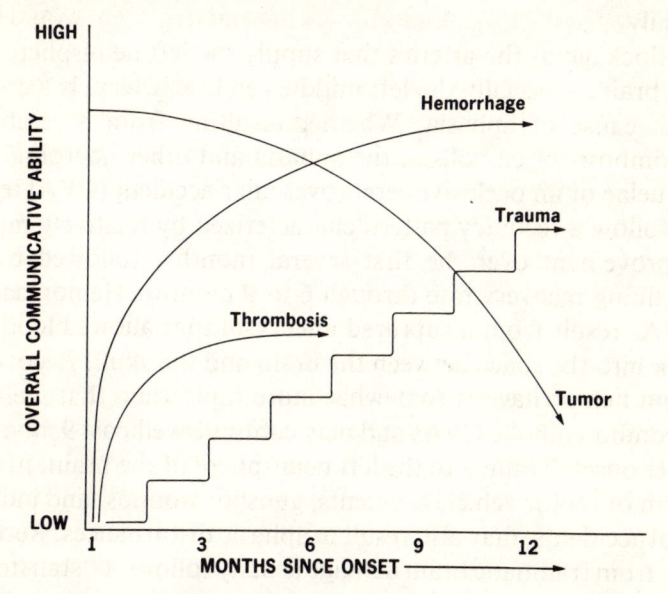

Source: Adapted from B. E. Porch, personal communication, 1981.

taneous recovery and is believed to occur with or without therapeutic intervention. It is attributed to such things as reduction in swelling of brain tissues, establishment of collateral blood supplies, and easing of pressure in the case of hemorrhage or trauma. Apparently, some brain cells have been only temporarily affected and regain normal or near-normal function during the process of spontaneous recovery, which is usually 3 to 6 months in duration.

The issue of spontaneous recovery has sparked controversy over the efficacy of treatment for aphasic patients almost since formal treatment has been attempted. In essence, some physicians and speech pathologists argue that once spontaneous recovery has taken place, it is impossible to improve the patient

damage can be very helpful to the nurse in planning patient care and in promoting realistic expectations for the patient and family.

Blockage of the arteries that supply the left hemisphere of the brain, especially the left middle cerebral artery, is a common cause of aphasia. Whether resulting from a cerebral thrombosis or embolism, the aphasia and other neurological sequelae of an occlusive cerebrovascular accident (CVA) tend to follow a recovery pattern characterized by relatively rapid improvement over the first several months, followed by a declining recovery rate through 6 to 9 months. Hemorrhagic CVAs result from a ruptured aneurysm that allows blood to leak into the space between the brain and the skull. Recovery from hemorrhage is somewhat more rapid than that seen in thrombo-embolic CVAs and may continue well past 9 months after onset. Trauma to the left hemisphere of the brain, in the form of motor vehicle accidents, gunshot wounds, and industrial accidents, may also result in aphasic disturbances. Recovery from traumatic brain damage usually follows a "stairstep" pattern, with periods of recovery interspersed among periods of minimal change. Finally, both malignant and nonmalignant tumors sometimes present themselves in the area of the left hemisphere of the brain, with resultant aphasic disturbances. Aphasia resulting from a tumor, unlike that resulting from the other causes mentioned, tends to become more pronounced as the tumor progresses. Removal of the tumor may have a variable effect on recovery, depending upon its size, its location, the presence of metastases, the degree to which normal brain tissue is sacrificed, and so on. Figure 1-2 represents general patterns of recovery following different types of brain injury (Porch, 1981).

NEED FOR REFERRAL

Many patients who suffer brain damage resulting in aphasia show some initial improvement during the immediate months following the episode. This phenomenon is referred to as *spon-*

types mentioned in the foregoing paragraphs. This results from more diffuse damage in the left hemisphere, involving both the anterior and posterior portions. A mixed aphasia is seen frequently, but usually shows a predominance of involvement of either the anterior or posterior type. Severely involved patients are sometimes referred to as exhibiting *global aphasia,* and may only be able to utter automatic or overlearned speech (swear words, counting, the days of the week, etc.).

The nurse may also encounter a fourth type of disorder, called *apraxia of speech,* that results from limited damage to a small area in the anterior portion of the brain. In this disorder, the patient appears to have no difficulty understanding or verbalizing, except that the motor sequencing of complex speech patterns becomes impaired—in the absence of any detectable muscle paralysis or weakness. For example, the following is a transcription of a patient with apraxia of speech attempting to say the word "photographic":

> phorto . . . phogo . . . phortogo . . . photogro . . . phortogaph . . . photograph.

In this example, the patient appears to have the ability to produce all the necessary sounds correctly, but is having difficulty arranging them in the proper sequence. Some clinicians believe that apraxia of speech is essentially the same as Broca's or nonfluent aphasia, while others treat is as a separate entity. In some cases, apraxia of speech is found as an isolated disorder, but it is frequently combined with the other problems mentioned earlier.

CAUSES OF APHASIA

Aphasia is caused by damage to brain cells that are represented by the portions of Figure I-1 entitled "Language Concepts and Motor Speech Patterns" and "Auditory Patterns and Language Concepts." Knowing specifically what has caused the

patient exhibits greater difficulty with speaking and writing, but has relatively good understanding of spoken and written messages. The speech may be characterized by problems in producing sounds in the proper sequence (such as "tevelision" for "television"), and it tends to be telegraphic —that is, to omit conjunctions, prepositions, and articles (as in "I went store home" for "I went to the store and then home"). Misarticulated words ("fen" for "pen") and perseveration are also commonly noted. This type of aphasia is variously referred to as *Broca's* or *anterior aphasia* (in reference to the location of damage), or *nonfluent* or *expressive aphasia* (in reference to the type of speech pattern). In view of the brain's organization, it is also common to find paralysis or weakness in the right arm and leg in the anterior (nonfluent) type of aphasic patient, since the left side of the brain in this region controls the right side of the body.

If the damage is more toward the back of the brain in *Wernicke's area* (Figure 1-1), the patient usually has a good deal of speech output, but has difficulty understanding incoming messages. Also, the speech may sound strange, since it is grammatically accurate but lacks content and meaning. Other speech characteristics commonly observed are circumlocutions —that is, attempts to talk around a topic or word that cannot be retrieved ("That's a, uh, you use it for this or that"). Word substitutions from the same category ("cup" for "glass," "knife" for "fork") and lack of inhibition (the inability to turn off the flow of speech) are frequently observed. This type of aphasia is often referred to as *Wernicke's* or *posterior aphasia* (location), or *receptive* or *fluent aphasia* (type of communication problem). Unlike the Broca's aphasic patient, the Wernicke's patient may have little or no motor involvement of the right side of the body, since the damage has occurred posterior to the motor region of the brain.

In many patients, the aphasia is a combination of the two

TYPES OF APHASIA

Just as there are many definitions of aphasia, there also exist many classification systems that have been developed in an attempt to categorize the various manifestations of communicative impairment observed within the overall condition of aphasia. Again, for the purpose of this discussion, the classification systems used in the majority of hospitals and rehabilitation centers are identified and compared in an effort to enhance communication between nurses and speech pathologists.

In general, the aphasic patient exhibits a specific type of disorder based upon the general location of damage to the left hemisphere of the brain. If the damage is toward the front of the brain in the region known as *Broca's area* (Figure 1-1), the

Figure 1-1. Schematic representation of the left hemisphere of the brain, showing approximate locations of Broca's and Wernicke's areas and other common landmarks.

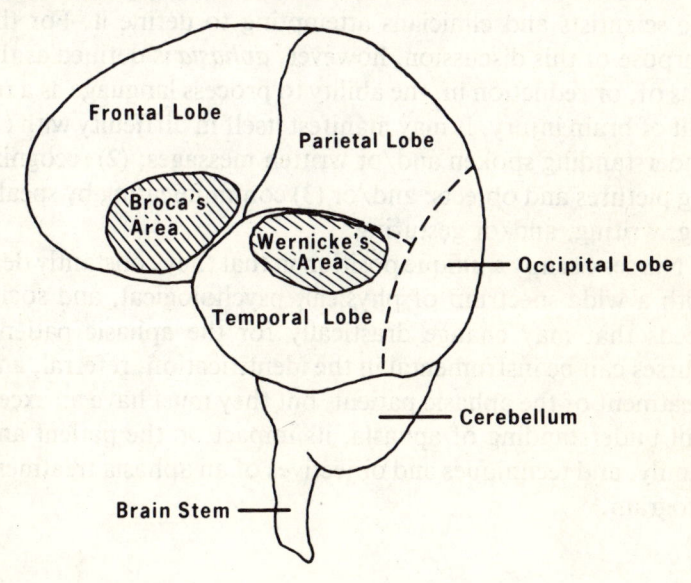

1
APHASIA

DEFINITION

The occurrence of brain injury is one of the most devastating problems encountered in today's medical settings. In many cases, brain damage means protracted hospitalization followed by a long period of outpatient rehabilitation. Aphasia frequently accompanies brain injury, and it is this disorder that is of concern here.

There are probably as many definitions of aphasia as there are scientists and clinicians attempting to define it. For the purpose of this discussion, however, *aphasia* is defined as the loss of, or reduction in, the ability to process language as a result of brain injury. It may manifest itself in difficulty with (1) understanding spoken and/or written messages; (2) recognizing pictures and objects; and/or (3) communicating by speaking, writing, and/or gesturing.

Nurses occupy a unique position in that they constantly deal with a wide spectrum of physical, psychological, and social needs that may change drastically for the aphasic patient. Nurses can be instrumental in the identification, referral, and treatment of the aphasic patient, but they must have an excellent understanding of aphasia, its impact on the patient and family, and techniques and objectives of an aphasia treatment program.

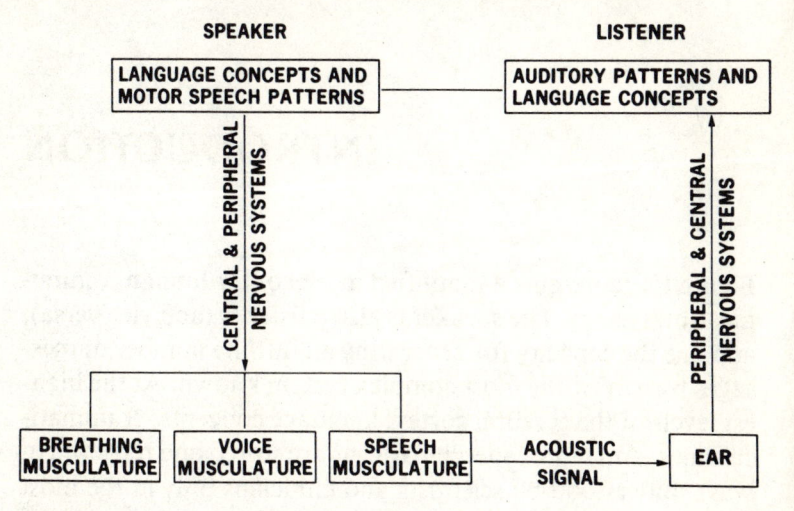

Figure I-1. Simplified model of human communication.

```
        SPEAKER                                    LISTENER
┌─────────────────────────┐              ┌─────────────────────────┐
│ LANGUAGE CONCEPTS AND   │              │ AUDITORY PATTERNS AND   │
│ MOTOR SPEECH PATTERNS   │──────────────│ LANGUAGE CONCEPTS       │
└─────────────────────────┘              └─────────────────────────┘
   │ CENTRAL & PERIPHERAL                       ▲ PERIPHERAL & CENTRAL
   │ NERVOUS SYSTEMS                            │ NERVOUS SYSTEMS
   ▼                                            │
┌──────────┐ ┌──────────┐ ┌──────────┐   ACOUSTIC   ┌─────┐
│BREATHING │ │ VOICE    │ │ SPEECH   │──  SIGNAL  ──│ EAR │
│MUSCULATURE│ │MUSCULATURE│ │MUSCULATURE│          └─────┘
└──────────┘ └──────────┘ └──────────┘
```

the chain have varying degrees of impact on communication, depending upon the type and severity of the interruption. The following chapters deal with disorders associated with interruptions in the chain commonly found in the adult population.

I begin with one of the most prevalent and tenacious problems—aphasia; continue through the dysarthrias, laryngectomy, glossectomy and maxillectomy, voice disorders, and hearing impairment; and conclude with specialized disorders that do not fit precisely in these other areas. It should be kept in mind that, while each disorder is discussed separately, two or more can and often do occur coincidentally. Again, as more of the communication chain is involved, a concomitant increase in resulting communicative impairment can usually be expected.

INTRODUCTION

Figure I-1 represents a simplified model of the human communication system. The speaker is also a listener (and vice versa), and has the capacity for generating an infinite number of messages by way of the most complex system known. At the highest levels of the cerebral cortex, language concepts, grammatical rules, and motor speech patterns are stored and retrieved in ways understood by scientists and clinicians only at the most superficial level. This information is transmitted through the central and peripheral nervous system to the respiratory, voice, and speech musculature.

Air from the lungs passes through the *glottis,* which is the opening bounded by the vocal folds. When the vocal folds are drawn together, the air from the lungs is forced through them, causing them to vibrate and make a buzzing sound. This sound passes through the *vocal tract* (formed by the pharyngeal, oral, and nasal cavities) and is modified by movements of the tongue, lips, and palate. Sometimes the air from the lungs is allowed to flow unimpeded by the vocal folds through the vocal tract, to be momentarily stopped or redirected by the speech musculature. These modifications create different sounds and groups of sounds, to which the listener has learned to attach meaning. The sounds impinge upon the listener's hearing mechanism and are transmitted via the peripheral and central nervous system to the highest levels of the cerebral cortex, where the language concepts, grammatical rules, and patterns are used to decode the message. Interruptions at one or more locations in

description of the disorder being discussed, its impact on the communication of the patient, concerns of the speech pathologist/audiologist in dealing with the disorder, and the nurse's involvement. Since it is acknowledged that the material is not all-inclusive, references and suggested readings are provided at the end of each chapter for the reader who is motivated to probe deeper; a glossary of terms is included at the end of the text to assist in deciphering the jargon of audiology and speech pathology. Where appropriate, lists of community resources have been included in the chapter text, and lists of aids and materials follow the references and suggested readings. The manuscript is designed to accompany audiovisual material presented in a workshop or in-service training program where numerous examples of each disorder can be studied, but is also written to stand alone as a reference text. Finally, the philosophical substrate underlying all of the material presented includes a commitment to establishing an atmosphere of understanding and cooperation among nurses and speech and hearing professionals for the delivery of high-quality care to the communicatively impaired patient.

Many people have contributed to this book, the most important being the communicatively impaired individuals with whom I have worked and from whom I have learned. I would also like to acknowledge the nurses who have sat patiently through workshops and in-service training programs, providing me with feedback, suggestions, and criticism for refining the material. Special thanks go to Carolyn E. Joneson, M.S.N., Laurel Gianchino, R.N., Judith Hudson, and Carol Suazo, as well as to my wife, Emilie, for her support and for her illustrations throughout the book.

PREFACE

It is unlikely that nurses dealing with adult patients on a daily basis do so without frequently encountering individuals with communicative handicaps. These may take the form of difficulties in understanding speech secondary to hearing loss, the inability to communicate because of a stroke, or the loss of voice following a laryngectomy. In many cases, nursing personnel have the most significant contact with such patients and play a crucial role in the rehabilitative effort.

While the special problems of the communicatively impaired patient have been receiving increased attention in the nursing literature, there has been little effort directed toward developing a single source to which nurses can refer on a day-to-day basis or use in a structured training program dealing with communicative disorders. This manual is an attempt to fill the void; it is an outgrowth of materials that have been developed and refined over a 10-year period of lectures, workshops, and in-service training programs presented in both the academic and hospital setting.

No attempt has been made to provide an exhaustive treatise on all aspects of diagnosis and treatment of speech- and hearing-impaired individuals. Nevertheless, experience and feedback from many colleagues in the nursing profession have prompted the inclusion of sufficient detail to enhance a better understanding of the overall characteristics of communication breakdown, rather than providing simply a list of "dos" and "don'ts." The material is presented in a format that includes a

FOREWORD

With the development of more sophisticated diagnostic and rehabilitative techniques, speech and hearing professionals have achieved marked progress in assisting those with communicative disorders. Nurses have been made aware of how to develop effective communication skills with only recent emphasis on special problems of the communicatively impaired. This manual can be an invaluable resource to the nurse who has been frustrated by an inability to communicate with the person handicapped in hearing and/or speech, who often is equally frustrated in establishing communication. Its easily understood content, supporting references, and suggested readings make for a usable text for the nurse who works with adults and their families in any health care setting.

Carolyn E. Joneson, M.S.N.
Former Associate Chief
Nursing Service for Education
Veterans Administration Medical Center
Albuquerque, New Mexico

on Ventilators 132   Locked-In Syndrome 134
Terminally Ill Patients 135   Summary 137   Suggested
Readings 137   Available Devices 138

GLOSSARY 139

INDEX 145

Postoperative Care 60 Rehabilitation 60 The
Patient's Family 73 Community Resources 74
Summary 74 References 75 Suggested
Readings 75 Other Resources 76

4. GLOSSECTOMY AND MAXILLECTOMY 78
Glossectomy: Definition 78 Normal Speech
Production 79 Changes in Speech and Swallowing
79 Preoperative Counseling 80 Postoperative
Rehabilitation 81 Maxillectomy: Definition
86 Effects on Speech and Swallowing 86
Preoperative Counseling 88 Postoperative Care 88
Rehabilitation 89 The Nurse as Counselor 90
Summary 91 References 92 Suggested
Reading 92

5. VOICE DISORDERS 93
Definition 93 Identification and Referral 93
Types and Causes of Dysphonia 95 Evaluation
and Treatment 99 Concluding Remarks 106
Summary 106 References 107 Suggested Readings
107

6. HEARING IMPAIRMENT 108
Definition 108 Normal Hearing 108 Types and
Causes of Hearing Impairment 110 Identification
and Referral 111 The Audiogram 112 Use of
Test Results 117 Hearing Aids 119 The Deaf 123
Community Resources 125 Summary 126
Suggested Readings 126 Materials for
Patients 127

7. OTHER AREAS OF COMMUNICATIVE
IMPAIRMENT 128
Right Hemisphere Damage 128 The Confused
Patient 130 Patients with Tracheostomy Tubes or

CONTENTS

Foreword, by Carolyn E. Joneson, M.S.N. ix

Preface xi

INTRODUCTION 1

1. APHASIA 3
   Definition 3   *Types of Aphasia* 4   *Causes of Aphasia* 6   *Need for Referral* 7   *Identifying the Aphasic Patient* 9   *Referring the Patient* 11   *Aphasia Evaluation* 12   *Use of Test Results* 16   *Rehabilitation* 18   *Case Summaries* 21   *Precautions* 25   *The Patient's Family* 26   *When a Speech Pathologist Is Not Available* 27   *Summary* 28   *References* 29   *Suggested Readings* 29   *Other Resources* 31

2. DYSARTHRIA 33
   Definition 33   *Types of Dysarthria* 33   *Causes of Dysarthria* 34   *Identification and Referral* 35   *Evaluation* 36   *Use of Test Results* 36   *Rehabilitation* 38   *Dysphagia* 44   *The Patient's Family* 48   *Community Resources* 49   *Summary* 49   *References* 50   *Suggested Readings* 51   *Other Resources* 51

3. LARYNGECTOMY 54
   Definition and Causes 54   *Before and After Laryngectomy* 54   *Preoperative Counseling* 58

Copyright © 1982 by Springer Publishing Company, Inc.

All rights reserved

No part of this publication may be reproduced, stored in a retrieval system, or transmitted in any form or by any means, electronic, mechanical, photocopying, recording, or otherwise, without the prior permission of Springer Publishing Company, Inc.

Springer Publishing Company, Inc.
200 Park Avenue South
New York, New York 10003

82 83 84 85 86 / 10 9 8 7 6 5 4 3 2 1

Library of Congress Cataloging in Publication Data

Ryan, William J. (William John)
   The nurse and the communicatively impaired adult.

   Includes bibliographies and index.
   1. Communicative disorders—Nursing. 2. Nurse and patient. I. Title. [DNLM: 1. Speech disorders—Nursing. 2. Voice disorders—Nursing. 3. Hearing disorders—Nursing. 4. Laryngectomy—Adverse effects. WY 158 R989n]
RC423.R9   1982   616.85′5   82-6088
ISBN 0-8261-3960-4   AACR2
ISBN 0-8261-3961-2 (pbk.)

Printed in the United States of America

The Nurse and the Communicatively Impaired Adult

WILLIAM J. RYAN, PH.D.

Foreword by Carolyn E. Joneson, M.S.N.

SPRINGER PUBLISHING COMPANY
New York

William J. Ryan, Ph.D. completed his undergraduate work in speech pathology and audiology at the State University of New York (Geneseo) in 1965. He attended Purdue University from 1965 to 1967 and spent the next 2 years in the U.S. Army. Upon returning to Purdue, he received an M.S. degree in 1969 and a Ph.D. in 1971 from the Department of Audiology and Speech Sciences. Dr. Ryan's professional career includes a 4-year appointment as Assistant Professor of Speech Science at the University of New Mexico, followed by a 3-year stay as Speech Pathology Section Chief at the Audie Murphy Memorial VA Medical Center in San Antonio, Texas. He is currently Chief of the Audiology and Speech Pathology Service at the Albuquerque VA Medical Center, a position that he has held for the last 4 years.

Dr. Ryan has authored or coauthored more than a dozen journal articles and numerous papers in the areas of aging, aphasia, laryngectomy, and central auditory processing, and has presented invited lectures in Canada and the United States on normal and disordered communication in the elderly. For many years, he has been active in the planning, development, and presentation of VA in-service training programs and workshops designed to assist other health care providers in dealing with communicatively impaired patients, and he frequently serves as a faculty member for the VA Regional Medical Education Center Program.

The Nurse and the Communicatively Impaired Adult

PORTAGE PUBLIC LIBRARY

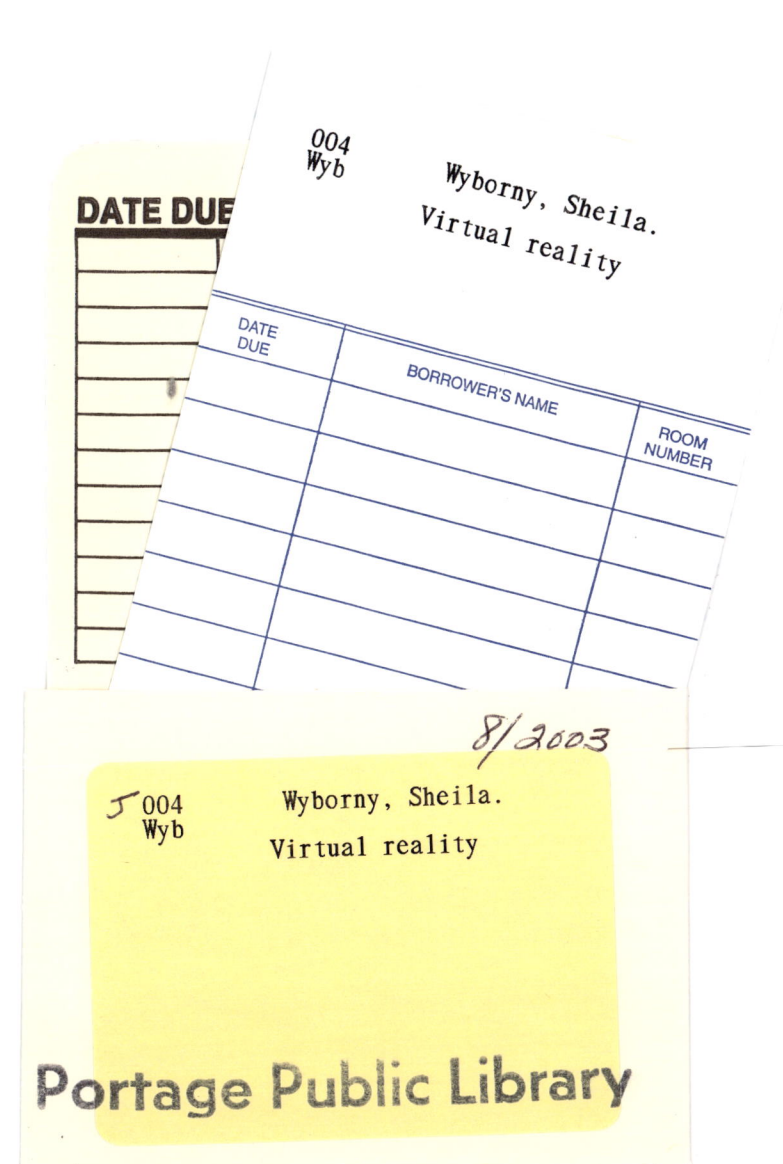

31814883500537   004 Wyb

Virtual reality

DATE DUE

DATE DUE	BORROWER'S NAME	ROOM NUMBER

004
Wyb   Wyborny, Sheila.
             Virtual reality

8/2003

J 004
Wyb   Wyborny, Sheila.
             Virtual reality

Portage Public Library

INDEX

Head mounted display (HMD), 6, 7, 15, 17, 18, 19, 20, 22, 37, 42, 43, 44
Heilig, Morton, 8, 10
HMD, 6, 7, 15, 17, 18, 19, 20, 22, 37, 42, 43, 44
Holodeck, 30

Illness, 21
Immersion virtual reality, 20, 30, 32, 39, 41
Interactive devices, 10, 18, 36, 44
Interactive virtual classrooms, 41–42, 44

Jobs, 28, 39
Joystick, 7, 19, 20

Law enforcement, 5, 39
Leisure, 32–33, 34, 44–45

Malls, virtual, 37–39
Mars, 25
Matrix, The, 33
Medicine, 5, 22, 23, 34, 35, 36–37, 43
Men in Black, 33
Military, virtual, 39, 41
Mobile computer system, 43
Mouse, 19, 20
Movies, 4, 18, 33
Music synthesizer, 18

NASA, 25
Nausea, 21

Parkinson's disease, 22
Pathfinder mission, 25
Pilots, 5, 28
Projector, 8

Real estate, 31
Reality Built for Two (RB2), 17
Reality simulator, 6–7
Rescue missions, 36–37
Robots, 22, 23, 25, 36–37

Schools, 26–27, 41–42, 44
Science fiction, 30
Sensorama, 10
Shopping, 37–39
Shutter glasses, 20, 36
Sketchpad, 12
Sound processor, 6
Space, 5, 45
Special effects, 30
Star Trek: The Next Generation, 30
Star Wars, 33
Stereoscope viewer, 7–8
Stores, 37, 39
Surgery, virtual, 22, 23, 36
Sutherland, Ivan, 10, 12, 13–14
Sword of Damocles, 13–14, 15
Synthesizer, music, 18

Tactile feedback, 34–35
Telerobotic systems, 36–37
Telesurgery, 22, 36
Television, 12, 15, 30
Theme parks, 20
Three dimensional images, 6, 8, 10, 15, 20, 26, 44
Total immersion, 39, 41
Touch, 34–35, 37
Tracking system, 43
Training simulators, 5, 27, 28, 39

Virtual reality
  commercial uses, 17
  defined, 6
  evolution, 6–18
  future, 34–45
  today, 4, 19–33

Waller, Fred, 8, 16
Weather forecasting, 24–25
Wheatstone, Charles, 7
Wired gloves, 7, 17, 18, 20, 22, 34, 37
World War II, 16

FOR FURTHER INFORMATION

Books

Baker, Christopher W. *Virtual Reality: Experiencing Illusion*. Brookfield, CT: The Millbrook Press, 2000.

Darling, David. *Computers of the Future: Intelligent Machines and Virtual Reality*. Parsippany, NJ: Dillon Press, 1996.

Grady, Sean. *Virtual Reality: Computers Mimic the Physical World*. New York: Facts On File, 1998.

Jeffers, David. *Cyberspace: Virtual Reality and the World Wide Web*. New York: Crabtree Publishing Co., 1999.

Jortberg, Charles A. *Virtual Reality & Beyond*. Edina, MN: Abdo & Daughters, 1997.

Website

Colorfully illustrated with history and current information about virtual reality.
http://archive.ncsa.uiuc.edu/Cyberia/VETopLevels/VR.History.html

INDEX

Animation, 18
Arcades, 10, 19, 20, 32
Architecture, 5, 31
Attack of the Clones, 33
Augmented reality, 43

Business, 20, 27, 31

Camera, 7–8
Cathode ray tube (CRT), 12
Cinerama, 8–11
Classrooms, virtual, 41–42, 44
Computer, 4, 5, 6, 10, 12, 13–14, 17, 19, 20, 39, 43
Computer aided drafting (CAD), 5, 31
Computer graphics, 10, 12, 43
CRT, 12
Cybersickness, 21

Desktop virtual reality, 20
Dizziness, 21
Doctors, virtual, 22, 23

Education, 5, 20, 25, 26–27, 34, 41–42, 44
Effectors, 7
Engineering, 43
Entertainment, virtual, 4, 8, 32–33

Fiber optics, 17, 41
Field trips, virtual, 26–27, 42
Flight simulator, 10, 12, 15, 28, 29, 39
Force feedback, 34
Full body suits, 18
Fun, 32–33

Glasses, shutter, 20, 36
Global communities, 40
Gloves, wired, 7, 17, 18, 20, 22, 34, 37
Graphics board, 6
Graphics, computer, 10, 12, 43

Hapatics, 34, 37
Hazardous waste, 36, 37

GLOSSARY

cathode ray tube a vacuum tube that contains electromagnets that form images when shooting electrons come in contact with the tube's phosphorescent screen.

Cinerama a system of synchronized cameras and projectors that produce images on an arc-shaped screen.

computer aided drafting (CAD) a computer program used for architecture or product design.

force feedback a system of tiny vibrators in a wired glove attached to a computer that can be stimulated to simulate contact with virtual objects.

hapatics the technology of providing virtual reality systems with the sense of touch, temperature, pressure, or any other tactile sense.

head mounted display (HMD) a headpiece with optical devices and sometimes small speakers that brings the user into the virtual world.

immersion virtual reality all-inclusive virtual reality technology that surrounds and immerses the user in the virtual environment by incorporating touch and sound in addition to three-dimensional graphics.

liquid crystal diode (LCD) a conductor of electricity that contains a liquid substance that becomes opaque when bombarded with an electrical charge.

shutter glasses glasses with liquid crystal lenses, which alternate rapidly between transparent and opaque and create the illusion of three dimensions.

simulator a device that creates test conditions as similar as possible to actual conditions.

telepresence the sensation of being in a particular place via computer-generated scenes.

virtual reality a computer-generated three-dimensional environment.

wired glove a glove that contains trackers and sensory devices that link the user to the computer and the virtual reality program.

ABOUT THE AUTHOR

Sheila Wyborny is a retired science and history teacher. She and her husband, Wendell, a broadcast engineer, live in Houston, Texas. They are currently building a home in a flying community, a neighborhood that is connected to a small airport. They look forward to having their airplane, and an airport, in their own backyard. Wyborny particularly enjoys the research stage of her writing projects, and looks forward to each new book. She also enjoys visiting schools and libraries and talking about her writing experiences.

Virtual reality programs that are in development today may enable us to better explore space in the future.

People will be able to stay at home and enjoy virtual hangouts with their pals. Rather than phone or e-mail each other, people will choose a place to meet in cyberspace. Friends might choose science-fiction locations, such as restaurants on other planets, or historical places such as King Arthur's round table or the New York World's Fair of 1939.

No one is yet sure how far virtual reality might go. One thing is certain, though—the technology will continue to change at an amazingly rapid rate. What is cutting-edge science today will be old news in a matter of months. Programs that are still in development today may be for sale in toy stores or computer centers in as little as a year's time. As a result, the story of the future of virtual reality may never really come to an end.

In the future, people will be able to learn about far away places through virtual reality.

will be much more complex and realistic, as well as more widely used, in the future.

Homework and research assignments will also be easier for students who cannot get to libraries. Research will be done on-line by computer, as many students do it today, but interactive three-dimensional models will also be possible. For example, with HMDs, these models will allow students to see the insides of cells and molecules for science projects.

Virtual reality has the potential to make education more meaningful for students. At the same time, however, students and their families will also be able to have more fully involved virtual experiences in their free time.

Biochemists study molecular interactions using virtual reality.

THE FUTURE OF VIRTUAL REALITY

Although virtual reality will continue to have many useful applications, people will still have leisure time, and virtual reality will ensure that there are many ways to enjoy it. People who want to take vacations will have the opportunity to tour an area as they consider a vacation there. They will be able to go on virtual tours of the Loch Ness region of Scotland or the Himalayas in Tibet without the time, expense, or physical danger an actual trip might demand. Even if a person does not want to take a real trip, though, virtual reality will make it an adventure to spend time with friends.

AUGMENTED REALITY

Augmented reality is related to but different from virtual reality. In augmented reality, computer graphics are superimposed onto transparent HMD visors,

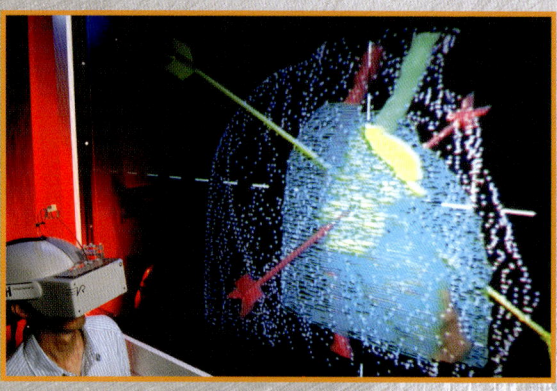

Images appear in front of, or on top of, real objects in augmented reality.

which make the images appear to be in front of or on top of real objects in the user's field of vision. The three elements of an augmented reality system are a head mounted display, a tracking system, and a mobile computer system. Although current systems are quite large, they will eventually be made into the lenses of glasses, and the images will move to coincide with the movement of the wearer's head and eyes.

Augmented reality has potential uses in many fields, including engineering and medicine. Since doctors cannot see inside the human body without bulky devices or invasive procedures, the HMDs with see-through visors can make ultrasound images appear on the affected area of the patient's body so doctors and surgeons can operate or give treatments. In engineering, schematic diagrams—drawings of the connected parts of devices—could be projected over wiring systems as electrical engineers construct complicated systems. Instruction manuals could also be projected directly over airplane engines, which would let mechanics see diagrams as they work.

students often have to travel great distances to get to school. Virtual programs will allow them to take part in classes with students from all over the world without ever leaving home.

Virtual reality allows students to learn any subject, such as music, from teachers all over the world.

Virtual reality field trips to places like Egypt will be available in the future.

Virtual field trips of the future will take students far beyond anything that is physically possible. As they wear HMDs and instrumented suits, students will be transported to the scene of historical events, such as the crowning of Egyptian kings. They will even learn about vision through virtual trips up the optic nerve to the brain. Although some similar programs are currently available, these experiences

42

of dollars and weight thousands of tons. Virtual reality will teach officers to guide surfaced submarines through harbors, which is very difficult. Officers have little time to learn these skills, but their careers and lives depend on whether they handle these very expensive vessels properly. Virtual reality training will provide realism without risk to actual submarines and will help officers learn important skills quickly.

Virtual reality simulators for submarines, like those for planes, may soon be available for training naval personnel.

VIRTUAL TEACHERS, TEXTBOOKS, AND HOMEWORK

Some of the most dynamic changes stemming from virtual reality technology will occur in education. Most parents want their children to have the best teachers and educational opportunities available, and virtual reality will make this possible. Certain kinds of virtual educational programs that use fiber optics and telephone lines are already in use. These are basically talking heads systems, however—students see head-and-shoulder images of instructors and communicate with them by telephone. Total immersion virtual reality will create interactive virtual classrooms instead. These will allow shared learning experiences and projects, as well as social interaction, just like a real classroom.

The best teachers in each subject will teach students through telepresence, regardless of where students are. In rural areas,

GLOBAL COMMUNITIES

Some computer experts foresee a day when virtual reality will create global communities. These would not be actual locations, but rather associations of people with similar skills, beliefs, or talents who would meet in virtual places. Some communities will be based on environmental issues, recreational interests, and shared experiences. These global communities will promote cultural understanding and facilitate the exchange of vast amounts of information.

As virtual reality systems become more lightweight and portable, these meetings of minds and cultures will be accessible to more people. Virtual reality conferences might take place at any time and may include participants from any country in the world. Language differences will no longer be a problem, thanks to virtual reality interface technology, which will link virtual reality programs with translation programs. Spoken language will be translated via computer and converted to either verbal or written text.

These global interest communities have the potential to be of great service to humankind. They could exchange business information, medical technology, or even information that could help prevent wars.

The system will also encourage managers of existing stores to adopt changes that will make their stores more efficient and profitable.

Virtual reality will also help shoppers. To avoid crowds and save time, people will don instrumented gloves at home and tour virtual malls and shops. They will be able to choose items, view them from all sides, and examine their texture and weight. If shoppers want to purchase the items, they will drop them into virtual shopping carts. Computers will total the purchases, debit the shopper's bank account, and send the goods to customers via real-world delivery services.

This kind of visual assessment will also be useful in law enforcement. Total immersion programs will provide vital training for police officers. The programs will include scenarios of crimes in progress, such as bank robberies, hostage situations, and traffic stops. Law enforcement personnel will gain valuable experience to help them protect themselves and the public in times of crisis.

The military also wants to find ways to keep both people and equipment safe. The navy is developing a virtual reality simulator to teach submarine officers navigational skills. Submarines cost millions

An instructor monitors his student as she operates a flight simulator. Virtual reality allows people in dangerous jobs to train safely without danger of injury.

Virtual reality may change how people shop.

disaster victims in locations where it would be too dangerous to send human rescuers. Although there are remote control devices that can perform some of these functions today, with

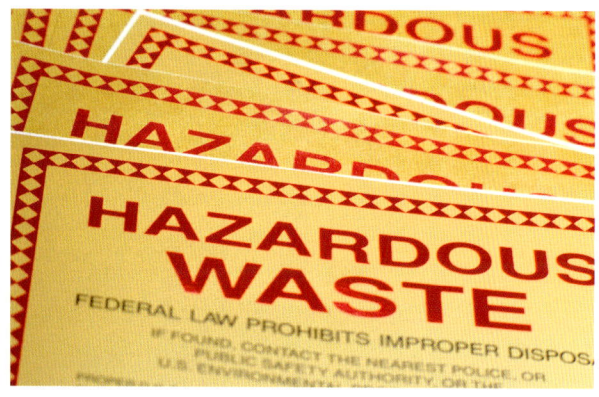

Someday, virtual reality may handle dangerous tasks like the disposal of hazardous waste.

the help of human directors wearing HMDs and wired gloves, future telerobotic systems will be able to do tasks that require highly refined coordination. These gloves will have hapatic systems to help the director detect texture, temperature, and resistance. With this refined sense of touch, the telerobots will be able to use tools and handle delicate objects without breaking them. They will perform risky tasks, such as painting tall structures and washing windows of skyscrapers. At the same time, inside those same buildings, virtual reality will continue to have an impact on business.

FROM VIRTUAL MALLS TO A VIRTUAL MILITARY

Virtual reality will provide ways to design, create, and test new products before they are actually made. This will help solve potential problems before they occur.

Virtual reality will one day help grocers plan store layouts. Grocery companies will be able to create virtual stores and organize space by product and brand. They will then be able to take virtual tours to see which displays work best for which foods and make adjustments as needed. Tests done in the mid-1990s showed that this "virtual stocking" will save grocery-store chains time and money.

SCIENTIFIC BREAKTHROUGHS: FROM VIRTUAL SURGERY TO VIRTUAL RESCUE MISSIONS

Medical students will soon use virtual displays to learn human anatomy. Today's future doctors work with diagrams and may occasionally dissect human cadavers that have been donated for research. Tomorrow's medical students, on the other hand, will have access to interactive virtual displays. Instrumented gloves and shutter glasses will enable these students to lift and examine the heart and other organs from virtual bodies.

Students will also be able to practice virtual surgery. Medical schools will be equipped with virtual reality operating rooms that will allow students to perform operations and see the results of both successful surgery and mistakes without any risk to real human lives.

In the future, virtual reality will handle other dangerous tasks without risk to people. For example, telerobotic systems will be used to clean up toxic waste, defuse explosives, and search for

Medical students can practice many different procedures using virtual technology.

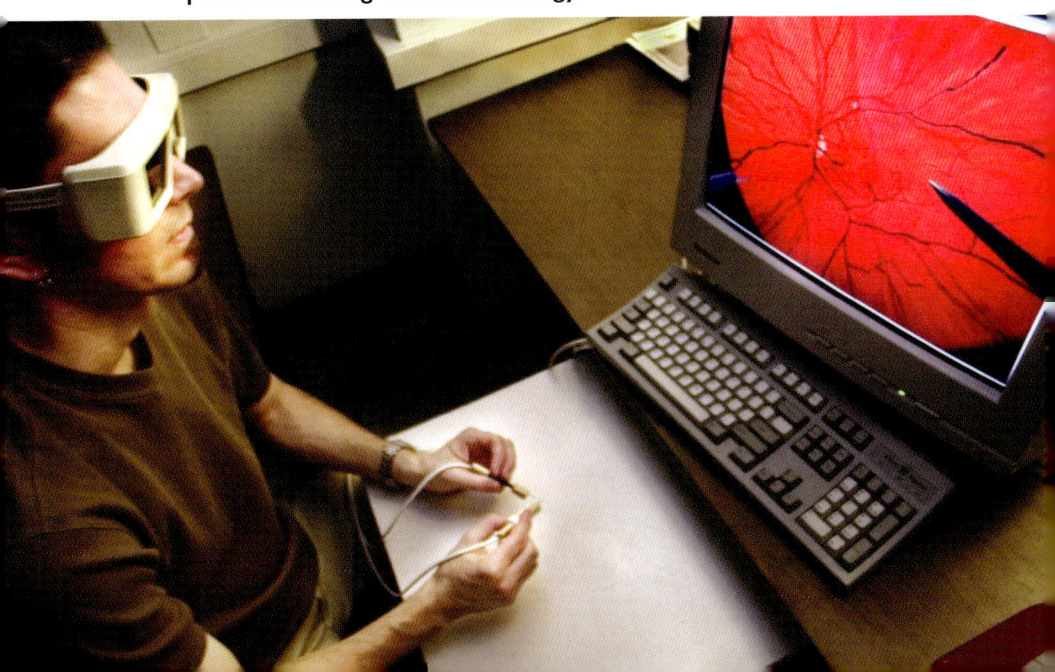

and temperature. Texture may be mimicked by means of small vibrators and tiny automated needles. Temperature is simulated by thermodes, cap-like devices slipped over the fingertips that use tiny pumps to produce hot or cold sensations. This enhanced sense of touch will be of great importance in the medical field.

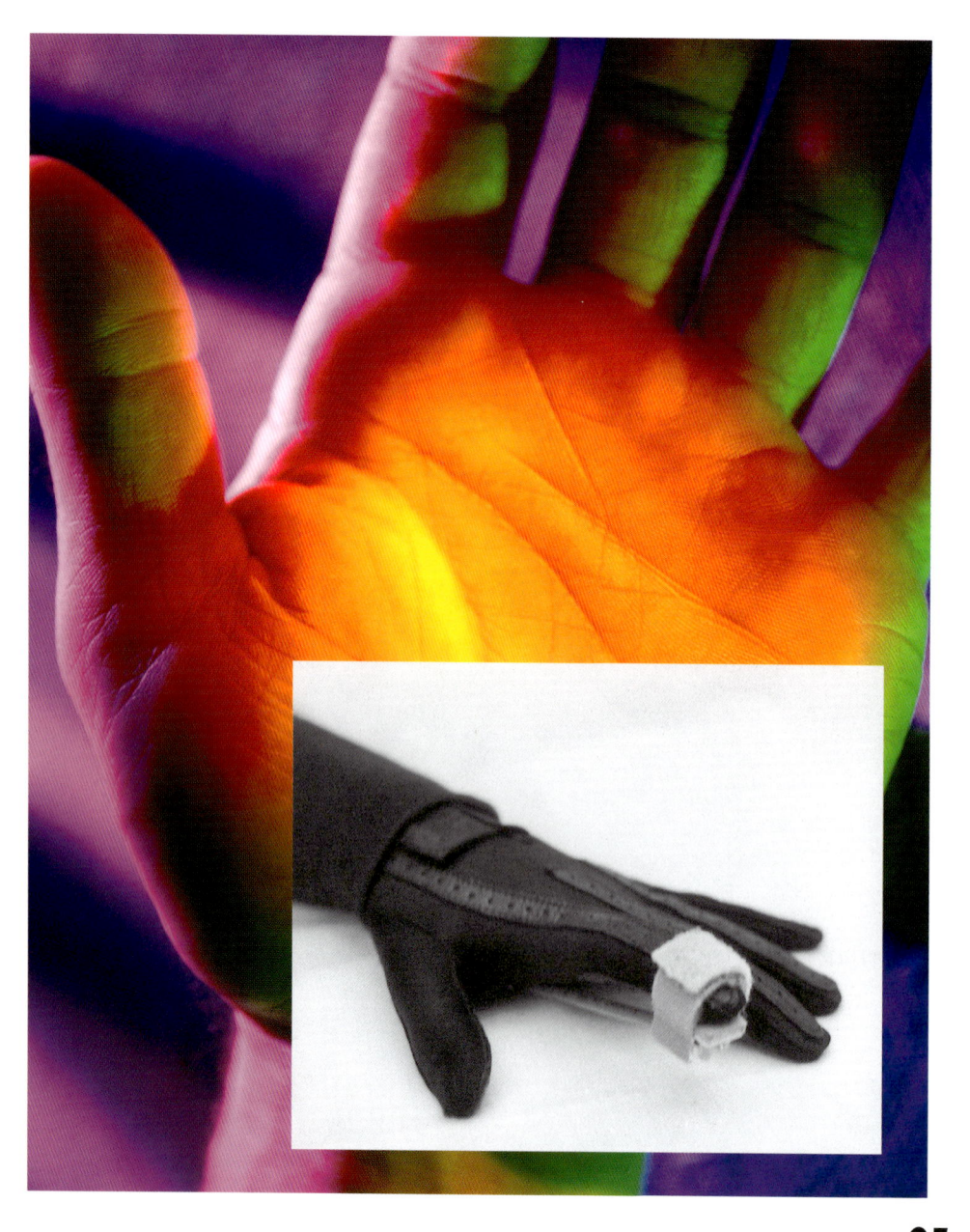

CHAPTER 3

LOOKING TOWARD THE FUTURE

Research is under way to develop virtual reality systems even more complex than those in use today. From medicine to education to leisure, researchers bring applications for virtual reality to an ever-expanding array of vital fields.

Not only do researchers want to find more uses for virtual reality, but they also hope to improve the technology that already exists. One way they seek to make the virtual reality experience more realistic is through an improved, refined sense of touch.

VIRTUAL FEELINGS

Hapatics is an element of virtual reality. The name comes from the Greek word for "touch." The technology of hapatics includes all the devices that simulate the sense of touch and let real people interact with virtual objects. Although wired gloves have helped users manipulate virtual objects for years, there has been little feedback or sense of touch. Devices that have been in development since the 1990s, however, can mimic the way virtual objects would feel in a user's gloved hand. Tiny air sacs in the fingertips simulate the pressure of the fingers as they make contact with solid objects. This type of hapatics is called "force feedback." Although this technology still needs to be refined, hapatics will one day go even beyond the simulation of picking up and moving virtual objects.

Tactile feedback is currently in the research stage. When fully developed, this technology will allow to user to feel both texture

Technology is still being developed that will allow users to feel texture and temperature through virtual reality.

In a scene from the movie *The Matrix*, actors perform stunts using virtual reality technology.

Sometimes, however, people want to view the virtual world but not actually participate in it. For them, virtual movies are a good choice. Virtual reality movies are on the cutting edge of technology. They amaze audiences with their scope and realism.

The Matrix and *Men in Black* were two films of the 1990s that made impossible stunts look real and showed interaction between live actors and virtual characters. In the past, actors called extras were hired to be part of scenes with large groups. Today, virtual reality can create the extra characters. These characters can even be aliens. The *Star Wars* sequel of 2002, *Attack of the Clones*, had huge battle scenes created with virtual reality.

Although it often seems as though technicians have gone as far as they can in movies, more complex virtual action on the movie screen is always possible. Like researchers in other fields, those who work in entertainment are very interested to find out just how far virtual reality can go.

HAVING FUN, VIRTUALLY

From simple desktop programs to sophisticated immersion systems, there are many ways for people to enjoy themselves in the virtual world. Users can choose the kind of virtual games and other activities they prefer.

One activity is the virtual arcade ride. Two to six riders enter a pod-like chamber with a screen and strap themselves into seats much like those on a roller coaster. The pod rests on a hydraulic system. Once the ride begins, the hydraulic system dips the pod to synchronize with the scene on the screen. Some of these virtual rides are so lifelike that people with back or neck problems are warned not to try them.

One of the most common uses of virtual reality is for entertainment in arcades.

VIRTUAL REALITY TRANSFORMS THE WORLD OF BUSINESS

From real estate sales to architecture, virtual reality has had a big impact on business. The real estate business is now much more efficient, thanks to virtual reality. Prospective home buyers can take virtual tours of available homes, which saves time and energy. The buyer can see not only the walls of rooms, but also the ceiling above and floor below.

It is even possible to tour homes and offices that have not yet been built. Today, buildings can be designed and viewed before the first brick is laid. Computer aided drafting (CAD) programs allow people to make virtual visits to structures, and tour rooms that do not yet exist.

Virtual reality has clearly become extremely useful to serious business professionals. Some virtual reality systems, though, are just for fun.

A technician works on interior design using virtual reality.

WHEN SCIENCE FICTION BECOMES REALITY

The *Star Trek: The Next Generation* television series and its successors featured special effects that seemed fantastic during the show's run from 1987 to 1994, but some of the devices that were first created in the minds of science-fiction writers have actually become reality. One such example is the Holodeck, a special room on the Starship Enterprise where crew members could indulge in some much-needed rest and relaxation between missions. The name Holodeck was derived from the word hologram, a projected three-dimensional image. The Holodeck went far beyond the projection of images, however.

   On the Holodeck, crewmembers could take virtual vacations. They could go anywhere in the galaxies, either in the present or through history. Not only could they visit these places, but they could also interact with the people they met there and even eat the virtual foods. Although technology has yet to develop this degree of interaction, the Holodeck bears a remarkable similarity to some of today's immersion systems, with the inclusion of touch and sound.

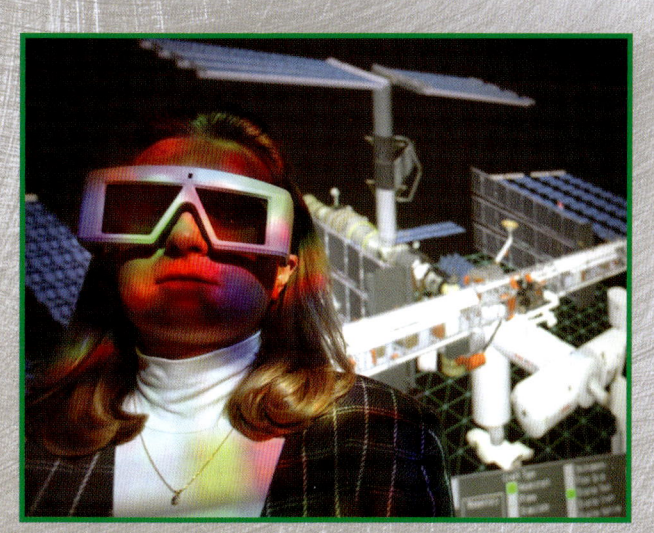

Stereo glasses give the user a three-dimensional view of whatever he or she looks at.

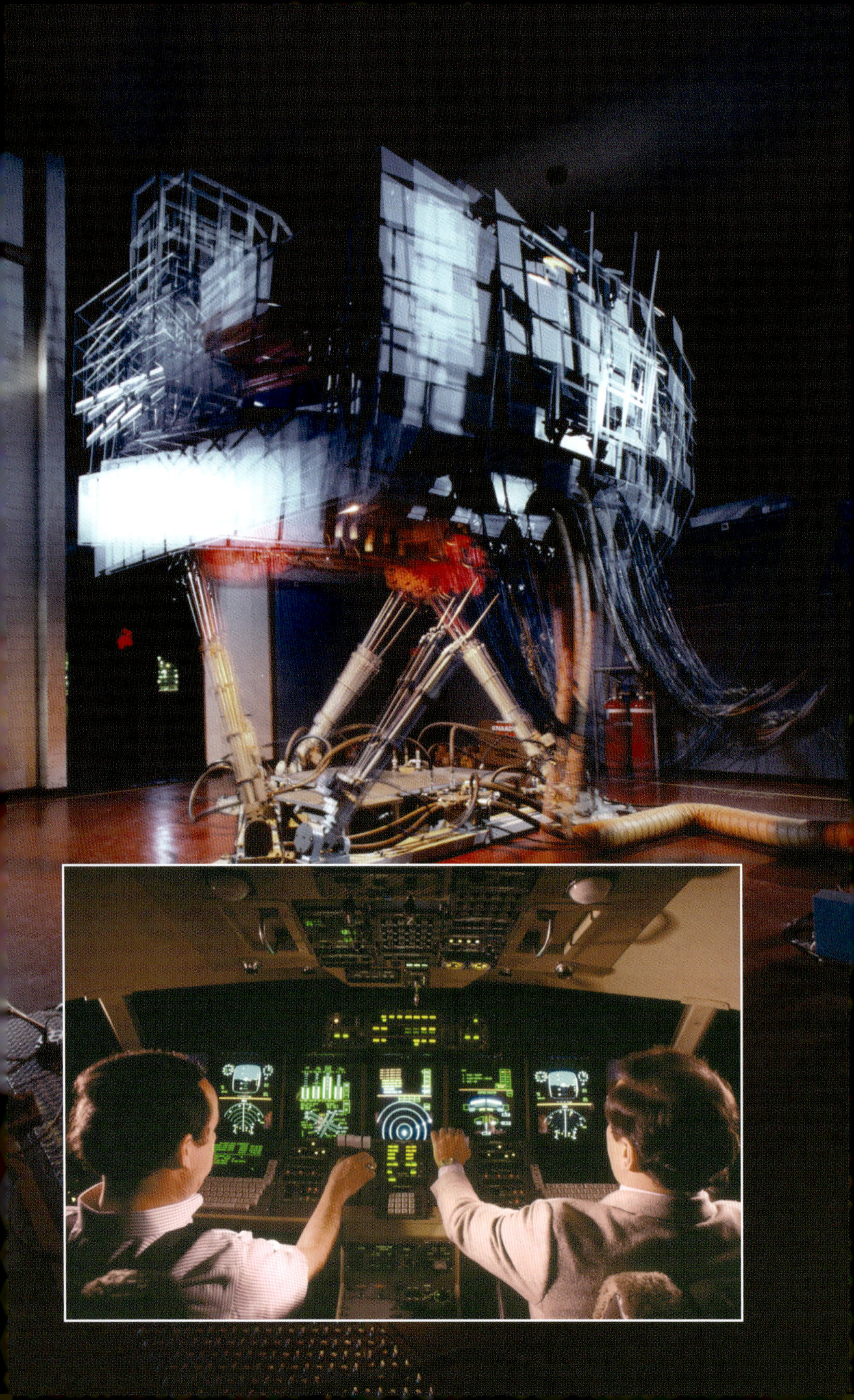

VIRTUAL TRAINING FOR REAL JOBS

Virtual simulators are designed to improve situational awareness—a person's ability to identify and respond to events. Training simulators have been developed for many purposes.

The University of Iowa Center for Computer-Aided Design has made a simulator to study drivers' reactions to varying conditions. The simulator tests the effects of medications on driving skills and response time, which can often mean the difference between life and death.

At the Naval Postgraduate School in Monterey, California, scientists have developed a type of treadmill and virtual reality device to help train foot soldiers. This special treadmill is omnidirectional, which means it can move forward, backward, and side-to-side. The treadmill can be used with an HMD to help soldiers learn skills for combat situations they might face on the battlefield.

Commercial airline companies also use virtual reality. Many major airlines have flight simulators that are realistic enough to train pilots for almost any emergency. From a mock cockpit, the pilot sees the computer-generated situation through the windshield. Hydraulics make the cockpit rise, dip, and roll in response to the way the pilot handles the instruments. Instructors can make the system simulate problems such as sudden bursts of wind, failed engines, and stalls. After they have experienced these emergencies in simulators, pilots are better equipped to deal with them if they really occur.

Because they create real life situations in the virtual world, these training simulators give workers valuable experience without any risk to lives or property. This ability to view possible outcomes before they happen has begun to be applied in other fields as well.

A flight simulator lets pilots (inset) practice flying without endangering themselves or others.

to plant and animal biospheres in a south Louisiana wetland or to a scientific base in Antarctica.

Educational opportunities for virtual reality continue to expand. Even as elementary and secondary schools explore new learning opportunities, businesses use virtual reality technology, such as simulators, to train employees or upgrade their skills.

Students can experience different places using virtual reality technology.

VIRTUAL SCHOOLS AND VIRTUAL FIELD TRIPS

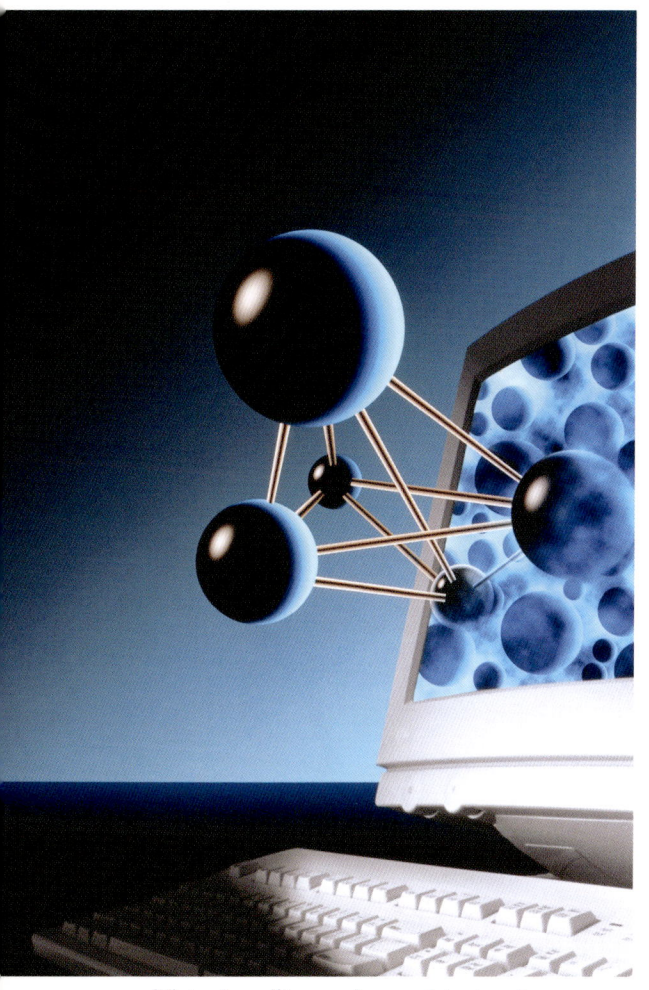

Virtual reality is an important training tool. At East Carolina University in Greenville, North Carolina, experimenters help schools use computer-generated models to create virtual worlds based on their home communities. Those models can be altered and studied to see how any proposed changes might affect business and traffic patterns.

In a social studies class, virtual models can help high school students see what would happen if a factory were built in a certain area of their town. In elementary math classes, virtual reality allows students to see the inside as well as the outside of geometric shapes and learn the relationship between solid three-dimensional shapes and flat two-dimensional shapes.

Virtual reality can also be used to create environments that would be difficult or impossible for students to visit. Information about the landforms, wildlife, and climate of an area can be used to create computer-generated field trips. Students can take virtual trips

Virtual reality can be used to teach students about three-dimensional shapes.

NASA used virtual reality with the robotic rover on Mars in 1997.

Over the years, special aircraft have flown into the center of powerful storms. The information they have gathered has been used in virtual reality systems. Computer programs have created model storms that let scientists study them and find ways to save lives and reduce damage.

Other scientists use virtual reality, too. The National Aerospace and Space Administration (NASA) uses virtual reality to take people to other planets. In 1997, the *Pathfinder* mission landed a small robotic rover on the planet Mars. The rover was equipped with cameras, and the pictures they took were sent to the World Wide Web as soon as they were received on Earth. This gave viewers a sense of telepresence: They felt they were there, even though they really were not.

Virtual reality in science has given average people an opportunity to view other planets and see weather systems up close. It has also proved helpful in other ways, especially in education.

Portage Public Library
VIRTUAL WEATHER REPORTS

Virtual reality is also used in weather forecasting. It allows television weather reporters to bring viewers right into storms and fair weather systems. With aircraft, weather balloons, and satellites, scientists gather weather data, such as temperature, humidity, and wind speed and direction, from points at a particular section of the Earth's surface and from altitudes up to 2 miles (3 km) above the Earth's surface. The information is fed into a computer, which analyzes it and creates a three-dimensional model of the weather system in that area. The computer also predicts how a system is likely to move and change.

Virtual reality can be used to predict weather.

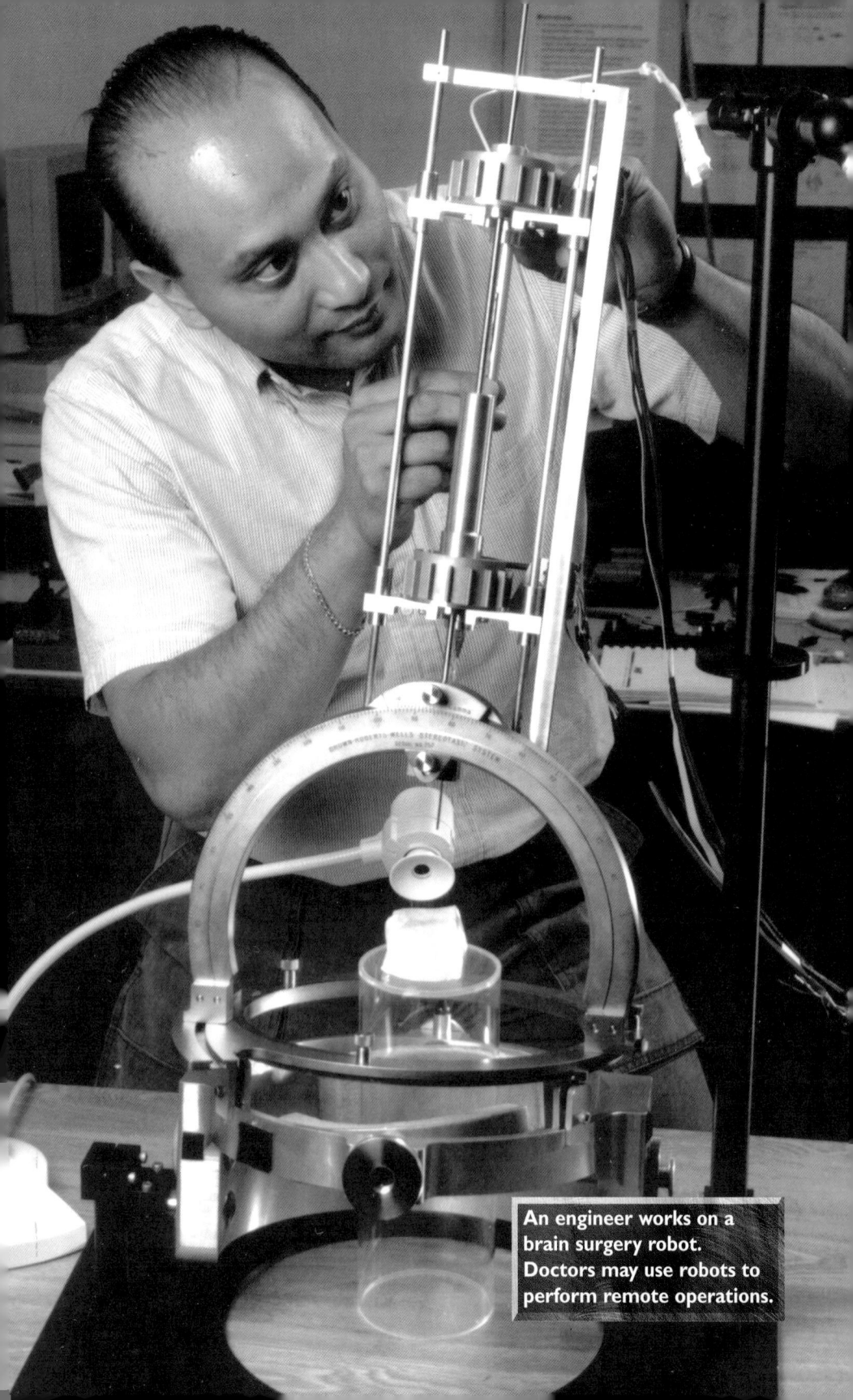

An engineer works on a brain surgery robot. Doctors may use robots to perform remote operations.

VIRTUAL DOCTORS SET UP PRACTICE

In medicine, virtual reality is used in many ways, from surgery to physical therapy. In 1997, a surgeon in Italy operated on a patient in Lisbon, Portugal, about 1,000 miles (1,600 km) away, in the first telesurgery. The operation was done with a $150,000 robotic arm controlled by the doctor, who wore a wired glove that could track his hand movements, while the patient was monitored by video cameras. The operation was a complete success.

Virtual reality has also helped people with Parkinson's disease, a condition that interferes with motor skills. Victims often shake uncontrollably and have trouble walking. Many Parkinson's patients are better able to walk if they have visual clues that tell them where to place their feet. An HMD has helped provide these visual aids. The HMD projects a series of lines onto the screen. The lines, which appear to be placed on the ground, tell Parkinson's patients where to put their feet.

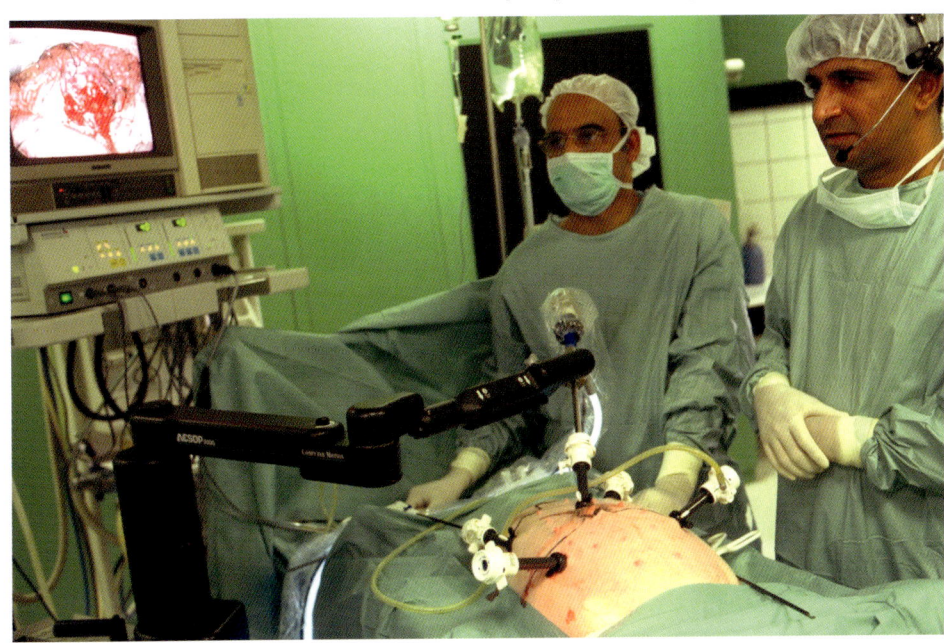

Doctors sometimes use virtual reality to perform surgery.

CALL THE DOCTOR! I'M CYBERSICK!

Virtual reality has created its own unique type of illness, one with very real symptoms. People who play a lot of virtual games can suffer some uncomfortable consequences from overindulgence.

In virtual reality, the eyes give the brain information about movement and balance that is different from what the body tells the brain. This conflicting information tells the brain that something is wrong. In response, the brain triggers nausea and dizziness, symptoms similar to those that accompany inner-ear disorders. Some virtual reality travelers suffer even worse effects, such as vomiting and injuries from falling. Cybersickness can be avoided if the length of visits to the virtual world is limited.

Frequent use of virtual reality can cause cybersickness, such as nausea and dizziness.

WAYS TO VISIT THE VIRTUAL WORLD

Immersion virtual reality involves most of the user's senses. It uses appliances such as HMDs to give a 360° view of the virtual world, and wired gloves to provide sense of touch and to track the player's place in the virtual world. Immersion systems can be very expensive. They need high-powered computers and appliances to run them. These systems are usually found only in arcades and theme parks, some military programs, and educational and research facilities. There are more affordable virtual reality systems, though.

Many virtual reality programs can be operated with the computer equipment found in the average home. These programs are called "desktop virtual reality." The computer's monitor acts as a window into the virtual world and allows the viewer to use a joystick, mouse, or keyboard to interact with virtual settings, people, and creatures. Desktop virtual reality can be enhanced with wired gloves or shutter glasses with liquid-crystal lenses, which turn the two-dimensional scene on the screen into three-dimensional images. Thanks to the availability of programs and appliances, virtual reality has already been adapted for many useful and vital applications. It has become a valuable tool in science, education, and business.

Many virtual reality programs can be operated on a home computer.

CHAPTER 2

TODAY'S VIRTUAL REALITY

There are several ways to experience virtual reality. Some programs require simple equipment, such as a joystick or mouse that the player uses to guide a character on the computer monitor's screen. More complex systems have large HMD systems suspended from overhead booms, and full-sized joysticks, like those found in helicopters, with buttons that can be used to "shoot" at an enemy. There are also sophisticated systems that link several cockpit-sized pods to a computer. Individual users sit inside the pods. The pods have screens that enable the player to see the virtual world, pedals to move the player's virtual character forward and backward, and joysticks to control other actions, such as hand-to-hand combat.

Virtual reality arcades are very popular. Today, there are many kinds of virtual reality programs, and players can choose their own level of involvement.

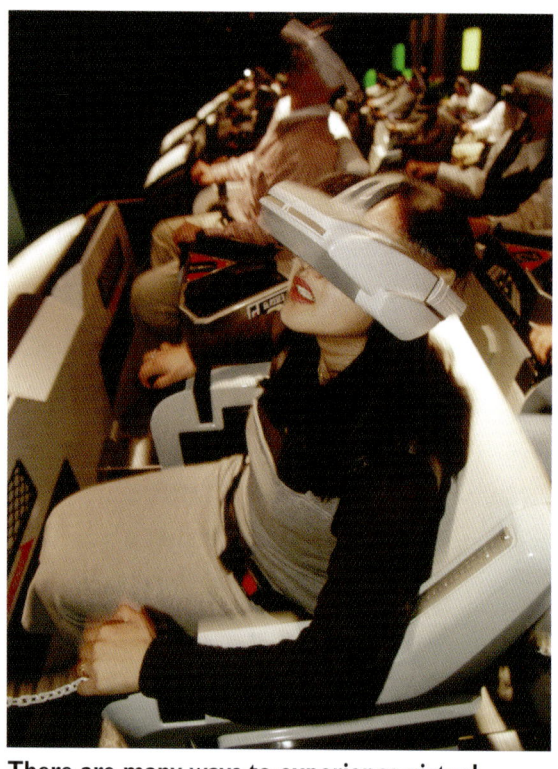

There are many ways to experience virtual reality, such as this ride in Japan.

REAL EQUIPMENT FOR A VIRTUAL WORLD

Interactive devices have gone far beyond the HMD and wired glove. In the 1980s, full body suits were developed that followed the motion of the user's body and controlled music synthesizers. These suits have also been used in the motion-picture industry to animate characters. As the equipment used to help people experience virtual reality develops, applications for virtual reality continue to multiply rapidly.

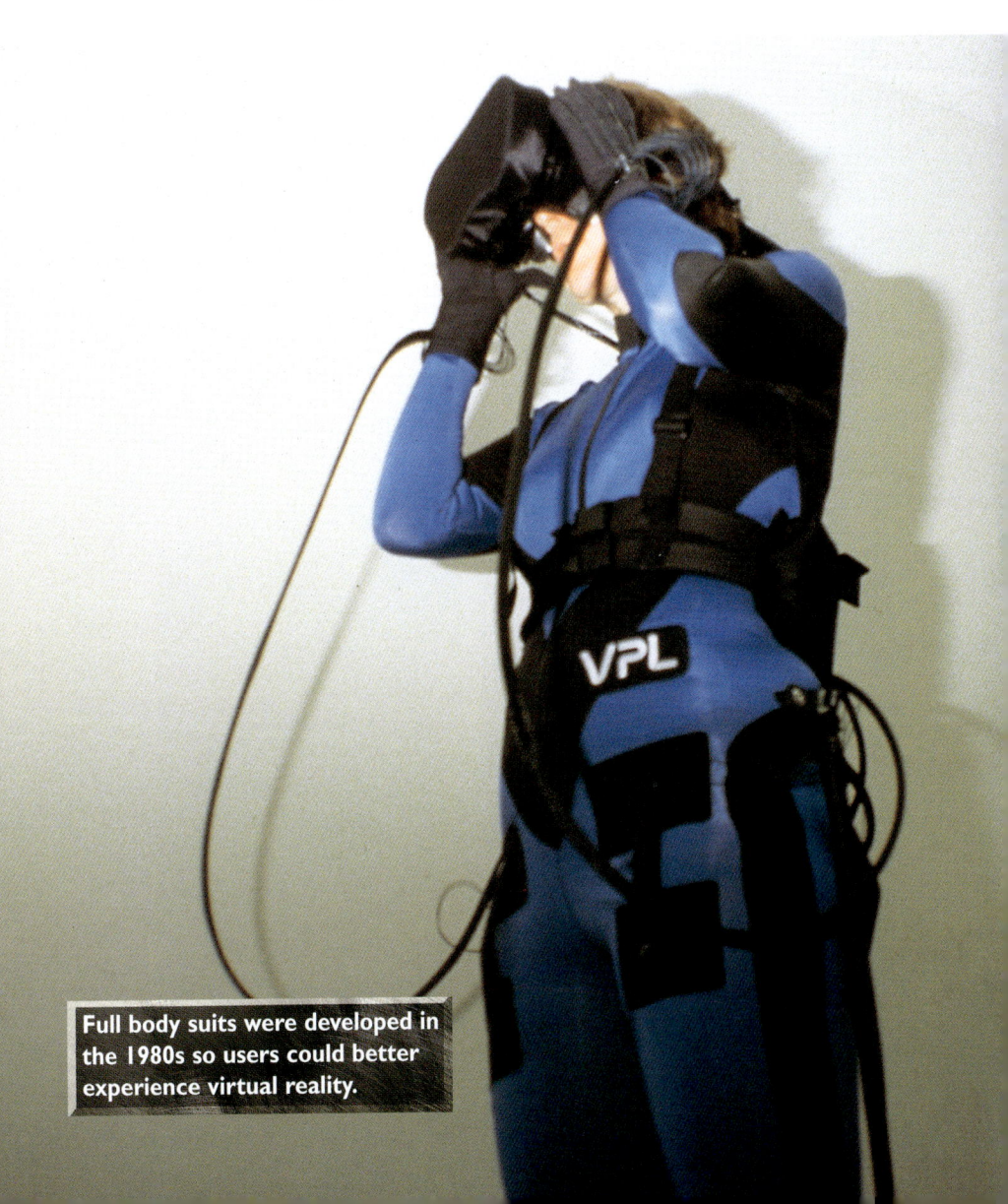

Full body suits were developed in the 1980s so users could better experience virtual reality.

VIRTUAL REALITY GOES ON SALE

Reality Built for Two (RB2), developed in the late 1980s, was the first commercial virtual reality system. It ran off an Apple Macintosh computer and came with a type of HMD and wired gloves that allowed the computer to monitor the position of the user's hand. Some wired gloves use fiber optic sensors to send tracking information, such as the bending of fingers, to the computer. Other use magnetic trackers to monitor the user's position.

RB2 had problems with image quality and body tracking. It could not easily follow the position of the user's hands. The system also had a tangle of wires and cables that connected the HMD and wired gloves to the computer. The biggest problem, though, was the price: $400,000 for a two-person system and $225,000 for the one-person version.

Despite its drawbacks, RB2 proved that virtual reality technology could succeed commercially, if people could afford it. It also began to create a new vocabulary of terms related to virtual reality. As the technology became more accessible, more people learned about the equipment needed to be part of the virtual world.

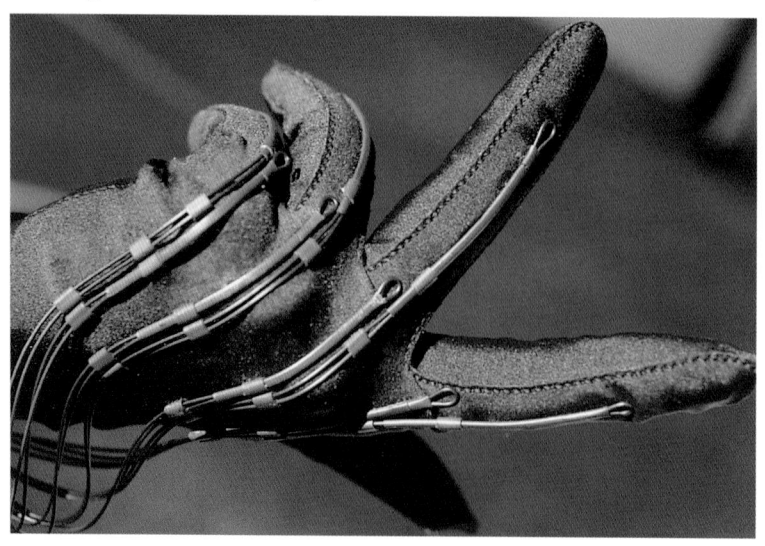

Wired gloves allow the computer to monitor hand movements.

BOMBS AWAY!

In World War II (1939–1945), the Allied war effort depended largely on bomber aircraft to weaken Axis power. This strategy required thousands of bomber aircraft and well-trained crews. Up to this time, training just one crew of four gunners meant that two aircraft had to be put in the sky: one to carry the trainers and the other to tow the target. To train enough bomber crews would have required half of the aircraft on active duty in Europe, would have consumed valuable fuel and ammunition, and would have posed major risks for loss of life in training accidents.

The Waller Gunnery Trainer was used to train pilots in World War II.

To train many crews quickly, safely, and efficiently, cinematographer Fred Waller developed a simulator in the early 1940s called the Waller Gunnery Trainer. The simulator had four gunners' stations and incorporated realistic effects. The guns had realistic recoil feedback, and the trainee could hear the sound of his guns and other aircraft noises through headphones. Waller's trainer was said to be the world's first virtual reality system.

A VIRTUAL HELMET

By the late 1970s, a head mounted display (HMD) more portable than the Sword of Damocles was developed. The HMD was a helmet-like structure that had screens for each eye. The military used the HMD to help jet fighter pilots monitor the aircraft's controls during high-speed air combat. These HMDs projected the control panel on the inside of the pilot's visor so the pilot did not have to look down. Because jet fighters fly extremely fast, if the pilot takes his or her attention from the windshield, even for seconds, disaster could result.

Although it was an improvement, the HMD was large and heavy. More development was needed to make it more versatile. Today's HMDs contain two tiny television screens, one in front of each eye. The images are slightly different on each screen, and viewed together, they appear as three-dimensional images. HMDs sometimes have small speakers to add sound to the virtual experience.

The efforts to develop better HMDs helped bring about useful applications for education, science, and medicine. Programs were developed for private use, too.

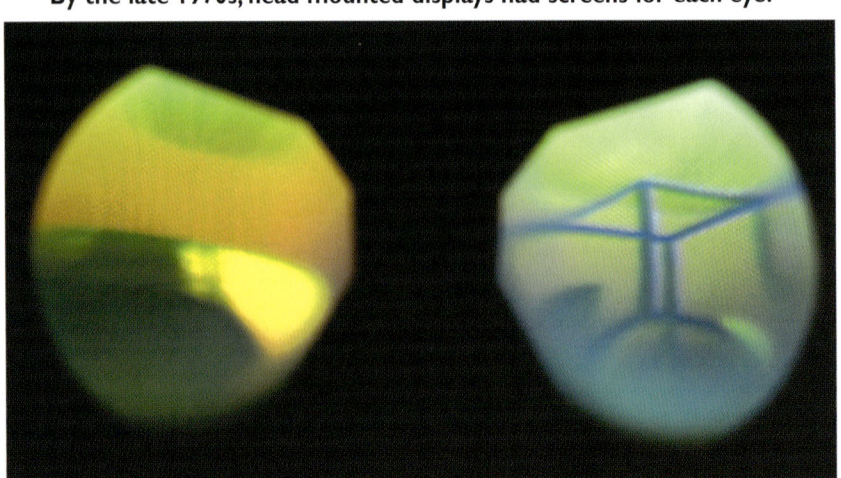

By the late 1970s, head mounted displays had screens for each eye.

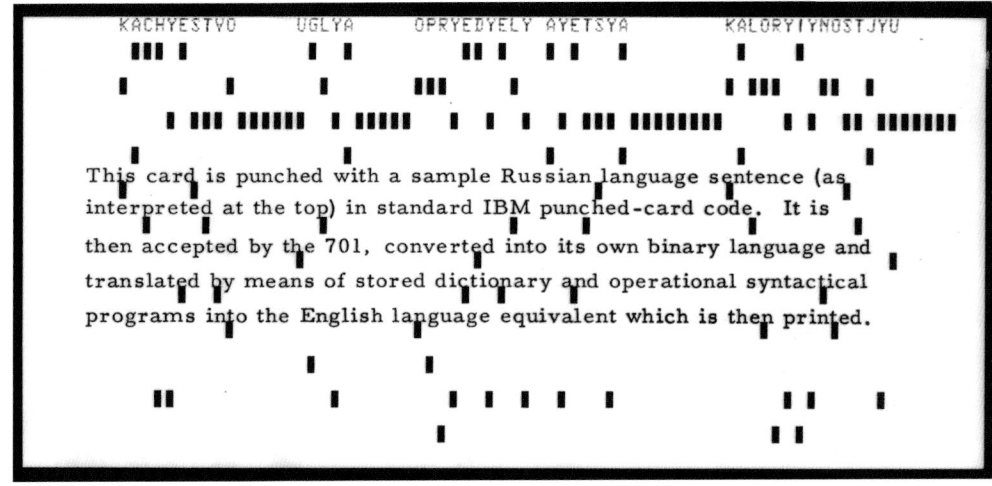

```
KACHYESTVO      UGLYA      OPRYEDYELY AYETSYA         KALORYIYNOSTJYU
```

This card is punched with a sample Russian language sentence (as interpreted at the top) in standard IBM punched-card code. It is then accepted by the 701, converted into its own binary language and translated by means of stored dictionary and operational syntactical programs into the English language equivalent which is then printed.

Computers used to accept data from punch cards like this one.

Sutherland used his ideas to create the "Sword of Damocles." This display suspended from the ceiling of the laboratory and had a helmet and goggle device that the user strapped on. It was named the Sword of Damocles after a Greek legend about a sword that hung from the ceiling by a single hair. This early virtual reality display created simple line images, like stick figures, that seemed to float in midair.

Before this time, data was carried on stiff paper punch cards with patterns of perforated holes that represented information that only a computer operator would understand. The Sword of Damocles proved that recognizable images could be created by computer-generated data. This would ultimately help prove Sutherland's claim that computer programs could be created for ordinary people.

The Sword of Damocles was quite expensive, however. It was also cumbersome to wear, and the types of images it could generate were limited. Still, many computer experts believed this technology would improve and develop a wide range of uses over time.

THE SWORD OF DAMOCLES

In 1965, Sutherland wrote an article titled "The Ultimate Display." In it, he expressed his belief that a computer could be developed for ordinary people to use. This kind of computer would help average people model and illustrate abstract concepts, such as mathematical and scientific theories, with computer graphics to better understand them. Computer programs would be learning tools.

Sutherland believed computers could be used to better understand mathematical and scientific theories.

Ivan Sutherland made use of the development of flight simulators like this one to create a way for average people to use computers.

commands, which required more advanced skills than most people had. Sutherland wanted to find a way for average people, not just scientists, to use computers. The system he developed was called a "sketchpad." This was the first step toward modern computer graphics. Sutherland's sketchpad used a light-emitting pen to draw simple designs on a cathode ray tube (CRT), which forms the display screen on televisions and computer monitors. The sketchpad could save almost any type of information—text or graphics—in the computer, and the information could be used later.

Sutherland saw the computer's monitor screen as a window through which viewers would one day see an almost real, or virtual, world. He believed researchers could make the characters, scenes, and action of this virtual world function just as they did in the real world.

A VISIT TO CINERAMA

Eighty-six-year-old Evelyn Reynolds remembers a visit she made to a Cinerama in Memphis, Tennessee, more than forty years ago:

When we first went into the theater, it looked normal. It was just a normal-sized screen. But then the music started, loud music, and a picture came on the screen. And in just a minute or so the announcer said, "Ladies and gentlemen, this is Cinerama!"

Then the curtains opened on both sides and you felt like you were sitting in the middle of it. It was more like a travel log than a movie. One part was about an airplane. And it was just like being in an airplane. You could see right over the front of it. I was all right as long as they were flying straight, but if they banked the plane, I had to close my eyes because it made me feel kind of sick. When I opened my eyes again, the plane would be straight. It was so real that I was scared, even though I knew I was in a theater.

Then they showed a scene of riding the rapids on a river. It was so real I could feel the water bumping under the boat, like I was riding in a car on a rough road. The water would swish up the side of the screen and I would dodge it. I felt like the water was going to get on me.

It was interesting, but too frightening. I wasn't ready to go back any time soon.

eyes and ears. Heilig wanted to engage the audience's senses fully, to give them something to touch and smell as well as see and hear. Heilig called his invention Sensorama.

Sensorama was a one-person arcade ride. The user sat in front of a small three-dimensional screen. Speakers, fans, and devices produced scents that went along with the movie. The problem was that the device could not withstand constant use, and broke down frequently. Slowly, Sensorama machines disappeared from arcades.

After the failure of Sensorama, there were few new developments in this kind of technology for a while. By the 1960s and 1970s, however, more interactive forms of three-dimensional technology were in development, thanks to an invention called the computer.

SUTHERLAND'S MODERN SKETCHPAD

In 1963, computer scientist Ivan Sutherland was involved with computer graphics experiments, then almost an unknown field, at the Massachusetts Institute of Technology (MIT). Sutherland made use of the radical computer design developments of the early 1960s, such as computer-generated flight simulators created by the Department of Defense. Before this time, computers could only run by use of complicated

Ivan Sutherland wanted to create a computer that the average person could use.

10

> Posters such as this one advertised the 1939 World's Fair. Cinerama was developed from an exhibit at this fair.

(5 cm) apart—roughly the distance between the eyes. The eyepiece portion of the viewer had prismatic lenses, which made the two views blend into one. The brain perceived the image in three dimensions, as it does with normal vision.

As a type of entertainment, the stereoscope enjoyed nearly 100 years of popularity. When motion-picture theaters developed in the 1920s, though, stereoscope viewers were largely forgotten. Even so, not long after the advent of motion pictures, people already wanted to see more than moving pictures on a flat screen. Plans were soon under way to give people what they wanted.

Movie special effects engineer and inventor Fred Waller was asked to create an audiovisual exhibit for the 1939 World's Fair in New York. The exhibit would be presented in a spherical building. To show a movie on an arc-shaped screen, Waller knew he would have to develop a special filming and projection technique. He put together a series of eleven synchronized cameras and a similar arrangement of projectors. Each projector beamed its film image onto a different part of the screen, which created a panoramic effect. Because the screen surrounded viewers on three sides, this system made them feel as though they were in the middle of the action. Over the years, Waller refined the system and narrowed the equipment down to five cameras and three projectors. With these improvements, Cinerama was born. During the 1950s and into the 1960s, the arc-shaped screens of Cinerama theaters sprang up all over the United States. Then, someone had an idea that went beyond Cinerama.

ENTERTAINMENT FOR ALL THE SENSES

Morton Heilig worked in the motion-picture industry in the 1950s, when Cinerama was at the height of its popularity. Although audiences enjoyed Cinerama, they could only participate with their

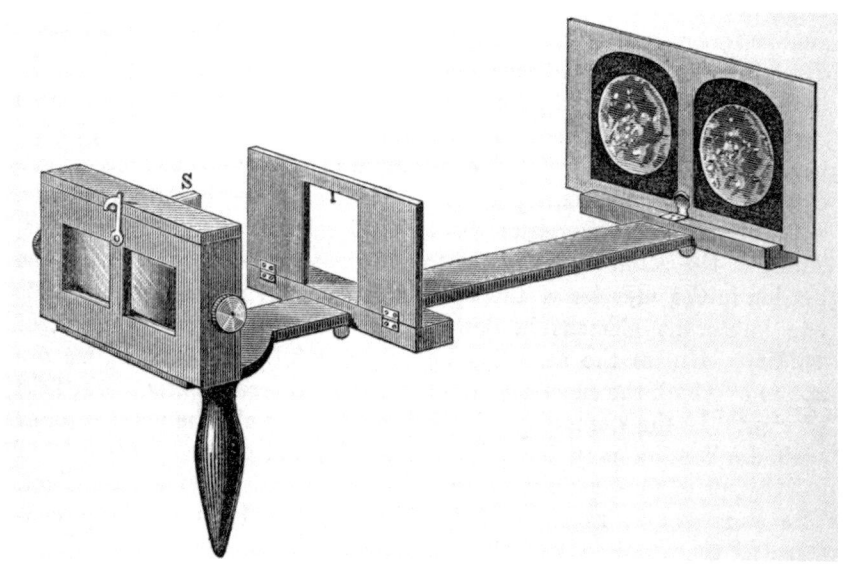

The stereoscope viewer was invented by Sir Charles Wheatstone in 1833.

consists of the input and output devices, also called effectors. These may include a head mounted display (HMD), joysticks, and wired gloves. The third part of the virtual reality system is the user, who controls what goes on in the virtual world.

Although today's virtual reality uses cutting-edge technology, the idea of virtual reality did not begin in the 21st century, or even in the second half of the 20th century. It began many years earlier, in the 19th century.

GIVING PEOPLE A VISUAL EXPERIENCE

In 1833, British inventor Sir Charles Wheatstone created a stereoscope viewer. It resembled diving goggles mounted on a rod about a foot long. On the end of the rod, a frame held a stereo picture. Stereo pictures were taken by a camera with two lenses, which produced two separate pictures a little more than 2 inches

CHAPTER 1

THE EVOLUTION OF VIRTUAL REALITY

Virtual reality is the science of creating three-dimensional simulated environments with computers. In the virtual environment, people can interact with simulated humans, animals, and objects. Some people describe virtual reality as a cartoon world that real people can visit.

The virtual reality system has three parts. The first is the reality simulator—a powerful computer that can run a virtual reality program so fast that there is no delay in the interaction between the user and the virtual world. The reality simulator contains the graphics board that produces the visual three-dimensional environment, the sound processors that create the sound effects, and the controllers for input and output devices that connect the user with the virtual environment. The second part of the system

A head mounted display is part of the virtual reality system.

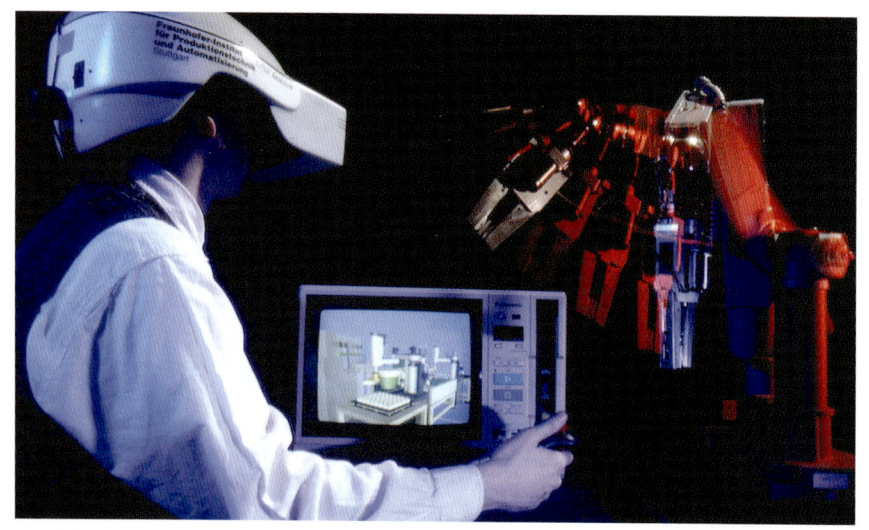

Virtual reality has had an impact on architecture and many other professions.

Virtual reality has a great impact on many professions. Architects can see their designs before the buildings are constructed, and doctors can treat patients who live in remote areas and cannot go to a medical center. Astronauts use virtual reality programs to train for space missions. Law enforcement personnel can face life-threatening situations without actually being in harm's way, and pilots can practice difficult flight maneuvers without any risk to aircraft or lives.

Virtual reality can help students prepare for their future occupations. Students use computer aided drafting (CAD) programs to learn product design skills. Medical students can examine, diagnose, and even practice surgery on virtual patients.

Like other kinds of technology, virtual reality constantly evolves and improves. Whereas the computers required for early virtual systems once took up the space of an entire room, some systems now fit in a space no bigger than a suitcase. Soon, systems will be able to fit into a small backpack.

As systems become more sophisticated, new uses for virtual reality are discovered. Many of the tasks done today were not possible just a few years ago. In the future, virtual reality will have even more applications that do not yet exist.

INTRODUCTION

Virtual reality as it exists today has been more than forty years in the making. The invention of computers brought the idea to life and enabled people to have experiences that might be difficult or impossible in the real world.

The technology of virtual reality continually changes the world in many ways. Virtual reality makes it possible for people to go places and do things without ever leaving their homes. It lets people travel not only to different places, but to different times as well. Through virtual reality, a person can experience simulated events in history and meet famous people from earlier centuries. Virtual reality is also a strong presence in entertainment, such as movies and games.

Through virtual reality, a person can experience what it is like to pilot an aircraft without really doing it.

TABLE OF CONTENTS

Introduction 4

Chapter One: *The Evolution of Virtual Reality* 6

Chapter Two: *Today's Virtual Reality* 19

Chapter Three: *Looking Toward the Future* 34

Glossary 46

About the Author 46

For Further Information 47

Index 47

THOMSON

GALE

© 2003 by Blackbirch Press™. Blackbirch Press™ is an imprint of The Gale Group, Inc., a division of Thomson Learning, Inc.

Blackbirch Press™ and Thomson Learning™ are trademarks used herein under license.

For more information, contact
The Gale Group, Inc.
27500 Drake Rd.
Farmington Hills, MI 48331-3535
Or you can visit our Internet site at http://www.gale.com

ALL RIGHTS RESERVED
No part of this work covered by the copyright hereon may be reproduced or used in any form or by any means—graphic, electronic, or mechanical, including photocopying, recording, taping, Web distribution, or information storage retrieval systems—without the written permission of the publisher.

Every effort has been made to trace the owners of copyrighted material.

Photo credits: cover, page 31 © Getty Images; pages 4, 10, 21 © Evans & Sutherland; pages 5, 7, 9, 13, 14, 15, 16, 19, 22, 24, 29, 32, 33, 38, 39, 42 © CORBIS; pages 6, 17, 18, 41 © VPL Research; page 13 © Hulton Archive; pages 20, 35, 37, 42 © PhotoDisc; page 23 © NASA Ames Research Center; page 25 © NASA Langley Research Center; page 26 © M. Agliolo / PhotoResearchers, Inc.; page 43 © H. Morgan / PhotoResearchers, Inc.; page 44 © G. Tompkinson / PhotoResearchers, Inc.; pages 27, 40, 45 © Corel Corporation; pages 30, 36 © AP Wide World; page 35 © CM Research

LIBRARY OF CONGRESS CATALOGING-IN-PUBLICATION DATA

Wyborny, Sheila.
   Virtual reality / by Sheila Wyborny.
     p. cm. — (Science on the edge series)
   Summary: Discusses the history, present uses, and future of the technology of virtual reality.
   Includes bibliographical references.
   ISBN 1-56711-789-9 (hardback)
    1. Human-computer interaction—Juvenile literature. 2. Virtual reality—Juvenile literature.   [1. Virtual reality. 2. Human-computer interaction.]   I. Title. II. Series.
   QA76.9.H85G725 2003
   004'.019—dc21
                                                        2002011926

Printed in China
10 9 8 7 6 5 4 3 2 1

SCIENCE ON THE EDGE

VIRTUAL REALITY

WRITTEN BY
SHEILA WYBORNY

BLACKBIRCH PRESS

THOMSON
GALE

San Diego • Detroit • New York • San Francisco • Cleveland • New Haven, Conn. • Waterville, Maine • London • Munich

Portage Public Library